ROUTLEDGE LIBRARY EDITIONS:
ENVIRONMENTAL POLICY

Volume 8

THE QUEST FOR WORLD ENVIRONMENTAL COOPERATION

THE QUEST FOR WORLD ENVIRONMENTAL COOPERATION
The Case of the UN Global Environment Monitoring System

BRANISLAV GOSOVIC

LONDON AND NEW YORK

First published in 1992 by Routledge

This edition first published in 2019
by Routledge
2 Park Square, Milton Park, Abingdon, Oxon OX14 4RN

and by Routledge
52 Vanderbilt Avenue, New York, NY 10017

Routledge is an imprint of the Taylor & Francis Group, an informa business

© 1992 Branislav Gosovic

All rights reserved. No part of this book may be reprinted or reproduced or utilised in any form or by any electronic, mechanical, or other means, now known or hereafter invented, including photocopying and recording, or in any information storage or retrieval system, without permission in writing from the publishers.

Trademark notice: Product or corporate names may be trademarks or registered trademarks, and are used only for identification and explanation without intent to infringe.

British Library Cataloguing in Publication Data
A catalogue record for this book is available from the British Library

ISBN: 978-0-367-18894-8 (Set)
ISBN: 978-0-429-27423-7 (Set) (ebk)
ISBN: 978-0-367-19353-9 (Volume 8) (hbk)
ISBN: 978-0-429-20197-4 (Volume 8) (ebk)

Publisher's Note
The publisher has gone to great lengths to ensure the quality of this reprint but points out that some imperfections in the original copies may be apparent.

Disclaimer
The publisher has made every effort to trace copyright holders and would welcome correspondence from those they have been unable to trace.

The quest for world environmental cooperation
The case of the UN Global Environment Monitoring System

Branislav Gosovic

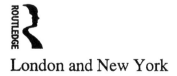
London and New York

First published in 1992
by Routledge
11 New Fetter Lane, London EC4P 4EE

Simultaneously published in the USA and Canada
by Routledge
a division of Routledge, Chapman and Hall, Inc.
29 West 35th Street, New York, NY 10001

© 1992 Branislav Gosovic

Typeset in Times by Michael Mepham, Frome, Somerset
Printed and bound in Great Britain by
Mackays of Chatham PLC, Chatham, Kent

All rights reserved. No part of this book may be reprinted or reproduced or utilized in any form or by any electronic, mechanical, or other means, now known or hereafter invented, including photocopying and recording, or in any information storage or retrieval system, without permission in writing from the publishers.

British Library Cataloguing in Publication Data
A catalogue record for this title is available from the British Library

Library of Congress Cataloging in Publication Data
Gosovic, Branislav.
  The quest for world environmental cooperation : the case of the
UN Global Environment Monitoring System / Branislav Gosovic.
  p. cm.
  Includes bibliographical references and index.
  ISBN 0–415–00458–6
  1. Global Environmental Monitoring System. 2. Environmental
monitoring—International cooperation. 3. Environmental protection—
International cooperation. I. Title.
TD193.G66 1992
363.7'063—dc20   91–41631
                                                                 CIP

To Maja and Mašan

Contents

Foreword	ix
Prologue	xi
Preface	xv
Acknowledgements	xxi
Abbreviations	xxv

Part I The roots — 1

1 UNCHE and UNEP — 3
   The 1972 Stockholm Conference — 3
   Institutional and substantive responses in the wake of UNCHE — 9
   The UNEP era: model and reality — 17

2 The origin and nature of GEMS — 35
   Environmental monitoring at the Stockholm Conference — 35
   Setting up the Global Environmental Monitoring System — 42
   What is GEMS? — 50

Part II GEMS in action — 59

3 Health-related environmental monitoring in GEMS — 61
   Nature and constituents of GEMS HEALTH — 62
   Indicators of progress — 67
   Factors at work — 73
   Assessment and prospects of GEMS HEALTH — 82

4 Climate-related environmental monitoring in GEMS — 91
   GEMS and human-induced climate change: the first steps — 92
   Early indicators of progress — 99
   Factors at play — 105
   Towards an integrated approach to global atmosphere monitoring — 113

5 Marine pollution monitoring in GEMS — 124
   The first steps and controversies — 125

viii *Contents*

	Pollution monitoring in regional seas: the Mediterranean model	132
	Marine pollution monitoring in GEMS: factors at play	140
	Towards comprehensive global marine pollution monitoring	143
6	Natural resources and integrated monitoring in GEMS	151
	Natural resources monitoring	153
	Integrated monitoring	166
	The prospects for the monitoring of natural resources and integrated monitoring	172
7	Monitoring long-range transboundary air pollution in GEMS	175
	Monitoring long-range transboundary air pollution in Europe	177
	The significance of EMEP for GEMS	184
8	GRID: Global Resource Information Database	193
	The origins of GRID	193
	The GRID pilot phase	196
	The implementation phase of GRID	197
	GRID and GEMS	199
9	GEMS: inching towards a global system?	205
	Assessing GEMS	205
	Whither GEMS?	211
	GEMS: the need for a fresh start	220

Part III What international organization for the twenty-first century? 221

10	Building blocks for a stronger UN	223
	International civil service	225
	Financing international organizations	236
	Institutional betterment as a permanent objective	243
	The challenge of interrelatedness	246
11	UN system at a crossroads	249
	Non-reform and change through unilateral action	249
	The need for a comprehensive reform of the UN system	258
	By way of a conclusion	271
	Name index	273
	Subject index	275

Foreword

One of the most important results of the United Nations Conference on the Human Environment held in Stockholm in 1972 was agreement on the need to establish more effective international cooperation in monitoring the environment and life support systems of our 'Only One Earth' and assessing potential risks to human life and well-being. When, following the Stockholm Conference, the United Nations Environment Programme (UNEP) was established, it created as one of its first programme initiatives the Global Environment Monitoring System (GEMS) in order to meet this need. This has since become a cornerstone of UNEP's programme, and the recently released report of the World Commission on Environment and Development recommends that GEMS be accorded even higher priority and much greater support in the period ahead.

Thus, Branislav Gosovic's study on GEMS is both timely and highly relevant. It combines a well-informed account of the history of GEMS as an integral component of UNEP's institutional evolution with a thoughtful and sometimes provocative analysis of what has gone well, what could have gone better and what can be done to make GEMS more effective in meeting the growing need for its services in future. The study derives special value and credibility from the fact that the author was able to observe the processes he describes in his role as a staff member of the UNEP secretariat. It thus reflects the insights as well as the inevitable biases of an insider. One does not have to agree with every point the author makes to welcome this as an extremely important and unique contribution to the history of international environmental cooperation. The lessons it points up will be invaluable to those charting the future development of GEMS as a central and indispensable element in the system through which that cooperation is effected. Also, the study will be especially useful when the related issues are revisited twenty years after Stockholm, at the 1992 United Nations Conference on Environment and Development in Brazil.

Maurice F. Strong

Prologue

In 1972, the United Nations Conference on the Human Environment (UNCHE), held in Stockholm, launched a new venture of international cooperation. Expectations were high and optimism pronounced. Global multilateral cooperation through the United Nations was still seen as a natural and promising road to follow. A mere fifteen years later, however, the accepted notion of multilateralism was being questioned and the UN was in the throes of a political, institutional, and financial crisis.

This crisis had as its immediate cause the actions of a few powerful developed countries. Their stated objective was to streamline the organization and eliminate the bureaucratic and financial inefficiencies. Their broader objective, however, was to reassert their influence and gain greater control over the work of the UN, which they felt was moving in an undesirable direction, and to neutralize what they considered to be an unwieldy majority of developing countries seeking change in the dominant international economic order, wealth distribution, and power structures.

The 1980s crisis was also a consequence and manifestation of long-simmering tensions and reflected organizational deficiencies, limited institutional capacity and mandate, and unsatisfactory performance as far as the expectations of many governments were concerned. It highlighted the failure to redefine and recast multilateral arrangements in order to keep up with changing realities, and the lack of agreement as to the functions and scope of international organization. The deeper roots of the crisis, however, were to be found primarily in the development gap between the North and the South, in great disparity in national economic and political power and governments' abilities to deal with domestic and international challenges, and in the diversity of their world views, interests, and priorities.

Whatever the primary motivations of those who triggered it, this crisis brought to the fore some of the key dilemmas of the UN, and the need for a positive process of change and institution-building in international organizations. Eventually and unexpectedly, it was supplanted by a more positive and hopeful atmosphere, not on account of any significant change in the UN system or in its ways, but because of changes in world political constellation, the positive attitude of the USSR towards the organization, and the newly perceived opportunities by earlier critics of how the UN system could serve their needs. Indeed, many of the underlying causes of institutional tension persisted or became aggravated.

The soon-to-follow Gulf events and the role that the UN was cast in served as a reminder of this fact, and cut short the spell of optimism. Also, the US-originated notion of a 'new world order' placed squarely on the agenda the questions of what kind of UN there should be and what role it should play in the coming age.

As the decade of the 1990s began to unfold, it was quite clear that the organization had embarked on a path of transition and change, with different visions and institutional paradigms competing.

This, and the recent ups and downs in the political fortunes of the United Nations, also highlights the need for the UN to build solid, two-way relations with its global constituency. As it is, among the public and the decison-makers there is a good deal of misinformation, misunderstanding and plain confusion about the UN. Some of this has its origins in persisting beliefs and somewhat romantic expectations of what the UN can and should do, and of what its powers and capabilities are; it is still seen by many as an entity independent of its membership and capable of institutional peformance on its own.[1] A lot, however, stems from deliberate, ideologically inspired hostility to international organizations or sheer dissatisfaction when they do not match the expectations of national policy. This hostility feeds on and also generates unfair judgements and half-truths that get readily diffused by global media which have not been known for their friendship to the United Nations.

The situation has not been helped by the fact that the inner workings of international organizations have been traditionally veiled. As they are precarious consensus bodies – bound by adversary positions of governments, vulnerable to pressures, defensive to criticism, and with a number of competing interests involved – it has not been easy to generate an in-depth and frank assessment of their performance and functioning. Institutionalized pretence and hypocrisy suit both the governments and the international civil servants. This has reinforced the inherent organizational and bureaucratic defensiveness. And it has contributed to a notable institutional lag and inertia. Furthermore, the efforts to study and understand international organizations have been made more difficult by the growing complexity and number of such organizations, making it hard to keep track of their proceedings – not to speak of their performance – even for those engaged in their work on a full-time basis.

The basic underlying premises and operational characteristics of international organizations need to be subjected to in-depth review and analysis, in the quest for fresh departures that would help the UN system to manage its problems and to become better adapted to new conditions and responsive to evolving needs.

In this process of re-evaluation, it is important that analytical and empirical

---

[1] This point is very nicely conveyed by the following quote: 'they (world organizations) are credited with an importance they do not possess; they are blamed for not doing what they are not given the means to do; faults that are often imaginary are ascribed to them, while their real faults go unnoticed; mythical explanations are invented to explain their ineffectiveness; and finally, there is very little recognition of the few significant results that they really do achieve'. M. Bertrand, *The Third Generation World Organization*, Nijhoff, Dordrecht, 1989, p.27.

reference points about the inner dynamics of international organizations are available to feed the debate, shape the thinking and assist in the design of new or modified institutions and approaches. This study of global environmental monitoring was undertaken with this as one of its principal objectives. Indeed, some of the very issues troubling the UN system and stirring up the passions of governments, international civil servants, and public opinion have also played an important role in the evolving story of the Global Environment Monitoring System (GEMS), which is presented in the pages that follow.

Preface

I

The 1970s were an important period in the evolution of the UN. A series of new issues was placed on its agenda; new institutions and mechanisms were created. Many of these concerns had to do with what came to be known as global interdependence, a concept formulated essentially to reflect the meshing of the destinies of countries and peoples, and their influence on each other through human-induced changes in the environment. This was a novel awareness, superimposed on the social, economic, political, and power relationships of the more traditional kind. The seeds were thus planted for social and institutional change, both nationally and as regards the modes of international cooperation. Challenges were mounting. More effective global responses and institutions were called for.

Not much progress was made in this direction, however. Indeed, the 1970s will also be remembered as a prelude to the full-blown UN crisis of the 1980s. A countertrend of frustrated expectations and fatigue with the UN began to make itself felt. Constraints on international cooperation increased, rigidities set in, and cleavages became more apparent. Power politics and unilateral pressures *vis-à-vis* the UN increased. The fragility and vulnerability of the international institutions were there for all to see.

As the 1980s progressed, the UN was caught in a bind. On the one hand, there were increasing demands, multiplying problems, an expanding agenda, and ever greater expectations; on the other hand, the organization faced its own inherent institutional limitations and rigidities which were compounded by a world economic and development crisis, an intensifying attack on its prerogatives and methods of work, a financial crisis, and erosion of its institutional, professional and political standing.

This is a study of the Global Environment Monitoring System (GEMS), a functional UN programme largely unknown to the general public, which was born and evolved during this unsettled period. GEMS is of interest not only because of its innate importance, qualities and objectives as regards the environment. It is also an activity that, in its own specialized and technical context, illustrates many of the underlying problems and dilemmas of international cooperation and global undertakings through the UN. It juxtaposes the constraints on international cooperation, the frustrations, and the slow progress with the need for, the potential of, and the promise inherent in such cooperation. And it sheds some light on the evolving drama of multilateralism played at the centre of the stage.

II

The 1972 Stockholm Conference on the Human Environment (UNCHE) was a precursor of many of the new issues and of the awareness that crystallized in the years that followed. Because of its systemic and multidisciplinary nature, it was difficult to accommodate the environmental issue within the classical, vertically divided institutional framework of the UN. The attempt to devise appropriate institutional follow-up mechanisms called for innovation and imagination. The United Nations Environment Programme (UNEP) was created to lead global environmental action. It embodied some new features that were designed to help it fulfil its difficult task, and to overcome some of the constraints inherent in UN bodies. It was also an institutional compromise, with many weaknesses and gaps. It was lodged within the UN system, which remained as it was and did not change to adapt to the requirements of environmental issues or of the heterodox new programme.

Earthwatch was one of the key codewords and concepts launched at the Stockholm Conference. It referred to environmental assessment. The word metaphorically depicts one of the basic functions of UNEP, namely, to stand guard and watch over the global environment. In the post-Stockholm period, Earthwatch has been used to refer to the environmental assessment component of UNEP.

The recommendations of the Stockholm Conference were grouped into two main categories of functions: 'environmental assessment', i.e., Earthwatch, and 'environmental management'. A residual category was also provided for, entitled 'supporting measures'.

- The *environmental assessment* category consists of evaluation, review, research, monitoring, and information exchange.
- The *environmental management* category consists of goal setting and planning, and international consultation and agreement.
- The *supporting measures* consist of education and training, public information, organization, financing, and technical cooperation. While this is a subsidiary category to the basic scheme, it is important politically, institutionally, and operationally as it refers mostly to activities aimed at assisting the developing countries.

This has served as the basic organizational scheme for UNEP's work ever since.[1] The distinction between assessment and management was straightforward:

- *Assessment* means to carry out research, to monitor (i.e., observe and measure), and evaluate the data thus obtained in order to establish facts and trends, to

---

1 See the Final Report of the Conference (UN doc. A/CONF.48/14/Rev.1, p.6, UN publication Sales No. E.73. II. A.14).

increase knowledge and forecasting capabilities, and to arrive at scientific conclusions that will be relevant and useful for decision-making and management;

- *Management* means to anticipate, decide, act, manage, and correct the course taken as necessary, relying on the scientific and quantitative inputs that are continuously generated by the process of environmental assessment.

Assessment was regarded as a relatively precise, quantifiable category. It was considered to belong to the domain of scientific and technical communities, and was supposed to generate the neutral, hard-data inputs and conclusions for management, a softer category involving a social process. The two were linked through a feedback loop, and together formed an integral whole.

According to the Stockholm scheme, the objective of environmental assessment, or Earthwatch, was to study the interaction between humans and the environment, to provide early warning of potential environmental hazards, and to determine the status of selected natural resources. Its four subsidiary functional categories were defined as follows.[2]

- *Evaluation and review*: to determine priority needs, based on the scientific and technical evaluation of findings and data, and to relate such specialized evaluations to broader concerns, as well as to bring to governments' attention significant trends, existing and potential problems, and possibilities for corrective and preventive action.
- *Research*: to add to the kind of knowledge specifically needed for understanding environmental processes and for interpreting related data. While such research would be carried out mainly at the national level, the international organization concerned would identify common needs, coordinate relevant activities, arrange for multinational support for research programmes designed to meet such needs, carry out periodic overviews, and undertake collection and analysis of data.
- *Monitoring*: to gather data on specific environmental variables; to evaluate such data in order to determine or predict major environmental conditions and trends and to give early warnings; and to integrate various national and international monitoring programmes on a global scale.
- *Information exchange*: to disseminate relevant knowledge and information within the scientific and technological communities, and to ensure that decision-makers at all levels have the benefit of the most comprehensive and accurate environmental information that can be made available.

The UNCHE recommendations regarding Earthwatch were formulated broadly, exhibited some terminological overlap, and did not distinguish between structure and function. This was not surprising given the institutional vagueness surrounding the outcome of the Conference, the novelty of the issues, and the lack of experience as to how the earthwatching would be carried out.

2   The Action Plan itself is rather vague when it comes to defining these subsidiary functions. A more explicit statement as to what was meant (and on which this text relies) is to be found in a background document for the Conference entitled 'International Organizational Implications of Action Proposals' (UN doc. A/CONF.48/11).

xviii *Preface*

Over the years after Stockholm, Earthwatch has continued to be ambiguous in concept and function, and as a programme.[3] This is, at least in part, a reflection of the ambivalence of UNEP's institutional role. It is called upon to be a global coordinator and mover of environmental action, yet is not endowed with institutional prerogatives, or material and staff resources needed to fulfil this role properly. From UNEP's standpoint, then, a vague and flexible notion of Earthwatch has been functional. It offers a symbolic, political, conceptual, and institutional umbrella. It is a unifying theme under which to promote environmental assessment, to group a series of programmes and activities and to coordinate the related work of the UN specialized agencies. Bearing a UNEP imprint, it contributes to institution-building and to its global image.

UNEP has tended to group a series of more or less important environmental assessment activities and projects under Earthwatch. For all practical purposes, however, Earthwatch has consisted of three global programmes, which were launched by the Stockholm Conference. Two of these, the International Referral System for Sources of Environmental Information (known as INFOTERRA) and the International Registry of Potentially Toxic Chemicals (IRPTC), are aimed at promoting the exchange of environmental information. Substantively and operationally they are not on a par with GEMS, the third major component of Earthwatch. In fact, in the minds of many people, GEMS is equated with Earthwatch. It is GEMS that does the earthwatching, is global in scope and is responsible for generating the monitoring data that represent the critical input for the assessment/management process.

III

GEMS is part of the complex heritage of the Stockholm Conference. It is embedded in the inherently ambiguous structure and practice of UNEP, about which little has been written. This background, which is essential for an understanding of the nature of GEMS, is dealt with in the first chapter. It accounts for the ambiguities in the definition of GEMS, which have frustrated many observers when trying to size up and evaluate this undertaking.

This study is primarily an institutional and policy treatise. It should be of interest not only to those in and outside government who are concerned with the environment and international environmental cooperation, but also to those who study, work in, or care about international organizations and the UN system in particular. It reviews how this global scheme was conceived and born, how it has evolved, and what factors have figured prominently in this process. While this is not a study of the technical aspects of GEMS, it should be also of some interest to the scientific

3   Definitions used are circular, refer to a collection of functions, and stay clear of the question of structure. For example, Earthwatch has been defined as 'a dynamic process of integrated environmental assessment by which relevant environmental issues are identified and necessary data are gathered and evaluated to provide a basis of information and understanding for effective environmental management' (doc. UNEP/GC/61). A later definition speaks of a process that brings together and analyzes environmental information in order to clarify issues and assist management (doc. UNEP/WG.30/3).

community. Indeed, scientists should be informed about the international setting where the schemes and global designs that they conceive are being implemented.

The discussion covers the origins and roughly the first fifteen years of the operational existence of GEMS, a period of continuing search, experiment, and efforts at self-assertion. This is sufficiently long for the purposes of identifying the basic and enduring factors at work, and drawing principal conclusions about GEMS. It is also a self-contained period between the two global UN conferences on environment.

In researching and organizing the materials and in trying to provide a common explanatory framework for the wide-ranging and varied activities involved, I explored several themes to the extent possible, given the limitations on my time and mobility. These themes, individually and in combination, have shaped the physiognomy and direction of GEMS and have largely determined the results achieved so far. They are:

- the underlying North–South tensions and differences between industrialized and developing countries in terms of interests, capabilities, levels of development, etc., as well as the East–West controversies and mistrust that affected a number of GEMS-related activities;
- the characteristic traits of the UN system, such as the difficulties in attaining coordination of activities, jurisdictional competition within and among organizations, the mismatch between objectives and the resources allocated for their attainment, and the gap between the ambitious blueprints and their practical implementation;
- the nature of the issues being addressed and the role of such factors as the state of scientific knowledge and level of technology, institutional and financial support available for follow-up activities, etc.;
- the dichotomy between, on the one hand, scientific objectives inspired by the need to know and understand the processes and phenomena under study and observation and guided by longer-term considerations and, on the other hand, the practical and management applications of environmental monitoring, motivated by the need for products that will be of direct and immediate use to governments, decision-makers, and managers; and
- the inherent tension between the practical requirements of implementation, which pull in the direction of decentralization, disaggregation, sectorialization, and simplification, and the essence of most environmental problems, which call for aggregation, integration, a systems approach, and a global perspective, all of which are difficult to achieve in practice.

The role of individuals is difficult to accommodate in an academic study of this kind; this is a pity because they represent an important factor in the evolution of international activities and their actions often provide a key element needed to explain the course of events. They can have a significant influence in an international activity like GEMS, where most governments have tended to participate superficially for lack of interest or technical capability, and because they do not consider that their vital interests are at stake. This situation leaves greater scope for individual action or initiative than is normally the case in international organizations, thus highlighting the importance of the background of the people involved, their qualities, skills and motivation, and how they play their roles, use opportunities, and cope with constraints.

IV

This is a study of the first halting steps of a global programme in the making. It is an attempt to present a comprehensive view of GEMS and to place it in the broader context of international cooperation.

It is not easy to write on a topic like GEMS, which spans different scientific disciplines and issues, and involves political and institutional variables that do not mix easily with the former. Also, it is not easy in a single volume to please different categories of readership that are interested in GEMS and its fortunes, or in details of given aspects of its operation.

Yet, the attempt had to be made to fill part of a void in the academic literature on environment and UNEP, and to leave behind a basic reference text that could be of some value to both the academic community and those involved directly in the process of international cooperation. Anyway, it was worth the effort to save from sure oblivion elements of a story that is bound to amuse students of history of international organizations in the decades to come, as they look back in wonder at the first unsure steps by the international community in confronting global environmental issues.

It should be stressed that this book was written and is supposed to be read as an entity, including the footnotes which often contain important comments and useful additional information. This is not an easy task, however, and many will skim the text, or pick and choose what interests them most. They will miss some of the threads of the argument and the empirical data which inform the analysis and the conclusions. However, every chapter is written so that it can be read on its own and make sense, provided a reader has glanced at the introduction beforehand. The final two chapters are essays apart which, starting from some observations derived from the study of GEMS, move into the broader realm of issues having to do with the very nature of international organization embodied in the UN system.

Acknowledgements

This study was initiated in 1976 when I was on sabbatical leave from the United Nations Environment Programme (UNEP). During that year I took part in the project 'International conflict and international regimes: the cases of oceans, the atmosphere, the environment and space', at the Institute of International Studies, University of California, Berkeley. The project was coordinated by Professors Ernst B. Haas and John G. Ruggie.

I continued with research and writing in the years that followed, sporadically and whenever I could summon energy and inspiration outside my regular work – first in UNEP, then in the United Nations Economic Commission for Latin America and the Caribbean (ECLAC), in the World Commission on Environment and Development (WCED), for a brief period in the United Nations Conference on Trade and Development (UNCTAD), and towards the end in the South Commission. The views expressed in the study are, of course, my own and do not reflect in any way on these institutions.

I wish to thank Maurice F. Strong, the architect of UNCHE in 1972 and also at the helm of the UN Conference on Environment and Development (UNCED) twenty years later, for having lent his name to the foreword of this book. While he may not share some of my views or agree with all of my arguments and conclusions, we share a commitment to and belief in the United Nations, and in the need for effective and visionary international cooperation if humankind is to secure a bright future for itself and for the planet Earth.

The study has greatly benefited from, and indeed was made possible by, my being associated with the UN and international life. I also drew on the wisdom and knowledge of my colleagues, both within and outside the UN system, though the responsibility for analysis, intepretation, and data used, as well as for possible errors of fact or judgement, is entirely my own.

My thanks are due especially to those who have worked on various aspects of GEMS, within UNEP and in other organizations. In particular, I owe gratitude to Francesco Sella, as well as to Harvey Croze, Michael Gwynne, Stjepan Keckes, Wayne Mooneyhan, and C. C. Wallen in UNEP; to Richard Helmer and Wilfried Kreisel in WHO; Rumen Bojkov and Albert Kohler in WMO; Neil Andersen and Gunnar Kullenberg in IOC/UNESCO; Harald Dovland and Peter Sand in ECE; and Gordon Goodman and G. Bruce Wiersma.

As concerns the various aspects of UNEP's institutional history, a great part of which I lived through myself, I am thankful for comments that I have received at various times from Paul Berthoud, Robert Frosch, Ashok Khosla, Vicente Sánchez, Maurice Strong and Peter Thacher.

Regarding the last section of the book dealing with the future of the UN system, these chapters went through many versions and changes of argument and style. As I worked on the succeeding drafts, I benefited greatly from the advice and reactions of friends familiar with the workings and predicaments of the United Nations. I wish to thank for their comments Maurice Bertrand, Jean-Jacques Graisse, Iqbal Haji, Peter Hansen, Ljerka Komar, Marc Nerfin, John Ruggie, Vicente Sánchez, and Michael Zammit Cutajar.

I am indebted to my academic colleagues Haas and Ruggie who encouraged me to start the study and gave me valuable advice as I went along, and to the former for having hardened me for scholarly projects of this kind through the dissertation experience with him at an earlier date. I hope that the end product meets with their approval, even though it is much too down to earth and institution-oriented for their taste and too light on the theoretical aspects of the emergence of international regimes and on planetary politics. This, however, is a rarefied and exclusive domain of academic debate, where those, like me, who are taking part in the daily process of international cooperation are reluctant to tread.

My thanks are also due to many friends in Nairobi, Santiago, and Geneva who, at different times over the years, helped with editing the manuscript and correcting my English. My very special gratitude is reserved for Alfred Lehmann who agreed to do the final editing and for Jean François Baylocq for his invaluable help with formatting the text in pageproof form, and processing many successive drafts.

This moonlighting study was completed at a high cost to my family and personal life. Over the years, it absorbed many weekends and evenings, and a great deal of attention and energy which I should have devoted primarily to my children Maja and Mašan. To them I dedicate this book. I can only hope that when they grow up they will forgive me for having used, in pursuit of my academic ambitions and internationalist inspiration, many hours and part of my vital energy that should have been devoted to them.

One of the hard lessons I have learned in the process is that projects of this kind should be left to full-time academics and researchers. Trying to do in parallel with one's job a study that would normally take a year or two to complete could easily take forever. This may well explain why books of this nature are seldom written by those who take part in the organizational processes that call for analysis. I endured because of my stubborness and also because I felt that this was a worthwile project that no one else was likely to undertake.

My inability to work in a sustained manner resulted in frustrating delays and the need for successive revisions to update the manuscript in order to keep up with the many new developments. In the end, this turned out to be quite functional. As the study dragged on, the story of GEMS became more complete, interesting, and relevant to international environmental action, and the environment turned into a major global policy preoccupation, largely driven by those very issues which figure prominently on the GEMS agenda.

Also, the question of the UN's future and reform moved into policy focus, causing a further delay in the completion of the manuscript; my initial objective to draw a few parallels between GEMS and the UN grew more ambitious under the impact of institutional ferment. Fortunately, the approaching 1992 UN Conference on Environment and Development gives me a decisive reason to end this undertaking, because, like all scholars, I hope that my work will not remain of academic interest only.

Echenevex, May 1991

Abbreviations

Abbreviation	Meaning
ACC	Administrative Committee on Coordination
BAPMoN	Background Air Pollution Monitoring Network
CCC	Chemical Coordinating Centre
CMEA	Council for Mutual Economic Assistance
DESCON	UN Desertification Conference
DOEM	Designated Officials on Environmental Matters
ECB	Environment Coordination Board
ECE	UN Economic Commission for Europe
ECLA(C)	UN Economic Commission for Latin America (and the Caribbean)
ECOSOC	Economic and Social Council
EEC	European Economic Community
ELAS	Earth Laboratory Applications System
EMA	Extended Measurement Activity
EMASAR	Ecological Management of Arid and Semi-Arid Rangelands in Africa and the Near and Middle East
EMEP	Monitoring and Evaluation of the Long-range Transmission of Air Pollutants in Europe (i.e. Evaluation and Monitoring of Environmental Pollution)
ENSO	El Niño/South Oscillation
EOS	Earth Observing System/'Mission to Planet Earth'
EOSAT	Earth Observation Satellite
EPA	Environmental Protection Agency
EPMRP	Environmental Pollution Monitoring and Research Programme
EPTA	Expanded Programme of Technical Assistance
ERAMS	Environmental Radiation Ambient Monitoring System
ERTS	Earth Resources Technology Satellite
FAO	Food and Agriculture Organization
FIGGE	First Global Experiment of GARP
FOGS	Functioning of the GATT System
GARP	Global Atmospheric Research Programme (of WMO)
GATT	General Agreement on Tariffs and Trade

GAW	Global Atmosphere Watch
GBO	Geosphere–Biosphere Observatories
GEEP	Group of Experts on Effects of Pollutants
GEF	Global Environmental Facility
GEMS	Global Environment Monitoring System
GEMSI	Group of Experts on Methods, Standards and Intercalibration
GERMON	Global Environmental Radiation Monitoring Network
GESAMP	Group of Experts on the Scientific Aspects of Marine Pollution
GESREM	Group of Experts on Standards and Reference Materials
GIPME	Global Investigation of Pollution in the Marine Environment
GIS	Geographic Information Systems
GMCC	Geophysical Monitoring for Climatic Change
GO_3OS	Global Ozone Observing System
GRAP	Global Risks Assessment Programme
GRID	Global Resource Information Database
HEALs	Human Exposure Assessment Locations
IAEA	International Atomic Energy Agency
IARC	International Agency for Research on Cancer
ICES	International Council for the Exploration of the Sea
ICSU	International Committee of Scientific Unions
IGBP	International Geosphere–Biosphere Programme
IGOM	Integrated Global Ocean Monitoring
IGOSS	Integrated Global Ocean Station System
IGY	International Geophysical Year
IIASA	International Institute for Applied Systems Analysis
IIED	International Institute for Environment and Development
INFOTERRA	International Referral System (IRS) for Sources of Environmental Information
IOC	International Oceanographic Commission
IPCC	Intergovernmental Panel on Climate Change
IRPTC	International Registry of Potentially Toxic Chemicals
ISLSCP	International Satellite for a Land-Surface Climatology Project
IUCN	International Union for Conservation of Nature and Natural Resources
JIU	Joint Inspection Unit (of UN)
KREMU	Kenya Rangelands Environmental Monitoring Unit
LANDSAT	Land Remote-Sensing Satellite
LEPOR	Long-term and Expanded Programme of Oceanic Exploration and Research
LIDÀR	Light Detection and Ranging
MAB	Man and Biosphere Programme
MAPMOPP	Monitoring of Marine Pollution caused by Petroleum
MAPS	Measurement of Air Pollution from Satellites
MARC	Monitoring and Assessment Research Centre
MARPOLMON	Marine Pollution Monitoring System

MEDPOL	Coordinated Mediterranean Pollution Monitoring and Research Programme
MKN	Steering Body for Monitoring the Environmental Quality of the Nordic Countries
MMA	Minimum Measurement Activity
MSC-E	Meteorological Synthesizing Centre – East
MSC-W	Meteorological Synthesizing Centre – West
NADP	(US) National Atmospheric Deposition Program
NASA	National Aeronautics and Space Administration
NGO	non-governmental organization
NIEO	New International Economic Order
NOAA	National Oceanic and Atmospheric Administration
OCA PAC	Oceans and Coastal Areas PAC
OECD	Organization for Economic Cooperation and Development
PAC	Programme Activity Centre
PCBs	polychlorinated biphenyls
PMK	Swedish National Environmental Monitoring Programme
RIOS	River Inputs into Ocean Systems
RS PAC	Regional Seas PAC
SCEP	Study of Critical Environmental Problems
SCOPE	Scientific Committee for the Study of the Problems of the Environment
SCPRI	Central Service for Protection against Ionizing Radiation (Fr.)
SMIC	Study of Man's Impact on Climate
SPM	suspended particulate matter
SPOT	Le Système Probatoire d'Observation de la Terre
SUNFED	Special UN Fund for Economic Development
SWMTEP	System-Wide Medium-Term Environment Plan
TECOMAC	Technical Conference on Observation and Measurement of Atmospheric Contaminants
TEMA	Training, Education and Mutual Assistance (of IOC)
TERS	Tropical Earth Resources Satellite
TIROS	Television Infra-red Operational Satellite
UNCED	UN Conference on Environment and Development
UNCHE	UN Conference on the Human Environment
UNCTAD	UN Conference on Trade and Development
UNDP	UN Development Programme
UNEP	UN Environment Programme
UNESCO	UN Educational, Scientific and Cultural Organization
UNIDO	UN Industrial Development Organization
UNISPACE	UN Conference on Peaceful Uses of Outer Space
UNSCEAR	UN Committee on the Effects of Atomic Radiation
USAID	US Agency for International Development
VCP	Voluntary Cooperation Programme (of WMO)
WCAP	World Climate Applications Programme

WCDP	World Climate Data Programme
WCED	World Commission on Environment and Development
WCIP	World Climate Impacts Studies Programme
WCP	World Climate Programme
WCRP	World Climate Research Programme
WGI	World Glacier Inventory
WHO	World Health Organization
WMO	World Meteorological Organization
WORRI	World Register of Rivers Flowing into the Oceans
WRI	World Resources Institute
WWW	World Weather Watch
WWF	World Wide Fund for Nature

Part I
The roots

The two chapters that follow provide some unavoidable organizational history, necessary if one is to understand the Global Environment Monitoring System (GEMS). The first chapter discusses the 1972 UN Conference on the Human Environment (UNCHE), where it all started, and the UN Environment Programme (UNEP), the executor of the Stockholm legacy and the parent organization of GEMS; the second chapter examines the beginnings and nature of GEMS itself.

1 UNCHE and UNEP

THE 1972 STOCKHOLM CONFERENCE

The Stockholm Conference on the Human Environment was a pioneering venture. Its novelty came primarily from the nature of the subject matter on its agenda. It spanned disciplines and jurisdictions. It could not be readily fitted within the framework of the organizations used by governments either to run their domestic affairs or to conduct international relations. Furthermore, the knowledge and appreciation of environmental issues was still rudimentary. Indeed, this was an area where international consciousness had been raised before the subject became a domestic concern in most countries. This was in contrast to other more traditional subjects dealt with by international conferences where, in devising their positions and strategies, governments could draw on their own practice, the existing domestic structures, well-defined interests, and an institutional memory. As a result, governments had to seek advice and guidance on how to approach this still uncharted terrain. They found it in the relatively narrow circle of those in the vanguard of environmental consciousness, namely, the scientists, academics and a few non-governmental organizations (NGOs).[1] Also, they found it in the secretariats of the Conference itself and of the specialized agencies of the UN system, which, as a result, enjoyed a significant degree of initiative and were the principal shapers of the proposals and consensus that eventually materialized.

At that time, many formulations of problems were still tentative and not widely accepted. Scientific knowledge and explanations were incomplete and often controversial. Moreover, there was a good deal of substantive and institutional uncertainty as to how to proceed and how to link environmental objectives with everyday decision-making and action, especially in the economic sphere.

The Stockholm Conference brought to the surface a number of underlying controversies. It yielded given formulations and decisions that have played an important role in shaping the perception of environmental issues and national and international responses in the period that followed.

---

1 A refreshing and unusual aspect of the Conference was the presence of enthusiastic and vocal non-governmental organizations and grass-roots movements. Also, among the delegates the usual government representatives and diplomats from the traditional UN circuit were not dominant. In fact, the heterogeneous mix of delegates' backgrounds and profiles reflected the lack of an institutional niche for environment in the domestic set-up of most governments.

4 *The quest for world environmental cooperation*

One of the difficult challenges faced by the Conference was how to define the conceptual and policy framework to be used in orienting environmental action. Inevitably, this issue was greatly affected by the underlying North–South tensions. Another hard question concerned the institutional design. A few remarks are devoted to each of these topics in the pages that follow.

Recasting the frame of reference

The national and international responses to environmental problems, as well as the various conceptual and institutional frameworks devised to deal with them, have been significantly influenced and shaped by the real and/or perceived conflict between environment objectives and what were considered as the more pressing social and economic goals and priorities of society.

Thus, before and during the Stockholm Conference, the developing countries' suspicion and in many cases lack of experience of environment issues affected the proceedings and the outcomes.

A good deal of the continuing controversy shrouding environment issues internationally can be attributed to the manner in which these early debates evolved and to the terms in which the issues were conceptualized. At the risk of oversimplifying, this setting will be summed up in the next few paragraphs.

Contemporary concern for the environment started to become vocal in the 1960s in the highly industrialized countries. In the mind of the common citizen, as well as of the decision-maker, the principal cause of this concern was the pollution of the atmosphere and of water which was beginning to affect visibly and immediately the quality of life. The new found interest in the environment extended also to the protection and conservation of nature, a concern spurred by the rapid and extensive penetration into the countryside of economic activities, human settlements, and transportation networks, and the resulting degradation of nature.

During this early period, the developing countries were hardly aware of or worried about environmental pollution. Domestically, for most of them, pollution represented an issue still beyond the horizon. Indeed, many considered pollution as something unavoidable, and even to be welcomed as a by-product of progress and of the economic and industrial development to which they were aspiring. They felt that on the global scale it was the responsibility of the developed countries, which were at high levels of economic development and had the necessary resources at their disposal, to control and prevent environmental pollution that was mostly of their own making, as well as to remedy the damage already done. As far as they were concerned and in view of their pressing social and development needs, the developing countries felt that worrying about pollution was a luxury which they could ill afford. If they were to act, they considered, they should be granted generous financial support by the international community, in addition to what they were already receiving in development assistance flows.

Regarding the protection and conservation of nature, the feeling in many developing countries was that the challenge for them was not so much one of preserving, protecting, and rehabilitating nature, as rather, and primarily, one of conquering and mastering it. They were aware of the vulnerability of their renewable natural resource base and of the need to exploit and manage it with care. They were annoyed, however, by the arguments of some non-governmental groups and scientists from the North who, among other things, called for strict conservation measures in the South, showing little sensitivity for their development objectives. It should not be surprising, then, that in the minds of many people in the Third World 'conservation' came to be regarded as 'anti-development'.

The budding environmental alarm of the late 1960s also extended to the exhaustion of some key natural resources. Some biologists and neo-Malthusian economists and demographers, in particular, were quick to point out the implications of rapid population growth and limited natural resources. As is often the case, panaceas were proposed. Two of these caught the attention of decision-makers and of the public, i.e., limiting population growth and limiting economic growth by adopting the zero-growth or steady-state development model for national economies.

Such ideas were not viewed favourably by the developing countries, especially as they came from the North which accounted for the lion's share of the global demand on the renewable and non-renewable natural resource base, and whose affluence, consumption and lifestyles were responsible for squandering energy and natural resources. They were seen as conflicting with their development aspirations and aimed at perpetuating their status of dependence and economic underdevelopment.

It was hardly surprising, therefore, that when the preparatory process for the UNCHE was launched in 1968, the developing countries were not among its enthusiastic supporters or keenest participants. Their reserve grew as the initial preparations proceeded. Given the very heavy emphasis being placed on pollution and environmental assessment and, in general, on environmental issues as seen through the eyes of the North, many came to perceive the Conference as an affair of the industrialized countries. In fact, public figures from some developing countries began to voice doubts that their governments would take part.

In order to secure greater interest and participation by the developing countries, and in the light of what had been learned during the initial preparatory process, the Secretary General of the Conference made a fresh attempt to define the conceptual framework. For this purpose he convened a panel of experts at Founex, in Switzerland, to clarify and define the relationship between development and environment. This was one of the milestones in the history of environment as an issue on the international agenda. It marked the beginning of a protracted effort to secure the interest and participation of the developing countries, to overcome the limitations of the initial North-based formulae, and to seek more universally appealing definitions of the environmental problématique linked to the principal concerns of development. The Founex report affected the orientation of the Conference and, ultimately, the direction and scope of international environmental action in the post-Conference period.[2]

2 The Founex report and selected background documents were reproduced in a volume entitled *Development and Environment*, Mouton, Paris and The Hague, 1972.

The panel attempted to clarify the interest and concerns of the developing countries with regard to the environment. It also laid the foundations for an integrated understanding of and approach to issues at stake. It formulated a comprehensive definition of the environment, arguing that environmental concerns are part and parcel of the development process. The experts considered that environment was not simply the biophysical sphere, it was also socio-economic structures; the two formed an interdependent and inextricable web. The debate on the environment was not concerned only with pollution and conservation, nor was the damage to the environment attributable solely to the process of development. In many cases, the damage was due to the very same socio-economic forces and causes that were at the root of poverty, underdevelopment, and inequity, and could be overcome only through the process of development, economic growth, and social change.

Thus formulated, environmental issues became more appealing and interesting to the developing countries. The conceptual framework proposed by the Founex panel also represented a more comprehensive definition of the complex relationships between environment and development, and the acknowledgement of the wide diversity of interests and environmental concerns in the global constituency to which UNCHE had to cater.

It implied an open-ended, functionally diffuse, development-oriented, and relatively politicized course. This was a rather different substantive and organizational proposition from the more specialized, technical, primarily environmental assessment approach, which was the favoured option of most developed countries.

The institutional question

At the Stockholm Conference, the question of follow-up institutional arrangements loomed large. The dominant view was that the subject matter of the environment – by its very nature multidisciplinary and trans-sectoral and defined to encompass the majority of development concerns on the international agenda – was so broad that it would not be sufficient to establish a single, specialized institution with functionally delimited jurisdictions and clearly marked fields of action. Theoretically, such an institution for the environment was possible. However, it was considered that it would be costly, large, and unwieldy, and would duplicate the mandates of other organizations within the UN family.

Two factors influenced the institutional approach adopted at UNCHE. First, the main UN specialized agencies felt that there was no need to create a separate UN body for the environment. Each having already staked out a claim to environmental aspects within its own field of concern, the agencies argued that collectively they could do a satisfactory job and that coordination should be resorted to. They lobbied against the idea of a new institution. In this they were given support by their constituencies, namely the respective national ministries, which had much the same views and were engaged in similar jurisdictional and institutional arguments on the home front. Second, in most governments there was no strong voice advocating the establishment of a major international agency for the environment. The reasons were the very uncertainty as to what was an optimal institutional approach and the fact that the environmental issue itself was rather low on the list of governments' priorities and did not have the backing of well-established domestic institutions to press its claims and champion its cause in the international arena.

In the absence of a clear-cut view and of policy motives and pressures for creating an organization endowed with a substantive operational mandate and the capabilities needed to cope with the task, an institutional compromise was chosen.[3] It was innovative to a degree, promised to deliver results, and corresponded to the substantive definitions and the conceptual framework that had crystallized in the preparations for the Conference; yet it did not disturb the institutional status quo in the UN system.

Its basic features could be summarized as follows:

- The new global organization would be a coordinating and stimulating body, encompassing various sectors and disciplines. It would energize and ensure an integrated and coordinated programme for the environment within the UN system, as well as within its member organizations, which would be directly responsible for executing its various components.
- Its secretariat would be small and would not be endowed with operational, substantive, research, or executive capabilities. By definition, these were to be provided by the specialized agencies and other bodies within the UN system in their fields of expertise and competence.
- In order to enable the secretariat to perform its functions and respond to the requirements of the broad mandate, the organization was to be equipped with a fund, adequately financed by voluntary government contributions.

The institutional blueprint provided the new organization with the breadth and scope of action required by its subject matter; this in itself was unusual in the UN system, which is vertically divided into the sectoral domains of various organizations. It held out a hope of harnessing all the components of the UN in a joint and coordinated environment programme. To reaffirm this intent, the new body was made into an organ of the General Assembly, with the title of United Nations Environment Programme (UNEP). Moreover, the post of the Executive Director was established at the level of Under Secretary-General (rather than Assistant Secretary-General, as had been proposed). The Executive Director was to be directly elected by the General Assembly upon nomination by the Secretary-General, rather than appointed by the Secretary-General and ratified by the General Assembly. This was supposed to confer higher status and a more independent political base to the Executive Director.

The coordinating but non-operational mandate, however, contained the risk of impotence and frustration. This was compounded by the proviso of a 'small'

---

3 Following the recommendations of the Stockholm Conference, the UN General Assembly took a formal decision on the institutional and financial arrangements. See GA resolutions 2997 (XXVII) and 3004 (XXVII). For the institutional paper on which the Conference based its decision, see 'International Organizational Implications of Action Proposals', UN doc. A/CONF.48/11. Also see docs. A/CONF.48/11/Add.1 and A/CONF.48/12.

secretariat, which represented an important concession to the specialized agencies. They felt that anything 'small' was less likely to interfere in their work, could not duplicate their activities, and could not undertake environment-related actions on its own.

The fund was the key ingredient in the institutional package. It was meant to counter some of these weaknesses. Its establishment was in itself quite exceptional. While the developed countries had traditionally opposed the creation of similar funds within the UN system, in the case of the environment they championed it. This showed their special interest in the new organization and their wish to see it succeed. They must have also anticipated that, as major voluntary donors to the fund, they would have an important degree of influence and control over the orientation of UNEP and its substantive programmes.

The fund was supposed to give flexibility to the new organization and widen its scope of action, free it from the straitjacket of the regular UN budget procedures, speed up its action and reaction, enable the secretariat to overcome some of the constraints of its small size and non-operational mandate, and give greater meaning to coordination. There was another view of the fund's role, however. The specialized agencies expected it to be simply a source of financing for their environmental activities for which no money was available in their regular budget, rather than a programme and policy support tool for the new organization.

This institutional blueprint – which rested on many assumptions that were difficult to realize and contained some internal contradictions – was not only to guide and shape the new organization on its bumpy road to self-fulfilment, but also to influence the very nature of international environmental cooperation in the years and decades to come.

The expectations

It is usual for the outcomes of UN conferences to be a compromise. When the participating governments leave at the end, they maintain their preferred views of what should be done in the follow-up phase and how it should be accomplished.

Although UNCHE was relatively all-encompassing in its outcome and intent, the industrialized countries stuck to their preferences for the assessment function and for pollution-related issues. The developing countries, though more interested in the subject matter than before, were not fully convinced. They continued to be suspicious of environmental assessment and were reserved on the subject of pollution. Development-related matters remained their dominant concern. They also seemed eager to obtain technical and financial assistance from the international community for launching environmental actions domestically, primarily those involving human settlements and natural resource management.

These two views were to collide in the immediate aftermath of the Conference, as each group tried to shape the fledgling organization in its own image.

Global conferences also engender expectations. Usually, a healthy dose of realism and scepticism is present, with observers and participants aware of how little international conferences can achieve and how difficult it is to implement their decisions in the real world. Stockholm was different.

Those who sensed or knew how difficult it would be to cope with the social and economic causes of environmental degradation and to mount appropriate and effective organizational responses were not really trying to be heard. UNCHE was a celebration, a happy occasion to launch a new and just cause, that of the environment. There was a great deal of enthusiasm, principally coming from the non-governmental community. Many of the supporters of the environmental cause chose to believe that things would be somewhat different this time, and that a well-organized and successful conference would lead to quick action and the hoped-for results. When implementation began, they were in for a disappointment from organizational politics, bureaucratic inefficiencies, complexities of intergovernmental discourse, and the slow pace of progress. Many quickly became impatient and disillusioned, voicing their criticisms and blaming the UN, UNEP, and the governments concerned for the failure to live up to the promise of UNCHE.

Probably no other international organization belonging to the second generation (the first being those established immediately after World War II), was launched with such high expectations. Unlike its immediate predecessors – the UN Conference on Trade and Development (UNCTAD) and the UN Industrial Development Organization (UNIDO) – which were created after a bitter conflict between the developing and the developed countries and were stymied in their work by continuing policy and substantive disagreements and opposition by the latter, UNEP came into the world basically unopposed (with the exception of some specialized agencies which lobbied aggressively to stop its creation). It had an innovative design, a fund to help it work more efficiently, and the wish of the industrialized countries to see it succeed.

In spite of all these relatively favourable prerequisites, making the institutional blueprint work according to expectations was a difficult proposition. This came to be recognized soon after Stockholm, both by the international officials entrusted with the task of organization and programme build-up, and by the governments and other actors who placed high hopes in the new institution.

INSTITUTIONAL AND SUBSTANTIVE RESPONSES IN THE WAKE OF UNCHE

One of the principal decisions of UNCHE, which assured continuity and gave a focus to global action, was its recommendation that the institutional machinery for environment be created. The Conference, however, only offered broad guidelines as to what this machinery should be like. Nor did it consider and decide on the work programme of the new body; the Declaration and the Action Plan that it elaborated were confined to providing the overall framework and the recommendations addressed to the UN system as a whole and to its member organizations.[4] This

---

4 For the decisions and report of the Conference, see UN doc. A/CONF.48/14/Rev.1, UN publication Sales No. E.73. II. A.14.

meant that the specific decisions regarding the programme content and the institutional modalities of UNEP still remained to be taken.

Two important initial decisions – one on the location of UNEP headquarters, the other on UNEP's programme priorities – were affected by North–South tensions. A third key consideration dealt with fleshing out the institutional structure recommended by UNCHE and adopted by the General Assembly, and devising practical modalities for its operation. Together, these three steps markedly affected the physiognomy and nature of UNEP and had an influence on the global international response to environmental problems in the period that followed.

The location of UNEP

When the General Assembly was approving the institutional package recommended by the Stockholm Conference, it also had to decide on the location of the new organization.[5] As in earlier instances, when UNCTAD and UNIDO were being established, the developing countries proposed that UNEP be located somewhere in the Third World. After considerable internal discussion among regional groups, in their collective stand as the Group of 77 the developing countries eventually agreed to endorse the candidacy of Nairobi as the host city. The matter was brought to a vote and adopted, with all the industrial countries abstaining and manifestly reserved about the move.[6] Thus, UNEP became the first global agency of the UN to have its headquarters in a developing country.

This was a particular disappointment to the industrialized countries, which still considered UNEP as 'their' organization. They felt that it should be located in their midst, where environmental pollution was most acute and environmental action and consciousness most advanced. More importantly, many argued that the institutional characteristics of UNEP had been designed on the implicit assumption that it would be located in Geneva. There, it would draw on the support of the existing administrative infrastructure of the UN Geneva Office, and would be in the proximity of the major UN specialized agencies. These agencies were to act as its specialized branches, which it would coordinate, and with which it would interact directly on a daily basis. A move to Nairobi, they felt, would be detrimental to

---

5   An insider's view of the events in the General Assembly relating to the establishment of UNEP is contained in an unpublished monograph by Diego Cordovez, 'The Making of the United Nations Environment Programme', 1975.

6   For the decision on the location of the UNEP secretariat, see General Assembly resolution 3004 (XXVII). For the voting on the draft resolution and the amendments involved, see the report of the Second Committee of the General Assembly (UN doc. A/8901, pp.24–37). When the principle of Third World location obtained the required majority, the developed countries attempted to postpone the decision on the location itself. They asked for a detailed feasibility and comparative cost study of different cities that had offered to host UNEP (among them, Mexico City, Madrid, London, Nairobi, New Delhi, New York, Geneva, and Vienna), to be submitted to the next session of the General Assembly. However, through skilled diplomacy and the use of procedural rules, the delegation of Kenya managed to bring the issue of the Nairobi location to an immediate vote. What favoured Nairobi, in particular, was the availability of good office space and conference facilities for immediate occupancy.

UNEP's effective performance of its organizational role. There is no doubt that the interest of these countries in UNEP was diminished by its move to Nairobi.

The developing countries, in contrast, seemed pleased with their political victory and became more favourably disposed towards the new organization to be located in their midst. They assumed that it would be more sensitive to their problems and needs. Also, because it would be physically removed from the traditional UN centres in the North, they felt that it would be less dominated by the concerns of the industrialized countries and their NGOs, a phenomenon evident during the preparations for and at the Stockholm Conference.

Locating UNEP in Nairobi represented an important institutional and policy departure in the UN system. It could not but have a series of practical consequences that have affected the evolution, work, and performance of this organization. Some of these need to be mentioned. For example, UNEP has experienced high staff turnover as well as difficulties in recruiting professional staff, owing to the reservations of many people, especially from the North, to working on a long-term basis in Africa. Also, the location meant having to build from scratch a complete organizational and administrative infrastructure. This organizational build-up and learning process was a lengthy and by no means simple task.

The location has also contributed to keeping UNEP outside the mainstream of the UN, and has had an unfavourable influence on its interaction with the specialized agencies and other supporting organizations due to communication difficulties and financial and time costs of the travel that was required. The extra travel placed an additional burden on the already small and overloaded staff.

Ironically, the Nairobi location of UNEP was not logistically favourable to the developing countries, in the sense that it has weakened their direct influence on UNEP and its work. This has occurred partly because the Group of 77 has not been able to constitute itself effectively in Nairobi as it had done in other UN centres, mostly because relatively few developing countries had embassies in this city, and those that did had as their main concern the bilateral relations with the host country. Thus, in the period between Governing Council sessions, for all practical purposes the Group of 77 has hardly functioned. In contrast, the developed countries, of both East and West, have permanent missions attached to UNEP and have functioned as regional groups on a continuing basis from the very beginning.

During the sessions of the Governing Council, the Group of 77 has also found it difficult to operate effectively. This is due, in part and in addition to the above, to the nature of the developing countries' representation on the Council, which often consists of delegates with technical background and without the necessary policy experience. Another factor has been the lack of a consistent position and real interest on their part in many issues on the UNEP agenda, and the preference for working through regional subgroups.

It is safe to assume, however, that had UNEP been located in Geneva, for example, the very fact that the Group of 77 functions there on a regular basis would have contributed to the adoption of joint positions and actions. Also, and as important, it would have linked UNEP's work more explicitly with the global development agenda. This was illustrated by the events at the first session of the

Governing Council, discussed below, held in Geneva in 1973 before UNEP moved to Nairobi. And it was confirmed almost two decades later during the debates leading to the decision to convene the 1992 UN Conference on Environment and Development. On that occasion, the developing countries were among the main advocates of holding the sessions of the Preparatory Committee in New York or in Geneva, rather than in Nairobi, precisely because in these cities they could deploy greater resources and link the preparatory process more easily with the development agenda and their particular concerns.

Deciding on programme priorities for UNEP

The Plan of Action adopted by the Stockholm Conference was a document addressing all levels of action. It included recommendations to the international community, governments, and intergovernmental and non-governmental organizations. The specific agenda and programme priorities of the newly created UNEP had yet to be elaborated, and the financial resources available in the Environment Fund had to be divided among these elements. This was the task confronting the first session of UNEP's Governing Council.

At Stockholm, the developing countries concentrated their attention and negotiating effort on the Declaration, as well as on the symbolic and policy issues and the general principles that it contained. They showed little interest in most of the Plan of Action, and were generally content to leave it to the developed countries and the secretariats of the Conference and of the specialized agencies to put the finishing touches on the draft worked out during the preparatory process. This was mainly a result of the fact that, during the preparatory process and at the Conference, the Group of 77 had virtually no platform on many of the specific issues under discussion, and did not really seem to care about them. Moreover, many of the delegates representing developing countries in negotiations on the Plan of Action were technical people or scientists who were not always sensitive to underlying policy issues related to development.

The Declaration, on the other hand, dealt with policy issues. Many of these were relevant or related to North–South controversies on trade and development current at that time – a well-known area of interest to all developing countries on which their common positions were clearly defined. Not surprisingly, it attracted more policy-oriented delegates and channelled the discussion in this direction.

Thus the developing countries – spurred by China, which had just joined the UN – reopened the discussion on a number of principles already agreed upon during the preparatory process. This eventually led to the adoption of a declaration more in line with their perception of the problems at stake. The difference in policy orientation between the Declaration and the Plan of Action is so marked that they hardly seem part of the same document, adopted by the same meeting. In fact, at the Conference one had the impression that two rather different meetings were taking place simultaneously. As it soon became apparent, it was the Plan of Action that was more important for follow-up; the Declaration, worded in very general terms and barely concealing fundamental policy and practical disagreements, made little impact and was simply shelved for occasional future reference.

It was only at the first session of UNEP's Governing Council that the developing countries turned their attention to the programme implications of the Stockholm Plan of Action, and tried to apply the policy orientation of the UNCHE Declaration to the choice of areas for priority action by UNEP.

There was a good deal of spontaneity and an element of chance in all of this. The developing countries, continuing the pattern evident at UNCHE, came to the first session of the Governing Council without a common position and expressed divergent views. In contrast, the developed countries were well organized and bent on channelling the work of the new organization into the areas that concerned them most, namely pollution and the functional task of assessment. The course of proceedings angered the developing countries and galvanized them into action as the Group of 77 which managed to present a common front in the final hours of the negotiations.[7]

Granted, the common position of the developing countries was little more than a reaction to what the other side wanted. It amounted to a reshuffling of priorities and a reallocation of resources in an initial draft resolution submitted by the President of the Council, which was largely inspired by the developed countries. None the less, it was also founded on a real difference of perception between the two groups of countries as to where the focus should be and it had important implications for the direction and orientation of the new organization.

In their statements, the developing countries put stress on drought, desertification, effects of pesticides and pest control, soil and water, water-borne diseases and health, human settlements, natural disasters, wildlife, economics and trade, industrial location, transfer of technology, additionality, and technical assistance and training. The developed countries, on the other hand, showed a marked preference for monitoring, information exchange, pollution control, transport of pollutants, oceans, atmosphere, energy, rivers, low-waste technology, recycling, standards and conventions, climate change, etc. The socialist countries of Eastern Europe – for whom this was the first opportunity to express their views as they had not taken part in the Stockholm Conference to protest against the German Democratic Republic's exclusion – exhibited broadly similar concerns to those of the western countries.

The developing countries were especially critical of the major emphasis placed on pollution monitoring, and felt that problems of particular interest to them were relegated to the background. Their view carried weight owing to their joint negotiating position. The developed countries were unhappy with the programme

---

7 This happened towards the very end of the session due to the action of the few delegates from the Geneva missions, seasoned in North–South negotiations in UNCTAD. They became progressively irritated by what they saw as a pre-scripted show by the UNEP secretariat and the few key delegations from developed countries. Up to that point, only the regional groups had been meeting. They pressed successfully for the Group of 77 to meet. The appearance of the Group of 77 was greeted with open displeasure by the developed countries. They argued that group confrontation was contrary to the 'spirit of Stockholm' and was an unwanted intrusion of the UNCTAD style and politics into the newly created organization.

priorities presented by the Group of 77. However, taken by surprise by the last-minute group action of the developing countries, they had little time to manoeuvre with the end of the session in sight. Rather than threaten the outcome of the Council and launch UNEP on a sour note, they opted for accommodation.

The priorities, as agreed to by the first session of the Governing Council, meant a considerable recasting of the direction and shape of the programme. It became quite different from what had been desired and anticipated by the developed countries and by the UNEP secretariat in the documentation it had prepared for the session. The developed countries were against ranking the areas of activity to which financial resources were to be allocated, but the view of the developing countries in favour of ranking prevailed. The first four priority areas accounted for 63 per cent of the apportioned funds. They were:

- human settlements, human health, habitat, and well-being;
- land, water, and desertification;
- education, training, technical assistance, and information; and,
- trade, economics and transfer of technology.

The programme favoured by the developed countries, namely Earthwatch, accounted for only 9 per cent of the total.[8]

This first decision of the Council was a benchmark in the early evolution and orientation of UNEP. It brought it closer to the mainstream of development concerns. It contained a more balanced and, by definition, broader picture of the interests of its global constituency. And it left the door open for the pursuit of a more integrated concept of the environment–development interrelationships formulated at Founex and implicitly endorsed by UNCHE. As such, it pleased the developing countries and the development-oriented constituency.

On the other hand, the decision alarmed the developed countries. They felt that it marked a drift by UNEP into the diffuse area of development, which was within the competence of other UN organizations. To them, it meant diluting the technical focus of the new organization and downgrading the environmental assessment component, particularly as related to pollution, which, they felt, should have been at the very core of international action, especially at the global level. It also implied spreading even more thinly the already limited Environment Fund resources and the staff of the organization.

Due to this programme shift, the new institution lost additional appeal for them, on top of their disenchantment about its location. Their initially declared interest and enthusiasm for UNEP, for doing things through UNEP and for acting on a global basis diminished. At the same time, their inclination to do things in their own forums was reinforced. Their regional organizations were not affected by the development controversies, rhetoric, and stalemate politics common to the UN;

---

8   For purposes of comparison, see the programme groups and apportionment of resources proposed by the UNEP secretariat (documents UNEP/GC/5 and UNEP/GC/8), and the decisions of the Governing Council 1(I) and 4(I), which list programme priorities for action and apportion funds for the fiscal year 1973 (UN doc. A/9025).

there business could be done efficiently by industrially advanced countries with comparable problems and with the necessary capabilities and resources to act.[9]

On a more general level, the events at the first session of the Governing Council marked the onset of a permanent tension in the fold of UNEP between broader development concerns, of which the environmental dimension is an integral part, and the narrower, 'environmentalist' interpretation of the problématique. Also, these events confirmed a certain dichotomy between assessment-related activities, on the one hand, and management-related ones, on the other.

The action of the first Council set the tone and provided the backdrop for the consensus politics between the North and the South for which UNEP came to be noted in the period that ensued. With the different groups of countries obtaining allocations of Environment Fund resources for their favoured activities within the programme, a lasting and relatively efficient basis for accommodation was formed. The fact that most of the time vital issues or critical national concerns were not involved, and that countries' positions were flexible, made this tacit division of the programme and of the financial resources possible.

Elaborating the institutional model for UNEP

Only the basic features of the new institution were outlined by UNCHE; the details of how they would be implemented in practice remained to be worked out by the fledgling secretariat of UNEP.

There was continuity with the Conference in that the Secretary General of UNCHE became the Executive Director of UNEP and the whole institutional compromise and model devised in Stockholm strongly bore his personal imprint. However, devising and implementing practical and effective solutions to fulfil the sweeping and often ambiguous roles of UNEP was a difficult undertaking. There were no precedents or successful models to go by. Moreover, while in theory institutional packages that appeared reasonable and likely to work could be put together, it was a different matter to make them function successfully in a situation of institutional status quo and general ambivalence on the subject of the environment.

The task that had to be faced was basically twofold:

- to get a firm grasp of a broad-ranging and dynamic subject by devising appropriate institutional solutions; and
- to try to overcome the various limitations imposed on the organization by its own restrictive terms of reference (in particular, the 'small, coordinating, non-

---

9 This latent but widely shared view was articulated on the eve of the Stockholm Conference. It was proposed that a small group of 10 or so leading industrialized and maritime nations (which are the main cause of environmental problems and have the means to correct them) should constitute themselves into a 'club for the preservation of the natural environment' and create an entity (International Environment Agency) charged with performing functions they would define and acting in the best interests of the world. See George F. Kennan, 'To Prevent a World Wasteland', *Foreign Affairs*, Vol. 48, No. 3 (April 1970).

operational and non-executive' secretariat), as well as by the nature of the UN system and the general institutional environment within which it had to operate.

This turned out to be a permanent challenge and preoccupation of UNEP's secretariat and Governing Council. Probably, more than any other organization in the UN system, UNEP has been engaged continuously in devising, revising, and explaining its role, performance, and various institutional schemes.

Immediately after the Stockholm Conference, the UNEP secretariat elaborated the concept of a three-level programmatic and catalytic approach to respond to the basic prescriptions of the Stockholm institutional model and to make the best of its own ambiguous position. The scheme was in line with the premises of General Assembly resolution 2997(XXVII). With some modifications and adjustments, it has guided UNEP over the years. It must be taken into account in trying to understand how it works and to explain its performance:[10]

- *Level One* of the programme approach refers to the information-gathering function, i.e., UNEP's acquisition of relevant knowledge and overview of environment-related matters, policies and actions (e.g., the state-of-the-art on a given subject; review of the state of the environment or of a specific environmental problem; inventories of organizations, of research in progress and activities undertaken, of financial resources allocated to environment-related actions, etc.). The principal purpose is to provide UNEP's secretariat and the governments with the vantage point necessary for the fulfilment of policy-making, coordinating, and programming functions; for example, by identifying critical and emerging problems and trends, progress made, gaps in action and research. It presumes effective collection, processing, evaluation, and continuous updating of pertinent information. Also, it presupposes that such information will be used as a basic building block in policy formulation and decision-making.
- *Level Two* refers to the environmental programme addressed to the UN family of organizations and to governments. It denotes both the policy-making and programming functions, as well as implementation and practical action. Based on inputs from the secretariat and acting as the central UN organ for the environment, UNEP's Governing Council is supposed to decide on objectives, strategies, policy guidelines, programmes, and programme elements. These should provide the framework and guidance for the UN system and for governments, which should then proceed to translate them into corresponding action. UNEP is not supposed to be directly engaged in implementation and action, except in those instances in which its direct involvement facilitates or encourages

---

10 Although the scheme is conceptually simple, it has caused some misunderstanding, particularly among governmental delegates. They were slow to understand or were not receptive to distinctions between programmatic levels and their organizational implications. For the first explanation of what eventually became the programmatic approach, see the speech of the Executive Director to the first session of the Governing Council (doc. UNEP/GL/L.10). The first systematic presentation of the concept is to be found in the programme report of the Executive Director to the second session of the Governing Council (doc. UNEP/GC/14). See also Governing Council decision 20(III) in UN doc. A/10025 and decision 47(IV) in UN doc. A/31/25. For a discussion of the catalytic role of UNEP see doc. UNEP/GC/82.

the action of others. This is where the so-called *Level Three* comes into the picture.
- *Level Three*, also dubbed the 'Fund Programme', refers to those components of *Level Two* where UNEP uses financial resources from the Fund to help catalyse – that is, initiate, promote and/or sustain – given activities aimed at contributing to or fulfilling particular objectives and goals. *Level Three* is in fact a euphemism for Fund-supported activities and projects, which have been the main tool of UNEP in asserting its organizational role and in advancing environmental objectives.

A good deal of confusion and controversy has arisen because of the non-operational and non-executive nature of the UNEP secretariat and because of the need to entrust the execution of projects to others. This is why the concept of UNEP as a 'catalyst' was launched. It has been used to convey the idea of UNEP gaining access to a given activity and influencing its content and direction, or helping to set up and launch a given action with the help of its financial resources. According to UNEP's definition, its financial support is temporary. It should last only until a given activity becomes self-sustainable, to be carried on independently by the organization in charge. In other words, at this point 'pure' *Level Two* takes over, with UNEP moving to other projects and activities, using its financial resources for further catalytic actions. In this respect, UNEP's understanding of its role has been at odds with the views of the specialized agencies. They feel that the resources from the Fund should be provided to them on a permanent basis. Indeed, the agencies have often pointed out, tongue in cheek, that for a chemical reaction to be sustained a catalyst needs to be present continuously.

UNEP's institutional story has basically revolved around efforts to make this logically simple yet logistically demanding superstructure operational. As such, it is of key importance in understanding how UNEP has gone about pursuing its substantive objectives and programmes, including the Global Environment Monitoring System (GEMS).

THE UNEP ERA: MODEL AND REALITY

The substantive and institutional package conceived at UNCHE and further elaborated in UNEP, and the expectations regarding its functioning and performance, were based on several interrelated assumptions. Institutionally, there were primarily two:

- that the UN organizations and the governments would accept the central coordinating role of UNEP in the environmental field, and that they would implement various elements of the global programme elaborated and agreed to in UNEP and allocate the necessary resources for this purpose; and
- that UNEP would be capable of satisfactorily performing the tasks inherent in the very nature of the subject and the various roles envisaged for the organization, in particular that of global coordination and leadership.

18 *The quest for world environmental cooperation*

Substantively, it was expected that national institutions, and those working in them, would be responsive and adaptable to the thinking and requirements generated by the environmental issues and their multidisciplinary, multisectoral, and often longer-term implications. Furthermore, in 1972 it was assumed that environmental issues would occupy an important place among national objectives, and that countries would assign to environment a corresponding weight in their international actions.

This section takes a brief look at how the above premises fared in practice.

Environment and coordination in the UN system

The structure of the UN system is based on specialization and sectoral responsibilities. The interrelationships among subject areas have been downplayed, efficient mechanisms to deal with them have not been created, and attempts to analyse their implications have been traditionally avoided or discouraged. The secretariats of international organizations, normally bent on preserving and strengthening their jurisdictional domains, only helped to reinforce these tendencies.

Environmental issues made an important contribution to the evolution of the UN by highlighting the need to devise integrated and multidisciplinary approaches for their management and solution. The terms of reference for the new environmental organization were based on this institutional logic. Environment, and UNEP, brought into the UN the need for a qualitatively new approach to problems on the international agenda. It represented the beginning of a long and slow learning and adjustment process.

Indeed, 1972 marked a watershed in the history of the UN. At least in theory, it was agreed that an integrated approach was called for to deal with complex, global issues such as the environment. An integrated approach meant looking at the very nature of the system, questioning its basic premises, and recognizing the main cause–effect linkages.

A systems or integrated approach to contemporary international issues troubles decision-makers and existing institutions, and presents challenges of intellectual grasp, training and education, planning, management, implementation, and institutional coordination. These tensions and contradictions have appeared at all levels of UNEP's work, as the organization has attempted to develop an overall and integrative approach and, in particular, ways of translating this approach into practice. One such attempt has involved novel forms of coordination within the UN system.

Traditionally, coordination in the UN system has been vertically divisive, avoiding duplication and protecting individual jurisdictional domains from encroachment by others. This type of coordination was ill suited to the multidisciplinary and integrative nature of environmental problems. Instead, a horizontal link-up was required that would introduce the environmental dimension into each sector and connect individual components into an overall strategy and programme. It implied a positive approach, cooperation, joint programmes and projects, new perceptions, and dynamic and imaginative people.

In theory, it is the governments that should give the impetus and provide the guidance for such coordination. In practice, they have rarely done so. The conflict surrounding environmental issues domestically has prevented most governments from assuming a clear-cut and consistent policy or institutional stance on environmental matters within the UN system. Governments' potential contribution to system-wide coordination is eroded, moreover, by the complexities and wide scope of the subject matter, the usual opacity of inter-agency politics, and the opposition of vested interests within the system. In practice, then, the responsibility for coordination has fallen mainly on the shoulders of the secretariats of international organizations, and coordination has been primarily an intersecretariat affair.

The specialized agencies play the key role. Their views and preference on institutional approaches have remained fundamentally unaltered since Stockholm. They did not see any compelling reason for the creation of UNEP; once it was there, as far as they were concerned its main function should have been to provide continuing financial support for their environment-related programmes and activities for which they were not obtaining regular budget financing. They wanted also UNEP to interfere as little as possible in their affairs.[11]

The UNEP secretariat, on the other hand, fully aware of the resistance and institutional constraints, tried a series of approaches to activate and impose new types of coordination. It formulated a number of statements on the problem as a means to persuade and proselytize. One of its early documents spells out its role as follows:

> UNEP is needed because sectors think and act sectorally and disciplines think and act within their own confines.... Difficulties have arisen from, and been marked by, the failure of sectors and disciplines to consider and act in concert with other sectors and disciplines. Someone is needed to consider, to be the focus for, and to lead action in matters that are, by the systems nature of the environment, cross-sectoral and interdisciplinary.[12]

Starting with the creation of the Environment Coordination Board (ECB) by General Assembly resolution 2997 (XXVII), numerous mechanisms have been

---

11 A historical footnote is of interest here. The UNCHE resolution on institutional arrangements speaks of the 'Governing Council for Environmental Programmes', rather than a singular 'Programme'. This was part and parcel of the compromise at Stockholm; it was an effort to appease the specialized agencies that wanted to obtain and/or maintain recognition for the identity of their activities. They also opposed 'a programme', which implied a single programme managed by the new organization. While the original text of General Assembly resolution 2997 (XXVII) contained the same wording, the endings 'al' in 'environmental' and 's' in 'programmes' were eventually dropped by the UNEP secretariat, as an 'editorial change' after consulting the principal sponsors of the resolution. Following a query from a delegation at the first session of the Governing Council, the Executive Director answered that the word 'environmental' translates badly into certain official UN languages, and that the use of indefinite plural 'programmes' deprives the new institution of the identity that is conveyed by the definitive singular 'programme'. It was also pointed out that the original wording caused the UN Secretary-General to nominate the first Executive Director without specifying the organization! Had he said 'the environmental programmes', it could have been interpreted as meaning that the Executive Director was directing the programmes of the agencies.

12 See doc. UNEP/GC/31, pp. 1–4. See also Maurice Strong, 'One Year After Stockholm: An Ecological Approach to Management', *Foreign Affairs*, July 1973.

devised and tried out in an effort to give meaning and content to coordination and joint action.[13] Embedded in the three-level approach, they included joint programming with individual agencies as well as joint thematic programming with several of them, System-Wide Medium-Term Environment Plan (SWMTEP), the adoption of resolutions and the elaboration of programmes at the intergovernmental level addressed to the UN system as a whole and its member organizations, regular meetings of Designated Officials on Environmental Matters (DOEM), etc. To promote some of these approaches, the financial resources of the Fund were used, for example, to make interaction possible among the agencies concerned, and to support projects which could serve as nuclei for larger programmes and networks.

In contrast to the notion of 'executing agency' used by UN Development Programme (UNDP), UNEP uses the concept of a 'cooperating' and 'supporting' organization. This is meant to highlight a difference: the supporting organization is not supposed to execute projects for or on behalf of UNEP; rather, it is engaged in a common programme and a joint endeavour. In line with this notion, the UNEP Fund is not supposed to finance more than 50per cent of the cost of a given project and does not pay for overhead costs. Overhead expenses cover the support services required by the extra-budgetary programme activities; UNEP's position is that, since it contributes to regular programmes of the agencies that are supposed to be financed from their regular budgets anyway, by definition there is no need for it to pay for overheads. These stipulations have prompted complaints from the agencies and other UN bodies, accustomed to executing projects on behalf of UNDP, which pays for their services and for the overhead costs.

In practice, coordination of the kind conceived by UNEP is difficult to attain, especially at the global level. Its scope and effectiveness hinge not only on the resources available, and on the attitudes and willingness to participate and learn of staffs of various organizations and government representatives, but also on the capacity of UNEP and especially its secretariat to lead, influence, persuade, manage, direct, and give substance to the coordinating task. The process has been a slow and difficult one, full of tension and frustration. Yet the partnership has held and progress has been made essentially through mutual dependence, i.e., agencies depending on UNEP's financial support and UNEP depending on their substantive and operational capabilities. It has also been forced upon all by the very nature of the subject matter, which requires different and new modes of cooperation. Throughout, the monies available in the Environment Fund have played a central role in binding the process, channelling the work of the agencies, projecting the influence of UNEP, and enabling it to launch various initiatives.

---

13 The creation of the ECB was a special concession to the new institution and an attempt to boost its coordinating role and importance within the UN system. Its existence was resented by the specialized agencies, in particular. They did not see any reason why the environment should have this privileged status, and why it should not be treated, like other subjects on the UN agenda, in the standard framework of the Administrative Committee on Coordination (ACC). They saw ECB as an attempt by UNEP to impose upon them some type of guidance and control. They pressed for its abolition, and deliberately downgraded the level of representation at its meetings. Eventually, they managed to have ECB abolished and its functions taken over by the regular ACC machinery.

UNEP: performing its roles

As noted earlier, 'small, non-operational secretariat' was a key phrase of the Stockholm institutional compromise, and was a concession to the specialized agencies and their supporters. To them, a small, non-operational secretariat was less of a threat and less likely to interfere in their jurisdictions and to undertake or duplicate activities that they felt they should perform. It turned out that this particular stipulation, when interpreted too literally and in a restrictive manner, was one of the weak links in the process and structure conceived at Stockholm.

As concerns the size criterion, everyone implicitly seemed to doubt that being small was a realistic proposition for the secretariat of a major global organization whose mandate was to range over virtually the entire spectrum of disciplines and jurisdictions. This was so even if its mandate were to be limited to a coordinating and catalytic function, which, although it required fewer people than the operational role, was none the less demanding and called for an adequate number of qualified staff. However, the governments of the developed countries were firm in their desire to avoid upsetting the specialized agencies and the institutional status quo. The notion of a small secretariat also met with general approval among governments who traditionally have not looked with enthusiasm at mounting financial costs for the upkeep of a continuously growing international bureaucracy.

To allay the fears that a secretariat of 30–40 people could not do a proper job and to reassure the environmental constituency, the concept of excellence was brought into the discussion. It was argued that a small, highly qualified and motivated secretariat could perform the coordinating and catalytic function with success. The Secretary General of the Conference is said to have argued that, with '18' top-notch professionals, the new organization could fulfil its mandate successfully – provided it were to obtain the envisaged degree of collaboration from the rest of the UN system, which was the basic working assumption of the whole structure. The number came from his business career, where at one point he had run a large and diversfied corporate structure with an 18-member board. While the analogy may not have been the most appropriate one in view of the fundamental difference in objectives, structure, and institutional prerogatives between a UN organization and a corporation, it had the merit of highlighting the crucial importance and potential of high-quality leadership and motivated staff in the United Nations endeavours.

On a more practical plane, however, those in charge of the Conference were quite aware of the limitations of a small secretariat. They sought and obtained concessions from governments that would be important in the period that followed. Taking advantage of the positive feelings towards UNEP in the wake of Stockholm, they secured approval for high-level posts for the new secretariat proportionately far in excess of the number of such positions in the secretariats of some larger organizations of the UN system. This was supposed to attract good, high-level people to the secretariat, thus helping it to fulfil the criterion of excellence. It was also meant to provide an adequate management capability for a secretariat that was bound to grow as time went by. This growth was to be attained – and here came the second concession – by using the money from the Environment Fund to pay for secretariat posts that might be needed in addition to the very few such positions provided for in the regular UN budget. It can be argued that this concession was contrary to one of the basic premises of the institutional package and correspondingly diminished the resources that would be available for other purposes. However, it left open the possibility of expanding the secretariat in order to cope with the organization's ambitious mandate and increasing needs.

Thanks to the Fund, then, the secretariat of UNEP has grown considerably over the years. However, most of this growth has taken place in the administrative and support structure for the establishment of self-sufficient organizational headquarters in Nairobi. The Programme Bureau – which, with its 50 or so professional staff members, is supposed to act as the core and brain of the organization – represents less than 10 per cent of the total and is smaller than the average department of a specialized agency.[14]

For all practical purposes, then, and where it really counts, the secretariat has remained seriously understaffed. A single programme officer would be placed in charge of a substantive issue (e.g., human health or ecosystems). He or she would be expected to perform many functions simultaneously, such as keeping up with the substantive and policy developments in the field, interacting with the specialized agencies, managing projects and preparing for intergovernmental meetings, and all of this on top of being involved in administrative affairs and in-house politics.

Excellence is meant to make up for smallness. In the case of UNEP, an additional challenge is to have staff who, in the words of the first Executive Director, should be skilled at operating 'on the interface between science and technology and public policy' and be able to 'initiate, influence, cajole, orchestrate, integrate and monitor'. The difficulty of finding the right kind of people is compounded by the fact that contemporary organizations, including educational institutions, do not offer sufficient breadth of knowledge and perception required by the subject matter of environment dealt with in a global setting. The right kind of people for the key jobs in UNEP are few and far between, and do not necessarily find their way into this organization. Those who did have usually experienced a good deal of frustration. It is quite revealing that, of the top-level people who have joined UNEP from and later returned to distinguished public or private careers, few have been successful within the organization. This, of course, can be explained to a degree by the very nature of UNEP and of the subject matter that it deals with, and the resulting frustrations of trying to fulfil the requirements of its substantive and institutional roles.[15] On the whole, professional excellence has not proven to be one of the strong

14 In 1982, ten years after UNCHE, the UNEP secretariat consisted of 427 staff members, of whom almost 300 were support staff. Of the total, 140 staff members were professionals, of whom only 44 worked in the Programme Bureau. Of the 44 officers in the Programme Bureau only 14 were charged to the regular UN budget; the others were placed against temporary posts established with the resources from the Environment Fund.
15 In the early years of UNEP, a special effort was made to bring in individuals from outside the UN system who had varied backgrounds and distinguished careers in different fields and disciplines, within and outside government. The effort was not always successful due to difficulties in persuading

points in UNEP's experience. The problem was accentuated by the smallness of the staff which did not allow any margin for error; once on board, there was not much that could be done about a staff member who did not possess the necessary quality or motivations.

As regards the 'non-operational' proviso, which likewise placed a substantial constraint on the institutional development of the new organization, the early leadership of UNEP succeeded in securing an opening for the secretariat to assume a more direct role in the implementation of environmental programmes. This was achieved at the first session of the Governing Council, when the governments were presented with and approved the concept of Programme Activity Centres (PACs).[16]

The PACs were designed to enable the UNEP secretariat to lead and manage various programmes recommended by the Stockholm Conference, especially those in areas where more than one specialized agency had an operational claim and/or in those areas where an institutional vacuum might exist. They were also meant to give such programmes continuity and institutional identity, to endow them with a degree of administrative autonomy within the UNEP secretariat, and to leave open the possibility of locating them away from UNEP headquarters and alongside specialized agencies. Specifically, they were to give a boost to the three fledgling Earthwatch programmes, i.e. GEMS, INFOTERRA (in those days called International Referral System or IRS), and IRPTC (International Registry of Potentially Toxic Chemicals), all of which were given a PAC status from the very beginning.

This played a very helpful role in their development. IRPTC was located in Geneva, while GEMS and INFOTERRA were set up in Nairobi. Their separate identity and considerable autonomy within the UNEP secretariat and, in particular, their direct control over their respective budgets gave the PACs an operational edge and flexibility, shielded them from intrasecretariat problems, and in general made it simpler for them to pursue their objectives.

As UNEP developed its operations, the Environment Fund played a key role. The creation of the Fund represented the linchpin of the institutional solution agreed to at UNCHE, and was meant to give UNEP the organizational clout and flexibility that it needed.

Approximately $100 million was available in the Fund during the first five years; this grew to $150 million in the following quinquennium, reflecting increasing inflation, rising costs, and the fall in the value of the US dollar. This may have been little compared to the real needs, especially after the available resources were divided between the various programmes, priorities, and activities, and financial support extended for the growth and operation of the UNEP secretariat. However,

some people to come to work in Nairobi, the pressures from the UN administration to rely on officials from within the system, and so on. While some outsiders brought in flair and innovative ideas, they were usually not at ease and were not particularly efficient working within the existing UN setting. The experienced UN civil servants were often more efficient as far as the workings of the system were concerned. However, they were not necessarily the most dynamic and innovative people. This problem was accentuated by the fact that the best of the UN élite were not ready to jeopardize their career development by joining an organization located away from the principal UN cities.

16 For the PAC concept see docs UNEP/GC/L.18 and L.20, UNEP/GC/14, pp. 38–46, and decisions 5(II) in A/9625.

it greatly helped to expand UNEP's capability and influence, it is more than most UN organs have at their disposal, and it has had a major impact on UNEP's organizational evolution and performance.

A high point in the history of the Fund occurred during the first session of the UNEP Governing Council when the rules of the Fund were adopted. Indeed, it can be said, with the wisdom of hindsight, that the adoption of these rules was one of the decisive moments in the organizational history of UNEP.[17]

Draft rules prepared by the UNEP secretariat were aimed at maximum flexibility *vis-à-vis* the governments as well as the UN administration at the Headquarters. They gave the Executive Director an opportunity to act and react promptly, to respond, innovate, catalyse, pursue ideas, proselytize and publicize. When the governments began to review the draft financial rules, a few were annoyed at what was being proposed. They would have preferred to see a UNDP model applied with an intergovernmental committee reviewing and approving the projects periodically.

Initially, it was the East European countries that argued for tight control of the Fund by governments. Western countries were behind the proposal of the Executive Director, reflecting their high esteem and trust in him following the Stockholm Conference. In general, they agreed with the US position voiced at UNCHE that the Executive Director should act like the general manager of a corporation, and the Governing Council like a board of directors. The latter, though it would have the ultimate say, would not be able to tell the manager in detail how to perform his job. The developing countries did not have a clear position on the issue.

As the session progressed, and as the shift in programme orientation took place following the pressure from the Group of 77, the western countries seemed to change their mind and to begin to see some merit in the earlier argument of their East European counterparts. The organization appeared poised to move along a different programmatic and political track from the one they had envisaged; also they began to suspect that the secretariat might be too sympathetic to the Third World views. At this point, the Group of 77 moved solidly behind the draft rules of the Fund. This was not only on account of their greater sympathy for the organization due to its changing programme orientation and the secretariat's behaviour. It was also because they came to believe that any kind of intergovernmental watchdog committee would eventually be dominated by the major donors, and that projects and activities of interest to them would have a better chance with a favourably predisposed Executive Director of UNEP in charge of the Fund. The shift in their position was crucial in the intricate negotiations that led to the adoption of the rules of the Fund. Thus, the North–South conflict played an important role in yet another key facet of UNEP's organizational history.

The basic argument was that a new and innovative type of organization was being created, and that it needed the capability and instruments to fulfil its mandate. The Executive Director required flexibility to act and respond. He was to be accountable for his actions to governments and to the Council. The notion of 'accountability' won out over the more traditional approach of 'previous authori-

---

17 See docs A/9025, and UNEP/GC/23 and 37.

zation'. Thus, with the approval of the Fund rules, a basic block was built into the UNEP structure, giving those in charge of the organization both a privilege and an opportunity.

Ever since, there has been a persistent tension between these opposing approaches, which embody rather different organizational perceptions. In practice, governments give very broad guidance to the Executive Director at the sessions of the Governing Council, by endorsing (or modifying) the programme actions and financial implications presented by him for approval, i.e., the allocations for a dozen or so budget lines representing major priority programme areas. The Governing Council grants the Executive Director a flexibility of plus or minus 20 per cent within each budget line, which enables him to adjust allocations to changing needs and situations. He is also given resources for programme support costs, which provide him with additional flexibility and a broader scope of action. The governments do not take part in the formulation, approval, or review of the projects or in the decisions as to how the resources within budget lines are parcelled out. This has been a sore point with some of them, unhappy at not being able to scrutinize and control the project approval and execution.

The Fund resources have made it possible for UNEP to be heard and to influence other organizations within the UN system, to initiate or support projects and programmes, to pursue independently lines of action and specific activities within its programmes, to play its catalytic role, to respond to needs as they arise and, in general, to avoid the straitjacket of the regular budget. Executive heads in the UN system who depend on the regular budget and who often lack even the flexibility to execute a study not proposed and approved when the budget was adopted, or to create a post outside the budgetary allocation, or to hire a consultant outside the resources provided for this purpose must envy the Executive Director of UNEP for this latitude of action and comparatively little financial accountability.

As noted above, the specialized agencies and other UN bodies did not endorse the notion of catalytic funding, and felt that the Fund should provide for their environmental programmes and projects on a continuing basis. The governments, as represented in the governing bodies of the specialized agencies, contributed to this state of mind. Rather than providing for such programmes to be funded from the regular budgets of these organizations, they preferred to have them supported by the Environment Fund. This has tended to reduce resources available in the Fund for new and catalytic activities by committing a growing portion of the funds on a more or less permanent basis to given activities beyond the four-year period that such support is supposed to last.

The developing countries also saw the Fund as a source of additional financing for their environment-related activities. These expectations were a reflection of real needs and have resulted in continuing demands on UNEP, in spite of the repeated clarification of Fund roles and capabilities. Indeed, the creation of the Environment Fund by the UN General Assembly was largely taken as an act absolving the development financing institutions of responsibility in this domain. In the absence of support from these institutions for environmental actions, such demands had to be responded to. While hardly covering even a fraction of the actual needs, this outflow of resources had the effect of reducing the Fund's scope of action. It also helped to perpetuate the uncertainty as to the Fund's roles, kept many unhappy because it could not deliver what they expected, and reduced the pressure on development assistance organizations to accommodate these needs.

The voluntary nature of the Fund must also be reckoned with, not only because of the effort needed to get the governments to contribute regularly, but also because it makes UNEP's management, programmes, and policies vulnerable to pressures of major donors. By promising greater contributions, by a slowdown in contributions, or through threats of reducing or ceasing to pay into the Fund, they can exert influence on the direction of the programme, on projects, on the policies of the secretariat, its staffing, etc. This has been the case in particular with the United States which, at one point, was contributing as much as 40 per cent of the total Fund resources. It did not hesitate from the very early days of UNEP to apply this type of leverage, an approach which it later generalized and extended to the specialized agencies and even to the regular budget of the United Nations.[18] These pressures led to important changes in UNEP's programmatic orientation, namely the toning down of its advocacy stand as concerns structural aspects of environment–development relationships and the negative consequences for the state of the environment in the South generated by the workings of the international economy.

The Fund has also to be managed, and moneys used and spent. This is no small task, and has had important effects on the UNEP secretariat. It has burdened the small staff and often complicated intrasecretariat processes and relations.[19] It has detracted from UNEP's role of strategy and policy formulation and substantive

---

18 The US announcement in 1976 that it would reduce its pledged contribution to the Fund created confusion and a slowdown in the planned programme. While the official explanation was that its move was part of a financial austerity plan caused by domestic economic difficulties, it was widely interpreted also as a sign of displeasure with the UNEP management. This displeasure began to mount after the Cocoyoc Declaration was issued by a seminar on Patterns of Resource Use, Environment and Development, jointly organized by UNEP and UNCTAD in 1974 in the wake of the adoption of the resolution on the New International Economic Order by the UN General Assembly. The US administration felt that UNEP had no mandate to be concerned with the broader development issues, which were already taken care of by other competent bodies, and should concentrate on environmental matters. In 1981, UNEP was shaken once more when the US Administration, reflecting domestic compulsions and its reduced emphasis on environment, proposed that the US contribution to the Environment Fund be terminated. The threat led to a freeze in activities and recruitment, uncertainty about the projects supported by the Environment Fund, and disturbances in the Programme.

19 The division of the secretariat into two equal parts, namely the Programme Bureau and the Bureau for the Fund and Administration, was envisaged originally as a scheme of checks and balances in the secretariat. In practice, it has not worked all that well and has been a cause of some internal difficulties. The Programme Bureau is supposed to deal mainly with substantive, programme, and policy issues; the Bureau for the Fund is supposed to administer and follow up the projects supported by UNEP, which are part and parcel of the larger programme framework conceived and managed by the Programme Bureau. As it soon turned out, the Fund money came to be the main tool with which UNEP could promote its objectives, and *Level Three* became the centrepiece of UNEP's action. The theoretical and practical distinctions between programme and projects became blurred, and the officers of the two Bureaux spent a great deal of time trying to sort out in-house and jurisdictional problems. To the Bureau of the Fund it appeared that the Programme Bureau was interfering with financial and administrative matters, while the latter felt that the former was appropriating too much control over the resources and was acting as a separate Programme itself.

coordination, and oriented it mostly to project funding and the activities at the micro-level.

In many instances, it has pushed the secretariat into the posture of a passive funding organization. There are two principal causes of this. One has to do with the over-extension of a small staff, i.e., too few people vs. too many projects and too many different subject areas. The burden is compounded by the attendant administrative complexities, which often consume much of – and in some instances most of – the time of the substantive staff, leaving it little chance to delve in depth into substantive issues or to follow the evolution and implementation of the various projects it may be responsible for. The second problem has to do with the gaps in the expertise and specialized knowledge required to deal with a number of projects and activities supported by the Fund. Lacking an adequate capability numerically and often qualitatively in all the fields on its agenda, UNEP is not in a strong position in its relations with outside organizations.

The Governing Council is the master of UNEP. Rather similar to the Stockholm Conference, the Council at its first session started with considerable excitement and publicity, high-level and expert participation, good representation of the scientific and NGO communities, a lively political exchange and substantively strong debate, and a promise of being fully in command and on top of its assignment. However, for a variety of reasons it did not live up to its promise.

Most important, in the immediate period that followed, environmental issues within the UN failed to become a priority concern of governments. This was reflected in the lowering of the level of representation and in the lack of interest in UNEP itself. Also, routine and organizational detail took away some of the original élan from this environmental gathering. Finally, on account of travel costs, the Nairobi location discouraged many of the non-governmental groups from attending, thus weakening an important dimension in the political process.

The Governing Council was engaged in a permanent search for a role and for self-assertion. In part, the problem was of a subjective nature; the Council, facing the UNEP secretariat, seemed unable to gain the upper hand and often appeared to act as a rubber stamp, endorsing ongoing and proposed programmes, and giving broad authorization for how the Fund resources ought to be spent. To some extent, this was a consequence of the Council's tactical disadvantage in that it was unable to have adequate insight into and control of the complexities and intricacies of a broad and varied Fund-backed programme. Because of the rules and procedures guiding project management and execution, the Council thus found itself on the margins of the project cycle and process, though this happens to be the chief mode of UNEP's influence and operation.

Several attempts to modify the procedures and to establish some sort of intergovernmental control over project formulation and execution, specifically by setting up a 'watchdog' committee, failed to get off the ground. Many countries were indifferent, while others did not want to disturb the laboriously negotiated institutional package approved at UNCHE and the first session of the Council. Still others thought that the trade-off was not a desirable one, and that such detailed interference by the Council would stifle the secretariat and the discretion and initiative of the Executive Director. This lack of operational involvement on the part of the Council in the project cycle of UNEP weakened the collective control over the vital link in the process. Yet, it left plenty of scope for those governments with political and financial clout and interest to intercede directly with the secretariat.

The Council also found itself frustrated in its attempts to perform its coordinating role because the governing bodies of other organizations in their work seemed to pay little heed to its recommendations. The guidelines and exhortations addressed to other organizations (i.e., *Level Two*) seldom got beyond UNEP reports. The UNEP secretariat could do little with such statements, either as directives or as a tool of persuasion. Few governments seemed to make a connection between what they decided in the UNEP Governing Council and what they instructed their delegations to do in the governing bodies of the specialized agencies.

In sum, then, the institutional model of UNEP embodied a series of favourable traits, combined with a number of constraints that tended to weaken its potential. The way the model has worked in practice – the manner in which the Governing Council and the secretariat have performed their roles and the way in which the Fund has been used – provide the essential backdrop for understanding the follow-up to the Stockholm Conference and international action in the domain of environment.

Changes in the broader setting

UNEP's trajectory and performance were also influenced by trends in the global setting, and the impact of these on the approaches to environmental problems.

The Stockholm spirit was not a true reflection of reality. There was a good deal of skilful publicity in the preparation and holding of the Conference, and in the promotion of its outcome, greatly assisted by the interested and friendly media in the North which otherwise tend either to ignore UN doings or to report on them in an unfavourable light. This helped to dramatize the issues and give them an aura of urgency and importance which they did not have among governments or among major interests and actors on the national and international scenes. Once the initial enthusiasm faded, and the frustrating and slow follow-up process was engaged in, those who expected quick and effective responses were in for a disappointment.

The period following the Stockholm Conference was a troubled one for development, environment, and the international community. It all started with the socio-economic fallout from the increase in the price of oil, which, while highlighting the relevance of environmental issues, in practice led to their being sidelined because of the serious perturbations in national economies which ensued. One of the difficulties was how to get governments to incur the necessary financial costs given the diffuse, non-quantifiable, and usually longer-term social and economic benefits of many environment-related actions, which could not be justified in terms of standard economic criteria and cost–benefit analysis.

The conservative governments that came to power in some key developed countries led to a reversal of the earlier domestic policies on environment, in that they opted for a markedly lower profile of the state in dealing with these problems and relied on economic incentives, and the 'spontaneity' of the market and societal processes. This was reflected also in a reduced role of these countries in the international discussions and action regarding environment, feeding on their negative attitude towards the notion of multilateralism and the UN.

The major slippage, however, was experienced in the Third World. To begin with, this overall economic downturn had an impact on the international approach to conditions in developing countries. It was quite clear at UNCHE that most of their environmental dilemmas, i.e., those related to desertification, tropical forests, environment-borne diseases, soil degradation, human settlements, were in fact hard-to-crack, structural and long-term development issues, requiring, in addition to national efforts, parallel international measures and support. Such assistance was anticipated by the Conference, or at least by the developing countries in interpreting its outcomes. However, this support failed to materialize. Indeed, the situation became worse as the international macro-economic context deteriorated and development cooperation foundered.

The debt and development crisis that swept most developing countries in the 1980s played the key role not only in further downgrading environment as a policy concern but also in worsening markedly the state of the environment in the South. The debt servicing outflows, deteriorating terms of trade, and structural adjustment policies imposed by the IMF resulted in large transfers of resources from South to North; the many consequences of this included mounting pressures on natural resources, a deteriorating urban situation, and worsening air and water pollution in most developing countries.

In sum, then, the period after Stockholm witnessed an aggravation and multiplication of environmental problems, and a lack of appropriate responses on the part of most societies, which were embroiled in serious economic and social crises, experiencing discontinuities in their development and facing a hostile international setting. Unavoidably, all this was reflected in UNEP, directly and indirectly. It hampered the anticipated progress in most domains. It contributed to UNEP's being marginalized and to erosion of its institutional position, the process already initiated when it moved to Nairobi and when the developed countries began to deal in their regional forums with many environmental issues to avoid the distraction of UN development politics. UNEP's position was not helped either by its continuing groping for an effective institutional role, or by the high expectations and promises that were usually not matched by the practical achievements of its secretariat, whose image suffered in the UN system and among its broader constituency.

The tenth anniversary of the Stockholm Conference reflected some of this frustration.[20] There was a broadly shared feeling that the situation was unsatisfactory and that UNEP was faltering in its assignment. This gave rise to an initiative

20 Among the UNEP-prepared evaluation reports on what had happened in the decade since the Stockholm Conference, see 'Review of Major Achievements in the Implementation of the Action Plan for the Human Environment' (doc. UNEP/GC(SSC)/INF.1), and 'The Environment in 1982, Retrospect and Prospect' (doc. UNEP/GC(SSC)/2). Both documents were prepared for the 1982 Session of a Special Character of the Governing Council, convened on the tenth anniversary of UNCHE. For the discussion at the session, its report and its recommendations addressed to the next decade of environmental action, see UN doc. A/37/25.

that culminated in a decision in 1983 by the UN General Assembly to establish an independent commission outside the framework of UNEP. Its mandate was to take a fresh look at the 'environmental perspectives' to the year 2000 and beyond, including the strategies and ways and means by which the international community could deal more effectively with environmental concerns.[21]

The World Commission on Environment and Development (WCED) was thus created in 1985 in response to this call by the UN General Assembly. Chaired by the then Prime Minister of Norway, Gro Harlem Brundtland, the Commission produced a report entitled *Our Common Future*.[22]

The WCED report does not say anything strikingly novel in terms of its key themes and conclusions, at least for those familiar with the long-standing environment–development debate. It reiterates and elaborates the basic themes of interdependence, identity of environment and development, and sustainable development, which were first sounded in Founex, then amplified in Cocoyoc and in the early works of UNEP.

While the report is very explicit as concerns the diagnosis and documentation of environmental degradation, it is rather timid and circumspect on analysis of its causes and hence weak and often contradictory when it comes to prescriptions of how to deal with the predicament. It is almost as if the report had undergone a censorship of sorts, whereby it deliberately avoids or soft-pedals some of the central issues, such as the social values and the dominant economic logic and mechanisms of contemporary society, the wasteful and irrational patterns of production and consumption and lifestyles, the effects of the structurally biased international economic order and of the transnational enterprises especially concerning the environmental situation in the Third World, the different responsibilities, capabilities and roles of rich and poor countries, and so on.[23]

This should not be suprising to those who know how reports of international commissions are written and pieced together as consensus documents, especially when these commissions and their secretariats assemble a divergent group of people, from the North and from the South, and ranging across the political spectrum. And with all of its members coming from the establishment, it is no wonder that the Commission did not have the strength or inspiration to probe such controversial issues which could annoy the power structure or which deviated markedly from the accepted orthodoxy.

The WCED report, however, has to be seen in the context of the period when it appeared. Although still a few crucial steps away from explicit recognition of the global web of socio-economic cause-effect relationships that drive the world and define the context of environmental degradation, the report represents a refreshing policy advance relative to the traditional, technocratic and single topic oriented

21 See UN General Assembly resolution 38/161.
22 The World Commission on Environment and Development, *Our Common Future*, Oxford University Press, Oxford, 1987.
23 For a critique of the WCED report and of its inner contradictions see H. Pietila, 'Environment and Sustainable Development', and T. Trainer, 'A Rejection of the Brundtland Report', both published in *IFDA Dossier*, No.77, May/June 1990.

analyses. The very fact that an attempt was made to treat the environment–development problématique comprehensively, and the result was given worldwide publicity, was of value in itself and an important step forward. The significance of the report lies in its timeliness, in its drawing on the experience and events from the intervening period since UNCHE, and in its presentation of empirical data to back up its alarming diagnosis and principal conclusions. As only a hardy few manage to plough through such documents, what really mattered was the selling to the public and decision-makers of the concept of 'sustainable development', as a rallying cry and a useful motto for follow-up action.

The basic policy message met with an increasingly responsive audience in all parts of the world, sensitized by the media, green political parties and NGOs, and alarmed by the accidents at Bhopal, Chernobyl (which, as it turned out, blew the lid off the environmental secrecy generally observed in the then socialist countries of Eastern Europe) and Seveso, by the African drought and famine, acid rain, tropical forest and species depletion, the ozone layer, the 'nuclear winter', and in particular the risk of global warming and the related climate change.

After the initial apathy in the decade after UNCHE, the events and processes in all parts of the world have contributed to push the environmental problématique to the top of national and global agendas, giving it policy and practical relevance that it did not have earlier. The growing and diversifying environmental impacts and the speeding up of the process of global environmental change, and no doubt the continuing study and improved understanding of issues, combined with the media involvement and the political pressure mainly from the grass-roots, played a key role in the evolving situation.

Willy-nilly, the world economy, peoples, societies, and the environment are becoming progressively and quite rapidly intertwined into a truly novel type of a global interactive system. The new system needs to be understood and managed consciously, which means that radically new and complex challenges confront the international community. Collective responses and new modes of cooperation are required, which will represent a profound innovation in the history of humankind.

In terms of such responses to environmental challenges, UNCHE was surely premature. It was on time, however, to initiate a slow process of policy and institutional maturation to prepare the international community to confront the challenge at a later date, when it will have no choice but to look in all sincerity at the many serious implications, costs and hard choices that environmental issues entail for socio-economic systems, for the world economy and for the relations between countries and peoples.

Such a point in time could be June 1992, when the UN Conference on Environment and Development is scheduled to take place in Brazil, as a major act in the global environmental saga initiated in Stockholm twenty years earlier.

In essence, the basic issues will be largely the same as at UNCHE, a replay of sorts, but for much higher stakes and with many situations and trends having become so serious that responses can no longer be delayed. Also, the tone of the rhetoric will become much shriller and the underlying controversies will surface as the time to act and to distribute the costs of such action approaches ineluctably.

32   *The quest for world environmental cooperation*

Symptomatic of the conceptual and institutional ambiguity and counter-pressures involved is the fact that areas of concentration as adopted by the UNEP Governing Council at its fifteenth session in 1989 are basically a volte-face from where it started in 1973 and a return to the priority agenda rather similar to the one favoured by the North at that time. They involve:

(a) protection of the atmosphere in order to combat climate change and global warming, depletion of ozone layer, and transboundary air pollution; (b) protection of the quality of freshwater resources; (c) protection of ocean and coastal areas and resources; (d) protection of land resources by combating deforestation and desertification; (e) conservation of biological diversity; (f) environmentally sound management of biotechnology; (g) environmentally sound management of hazardous wastes and toxic chemicals; and (h) protection of human health conditions and quality of life, especially the living and working environment of poor people, from degradation of the environment.[24]

True, the listing largely reflects the environmental impacts side of the global picture and encompasses all the key concerns as perceived today. On the whole, however, it seems to overlook or at most pays lip service to the developmental side of the equation. Yet, this is where both the roots of the problems and their solution are to be looked for.

The broadly conceived agenda of the 1992 Conference, as agreed to by the UN General Assembly in 1989, could not avoid these fundamental issues. It focused on the environment–development link and decided that environmental issues need to be addressed within the development context. Thus, significantly, and in contrast to the 1972 Conference which was on the 'Human Environment', the 1992 Conference is on 'Environment and Development'.[25]

Also, for the preparations of the Conference the General Assembly established an *ad hoc* secretariat in Geneva, headed by a Secretary General, and channelled the preparatory process essentially through Geneva and New York. The UNEP secretariat thus found itself on the sidelines of the preparatory process, which under normal conditions should have been its institutional prerogative.

This was due to several mutually reinforcing factors. One of them had to do with the wish of the Group of 77 to keep the preparatory process in places where it was better organized and could deploy its collective strength, and close to the foci of development debate. Another factor was undoubtedly UNEP's backslide over the years into the more narrowly defined domain of 'environmentalism', its shying away from basic development issues and North–South controversies, and its concentration on methodological paraphernalia. All this was done largely to please the North. In the process, UNEP distanced itself from the development mainstream and probably lost some favour with the development constituency in the South that it enjoyed earlier.

In parallel, some developed countries have been largely dissatisfied with UNEP,

---

24   See resolution 15/1 in UN doc. A/44/25 and pp. 13–14 in this chapter.
25   See General Assembly resolution 44/228.

but for different reasons, including the way it had evolved and the style of its management. For years they have been looking for ways to energize or sidestep UNEP and to seek more satisfactory mechanisms for dealing with the environmental issues. Indeed, it was their initiative that led to the creation of the World Commission on Environment and Development. This convergence of impulses yielded a broad consensus that the preparatory process should take place outside the existing structures of UNEP.

Global environmental politics are heating up worldwide. For certain, the 1992 Conference, both in its agenda and in its venue in Brazil, has all the elements for a confrontation and, with luck, flexibility and vision, accommodation between the North and the South, essentially on how to share and manage the global resources. The first exchanges in the preparatory process, and the framework arrived at through the negotiations among governments, unfortunately do not give one much reason to cheer about, and indicate persistent tensions and rigid positions.

Twenty years after UNCHE, the developing countries and their peoples have become aware of the importance of environment for their national development, and for their relations with the North. They are also better aware of the economic implications, i.e. direct and indirect costs of non-action, as well as the high costs of most measures that are required.

At the same time, the North is much more clear in its mind that the situation in the South holds one of the keys to the future of the global environment. It is convinced that the population numbers in the South, their needs and aspirations, when combined either with the existing trends rooted in underdevelopment and the development crisis or with improved development conditions based on current modes of production, technologies and patterns of consumption, yield rather unsustainable global scenarios for the future.

Indeed, for the North, the situation is one of being between the devil and the deep blue sea. If it ignores the South and allows it to stew in its own juice, the environmental pressures of poverty and underdevelopment will not only undermine the very sustainability of development in these regions, leading to social, economic, political and environmental convulsions, but will also affect the global environment with consequences that no one will be able escape. If the South develops much along the lines that the North has pursued itself and which the South would like to imitate, other global environmental variables will be also affected, even more dramatically, in particular those related to greenhouse effect and the ozone layer.

If some kind of a globally sustainable, long-term strategy is to be followed, this would entail major costs and require significant social, economic and political changes, including those in international economic relations, which at present may still be largely unpalatable to the traditional establishment and power structures in the North. And, as if this were not enough, the impacts of political and social changes in USSR and Eastern Europe, and the serious environmental pollution in this region which requires urgent and costly attention, make additional demands on the advanced and prosperous countries of the North and could divert some resources which otherwise could have gone to the South.

Deeply worried, yet without a global strategy and vision how to handle the evolving situation and disinclined to consider sacrifices on their own part, the developed countries have continued the usual 'single issue' campaigns regarding the situation in the South. This approach, notable for its refusal to admit linkages and for dealing with issues in isolation, has succeeded in projecting a public image in the North of developing countries as environmental pariahs.

Spurred on by 'environmental fundamentalism' arising from the grass-roots and the media in some developed countries, the North has been pressing for 'environmental conditionality' in multilateral and bilateral development assistance projects. For most developing countries, after the disastrous development experience of the 1980s, this has been like rubbing salt into the wound and has given rise to complaints of 'environmental colonialism'.

The underlying issues are politically explosive. Also, they are highly complex. They need vision and generosity to be managed and sorted out. They need resources, organization and new technologies. They need the recasting of traditional economic and social paradigms and a lot of fresh thinking starting from a global viewpoint. In all these domains it is the North that holds the keys to success.

As for the South, the very fact of its being there and its place on the global environmental map makes it into a major actor with a principal role in any attempts to secure sustainable development on planet Earth. This endows the developing countries, provided they manage to organize themselves properly, with an important source of collective bargaining strength *vis-à-vis* the North, which they could use across the board in their relations with the developed countries.

Whatever the future course of events, the policy debate, international negotiations, and action will need to draw on and be informed by advanced knowledge and understanding of the Earth as a system. In this, such international mechanisms as the Global Environment Monitoring System have a vital role to play. Ultimately, it is the quantitative data and indicators that will tell us about the state of the planet, that will influence, if not determine, our political, economic, social and emotional responses, and that will be at the very base of international regulatory regimes and dispute settlement mechanisms

In the pages that follow, the GEMS story is traced over the years, as it evolved on the margins of international attention. Many of the seemingly unimportant details have their function and relevance to the broader attempts to build a superstructure needed by the international community to deal with environmental problématique. What has been lacking is a conceptual framework and the threads linking the discrete pieces into a coherent and informed long-term global undertaking.

With each actor engaged in the pursuit of his own endeavour, project or measurement, and with UNEP largely unable to maintain its hold on the whole undertaking, there was no one to provide this overall vision. Indeed, this was a major lacuna of global environmental monitoring, and one of the purposes of this study is to try to piece together a systemic explanation for what has been done so far, as a base for a rationale for what might be done in the future.

2 The origin and nature of GEMS

Setting up a system of global environmental monitoring was presented and considered as one of the most important tasks facing the international community in its effort to attain the objectives and specific goals agreed to at the Stockholm Conference. The Global Environment Monitoring System (GEMS) was supposed to do the *earthwatching* and to provide the monitoring data needed for assessment of the state of the environment and for its environmentally sound uses and management.

This chapter reviews how GEMS came into being and how it was shaped through the interaction between governments, the scientific community, and the secretariats of international organizations.

ENVIRONMENTAL MONITORING AT THE STOCKHOLM CONFERENCE

Although environment-related monitoring was brought to the attention of the international community only at the time of the Stockholm Conference, the notion and practice of such monitoring were well established. Systematic data-gathering and observation networks related to weather were more than a century old. Certain geological and oceanographic phenomena had been monitored regularly; for example, the International Council for the Exploration of the Sea (ICES) had established a data centre on physical aspects of oceanography for the Baltic and the North Sea as early as 1902. Systematic national observations of health indices, soils, forests, and so on were carried out routinely, especially in economically advanced countries.

Moreover, global monitoring networks were already functioning. For example, the monitoring of radionuclides resulting from atmospheric testing of nuclear weapons was begun in 1955 under the aegis of the UN Scientific Committee on the Effects of Atomic Radiation (UNSCEAR). Relying initially on data supplied by the US and UK stations positioned around the world, it represented the first coordinated effort on a global scale to monitor humanity's impacts on the environment. The World Weather Watch (WWW) was initiated in 1963 in the framework of the World Meteorological Organization (WMO), and boasted a global network for monitoring, processing, and reporting a set of weather-related parameters. Also,

36 *The quest for world environmental cooperation*

a modest yet global in scope network for measuring the ozone distribution was initiated during the International Geophysical Year (IGY) in 1957.

Spurred by the approaching Stockholm Conference, and the discussion and proposals made during the preparatory phase, the UN specialized agencies concerned began to plan environmental monitoring corresponding to their respective programmes and fields of specialization. This was meant as a way to respond to the requirements of this new subject. It was also an attempt to stake out territorial claims in what appeared as an attractive operational activity, pending the decisions of the Conference and the establishment of the new UN organization for the environment.[1]

Although most of these activities were still on the drawing board – with the exception of WMO's Background Air Pollution Monitoring Network (BAPMoN) which had a few of its baseline stations already in operation – they had definite substantive and institutional implications. They represented concrete proposals that were submitted to the Conference for its endorsement and inclusion into a global plan of action. Steps were taken and programmes initiated, which certain institutions identified with and fought for. This was done from the perspective of single environmental medium and existing organizational jurisdictions, and before there was a chance to explore systematically the requirements of an integrated global system of environmental monitoring.

Environmental monitoring has been defined in a variety of ways. The essence of all definitions is that it represents a scientifically designed and continuing observation and/or measurement of given indicators or environmental variables, in accordance with a fixed schedule in space and time, and using comparable methodologies and standardized procedures. Environmental monitoring can involve such technical operations as hourly observations of trace gases, the daily measurement of water quality indicators, annual or seasonal surveys of forest cover, and the periodic sampling of heavy metal concentrations in seaweed or food. Supplementary data collections can be activated during specified periods, such as natural disasters, epidemics, high pollution episodes, and random sampling of biota can be undertaken when necessary.[2]

---

1 For an overview based on the reporting by individual agencies, see 'The Report on the Current Activities and Plans in the UN System on Monitoring on the Eve of UNCHE', doc. A/CONF.48/IWGM.1/Inf.1, July 1971. Most prominently, as regards monitoring in the atmosphere, the report mentions the WMO's Background Air Pollution Monitoring Network, which was intended to monitor atmospheric turbidity, gaseous constituents and solid particles, and chemical constituents in precipitation; and the Environmental Air Pollution Network of the World Health Organization (WHO), intended to monitor sulphur dioxide and particulate matter at impact sites in urban and industrial locations. The report also lists: human health-related monitoring of inland waters – WHO; monitoring of pollutant levels in Man and Biosphere reserves – United Nations Educational, Scientific and Cultural Organization (UNESCO); monitoring of food contamination by chemicals – WHO and the Food and Agriculture Organization (FAO); monitoring of radioactive hazards in the environment – WHO, International Atomic Energy Agency (IAEA), UNSCEAR, etc. Moreover, it gives an overview of the complementary programmes that can provide relevant background information and supporting facilities for environmental monitoring, such as World Weather Watch of WMO and UNESCO's International Hydrological Decade.

2 See *Global Environmental Monitoring System*, ICSU/SCOPE Report No.3, 1973, p.13. Useful distinctions have been made among 'baseline monitoring', i.e., to establish the levels and distribution

The monitoring data thus generated can contribute to improved scientific knowledge and understanding. It can also serve as an input into the decision-making and management process as related to the environment, and it is primarily this function that governments had in mind.

A flexible and vague working definition of environmental monitoring, used in the early stages of the preparations for the UN Conference on the Human Environment (UNCHE) – 'a system of continued observation, measurement and evaluation for defined purposes' – provided a sufficiently broad framework to accommodate all the purposes of environmental monitoring and uses of the monitoring data that were likely to emerge.

These purposes were elaborated upon by an Intergovernmental Working Group on Monitoring, which met during the preparations for UNCHE.[3] The basic objective of global environmental monitoring was to generate 'the information necessary to ensure the present and future protection of human health and safety, and the wise management of the environment and its resources'. The data and information gathered were to increase the knowledge on natural and human-caused changes in the environment and their effects on people's health and welfare, as well as to improve scientific understanding of the global environment and of how a dynamic balance is maintained in ecosystems. Systematic, worldwide monitoring would make it possible to determine short-term and long-term trends in the environment and to supply decision-makers with early warnings about significant or detrimental environmental changes or natural disasters. Monitoring was also to serve, in given instances, for 'checking the effectiveness of the established regulatory mechanisms'.[4]

In setting up global environmental monitoring, it was agreed that, to the maximum extent possible, the system should be built on the basis of the existing national and international monitoring programmes, and that at the same time the coordination mechanisms in the UN system ought to be improved. This compromise formula reflected the existence of different views as to the best approach

---

of pollutants at a given point in time; 'trend monitoring', i.e., repeated measurements to obtain spatial and temporal trends; and 'target monitoring', i.e., monitoring target exposure to pollutants.

[3] For the report of the Group, see doc. A/CONF.48/IWGM. I/8. The Group was asked by the Preparatory Committee for UNCHE to define the purposes of monitoring and to review and assess the adequacy and compatibility of the existing regional and worldwide monitoring activities. It was also to assess the value of using, coordinating, and extending existing systems; of supplementing them with remote sensing; and of establishing new monitoring systems. Furthermore, the Group was to consider which bodies should receive monitoring data and evaluate them on a worldwide scale, and to which body the overall evaluations should be submitted for remedial action. See doc. A/CONF.48/PC.9, para. 46. An Intergovernmental Group on Marine Pollution dealt with the question of monitoring the marine environment. For its reports see docs A/CONF.48/IWGMP. I/5 and IWGMP. II/5.

[4] During the discussion, it was proposed that environmental monitoring should also include checking compliance with agreed standards and norms. While this was a logical application of monitoring, and was apparently acceptable to everybody when it came to domestic affairs, that was not the case with the international setting. To take into account this reservation, one of the basic documents for UNCHE explicitly states that monitoring 'does not connote the policing or surveillance of compliance with regulations or standards' (doc. A/CONF.48/11, p.12.) Yet, as is usual, the door was left open for a broader interpretation when the same document noted that information obtained through monitoring 'will be a valuable indication of the effectiveness of control measures'.

38 *The quest for world environmental cooperation*

for system design and implementation. Some governments, with the strong support of the specialized agencies, were in favour of relying fully on the existing structures and programmes; others, thinking along similar lines as the Conference secretariat, leaned towards a central coordinating mechanism, which would help to overcome the single-medium, compartmentalized approach of the specialized agencies and would give an integrated character to this global endeavour.[5] This difference of opinion anticipated one of the important issues to be faced in the post-Stockholm period and in efforts to set up an effective and rational global environmental monitoring system.

The preparatory process was a useful one. It provided the Conference with the initial considerations regarding the setting up of global environmental monitoring, and with a set of specific proposals for initiating various types of activities.[6] It also gave an indication of things to come and issues to be faced during the implementation stage. As mentioned, this involved the question of data uses and the need to reconcile the existing jurisdictional and subject-matter divisions with the need to build an integrated and coordinated global system. It also involved the problem of developing countries' participation. They hardly took part in discussions on environmental monitoring during the preparatory process for the Conference; they were not represented among the scientists who prepared the SCOPE report and only two of them – Argentina and Brazil – took part in the Intergovernmental Group on Monitoring. They showed little interest in the subject, especially as pollution monitoring was the dominant concern.

The UNCHE Action Plan addressed environmental monitoring in a comprehensive manner. Virtually one in every four recommendations had a monitoring component, in keeping with the notion of monitoring/assessment being a functional task. Table 2.1 gives an overview of these monitoring-related recommendations.

5 In its proposal, the US suggested that the system should be built 'in so far as possible' on the basis of the existing monitoring activities. The US was one of the countries that felt there was a need for a central programmatic and institutional framework to initiate and coordinate monitoring actions. It argued that any global monitoring programme would fail without appropriate institutional arrangements. (See doc. A/CONF.48/IWGM. I/Inf.10.) The UK was of the opinion that no new international agency should be established to conduct monitoring, and that matters of coordination could be resolved through continuing discussions to define responsibilities between existing programmes and agencies. (See doc. A/CONF.48/IWGM. I/Inf.6.) In the end, there was no overt controversy on this issue when the discussion took place in the working group, and the matter was resolved as indicated above. The group recognized the need for a coordinating mechanism on monitoring within the UN framework and recommended a study to explore how coordination could best be achieved (doc. A/CONF.48/PC.11/Add.1, p.5).

6 The Conference had at its disposal the objectives and guidelines for global environmental monitoring proposed by the Intergovernmental Working Group (doc. A/CONF.48/IWGM. I/8). It also considered the basic conference paper, prepared by the secretariat, on the identification and control of pollutants of a broad international significance (doc. A/CONF.48/8). The recommendations contained therein were eventually adopted by the Conference without major alterations. Some countries also submitted proposals on monitoring, most notably the US (doc. A/CONF.48/IWGM. I/Inf.10). An important study on global environmental monitoring was prepared by the Scientific Committee for the Study of the Problems of the Environment (SCOPE), a body established within the framework of the International Committee of Scientific Unions (ICSU). This represented the scientific community's contribution to the Conference on the subject of environmental monitoring. See *Global Environmental Monitoring*, ICSU/SCOPE Report No.1, Stockholm, 1971.

Table 2.1 Environmental monitoring in the UNCHE action plan: some principal outcomes

Recom-mendation number	Parameter to be monitored	Medium	Lead agency	Purpose
	Pollution-related			
79	Atmospheric constituents and properties affecting meteorological properties and climate	Atmosphere	WMO	To monitor their long-term global trends, especially distribution and concentration of pollutants, in order to understand and predict climate change
55, 87, 90	Pollutants in marine environment	Oceans, rivers	FAO, WMO, IOC, UN	To monitor high-priority pollutants in ocean water, sediments and organisms; to establish a registry of major rivers according to their discharge of water and pollutants into oceans
76, 77	Environmental agents that may be a health risk	Air, water, man	WHO	To monitor these agents at impact sites; to link results to the health of the local population; to provide data for the early warning and prevention of their deleterious effects on humans and for assessment of their potential risks, with special regard for carcinogenicity, mutagenicity, and teratogenicity
78	Contamination of food by chemical and biological agents	Food	FAO, WHO	To monitor trends of food contamination and pollutant levels in food unsafe for human intake
80	Effects of pollutants on ecosystems	Eco-systems	UNESCO	To monitor hazardous compounds in biological and abiotic material at representative sites, their movement and residence times, and their effects on ecosystems in general, including the reproductive success and population size of selected species
29	Pollutants in wildlife	Wildlife	UN	To monitor effects of pollutants on wildlife, as indicator of future disturbances and impacts on humans
57	Environmental effects of energy use and production	Not specified	UN	To monitor carbon dioxide, oxidants, nitrogen oxides, heat, particulates, oil, radioactivity

Table 2.1 (Continued)

Recom-mendation number	Parameter to be monitored	Medium	Lead agency	Purpose
75	Radioactive materials	Not specified	WHO, IAEA	To establish a registry of releases into biosphere of significant quantities of radioactive materials
73	Sources, pathways, exposures and risks of pollutants	Not specified	Not specified	To acquire knowledge for the assessment of pollutant sources, pathways, exposures, risks
74	Potentially most harmful man-made chemicals	Not specified	UN	To monitor production figures, and pathways from factory via utilization to ultimate disposal or recirculation
Natural resources-related				
20	Soil capabilities and degradation	Land	FAO, UNESCO	To monitor and map trends in different aspects of soil degradation, soil capabilities, soil regeneration, etc.
25	Forest cover	Land	FAO, UNESCO	To monitor forests, the balance between world forests, biomass, and environment, and changes in forest biomass making impact on environment
39, 40	Genetic resources, especially those endangered by depletion or extinction	Genetic resources	FAO	To keep inventories of genetic resources
46, 49	World fisheries and living aquatic resources	Oceans	FAO	To expand monitoring of world fishery resources and to include additional biological parameters
30	Animal species endangered by their trade value	Wildlife	FAO	To monitor current situation

All the monitoring requirements identified in the preparatory process found their way into the Action Plan. Special emphasis was placed on monitoring air pollution and its impacts on climate; air, water, and food pollution affecting human health;and marine pollution. Emphasis was also given to monitoring the natural resource base.

In more general terms, the majority of the recommendations were sector-oriented, endorsing the existing monitoring programmes or plans of the specialized agencies, and/or proposing their expansion to cover additional variables. Those areas where no agency had assumed a clear responsibility, such as the environmental impacts of energy production and use, or pollutant pathways, were to be coordinated by the UN Secretary-General, a euphemism for the new environmental machinery to be created. These problems required integrated approaches.

By addressing both pollution and natural resource monitoring, the Conference in fact provided the base for a comprehensive system of environmental monitoring. In addition, it noted the need to involve the developing countries as full participants and to provide them with financial and technical assistance in order to boost their monitoring capabilities.

In its deliberations and conclusions, the Stockholm Conference stayed clear of the controversies involving institutional arrangements, the need for an integrated approach to environmental monitoring, and the linkage between monitoring and environmental standards regimes.

The Action Plan contains only one mention of monitoring as a global system (in the context of Earthwatch), without elaborating upon its objectives and features. Moreover, while entrusting the implementation of monitoring tasks and sectoral coordination to the lead agencies, it does not address the question of how to tie these monitoring programmes and activities into a global system, or the issue of a central coordinating body for monitoring. Indeed, the Action Plan speaks of monitoring 'systems', rather than a 'system'.

In view of the strong sectoral pulls and pressures from the specialized agencies, the weak position of the UNCHE secretariat to press for suitable and well-defined institutional answers, the uncertainty about the best way in which to proceed, and the ambivalence among the governments concerned, it was best to leave the institutional issue alone.[7] At any rate, these matters had to be considered at a strategically more opportune moment, when the new institution for the environment was in place.

While this approach was necessary and reflected the conditions that prevailed and what was possible at the time, it meant that the Conference endorsed the already powerful centrifugal, sectoral tendencies by recommending that specific actions be undertaken by different agencies. It also meant that an opportunity was forgone to formulate a clear-cut policy and statement of principle regarding an integrated

---

7 For an indication of this, see the document submitted by the Conference secretariat to the Intergovernmental Group on Monitoring. It states that the Group might find it 'wise to avoid as far as possible the institutional matter, such as whether within the UN system any such integration (i.e. of monitoring activities and action) should take place by intergovernmental or interagency machinery (doc. A/CONF.48/IWGM. I/2, p.5).

global environmental monitoring system that could serve as a reference and guide in the years to come.

SETTING UP THE GLOBAL ENVIRONMENT MONITORING SYSTEM

The establishment of a global environmental monitoring system was considered to be one of the key and most urgent tasks for the new organization, both by its secretariat and by industrialized countries.

The majority of the questions having to do with the establishment of such a system were passed over by the Stockholm Conference, and had to be addressed within the framework of UNEP as the first step in the implementation phase. Inevitably, this was also part and parcel of UNEP's effort to shape its own institutional and substantive roles and to devise approaches for putting the UNCHE legacy into practice.

The first issue of whether or not environmental monitoring should be a distinct programme was discussed and resolved immediately after the Conference within the UNEP secretariat, together with its close advisers. The view that monitoring should be simply an integral component of each area of the emerging programme, and of the projects where necessary, although recognized as conceptually and substantively sound, was rejected on practical grounds. It was concluded that, for institutional and policy reasons, environmental monitoring should be a separate activity and should be given its own programme and institutional identity and visibility, as the Global Environment Monitoring System, or GEMS.[8]

Before the system could get going, it was necessary to consider and agree upon the place to be given to environmental monitoring in the programme of UNEP, the variables and processes to be monitored, and the institutional features of the system.

GEMS at the first session of the UNEP Governing Council

The question of GEMS inevitably became entangled in the confrontation between developed and developing countries at the first session of the Governing Council. Some developed countries felt that environmental monitoring should occupy a dominant position in the fledgling programme of UNEP. The secretariat also placed a marked emphasis on environmental monitoring in the programme document it submitted to the Council.[9] The developed countries furthermore proposed that in the initial phase of GEMS efforts should be focused on pollution monitoring only.

---

8 The acronym GEMS began to be used officially in the context of the consultations of the Interagency Working Group on Monitoring, established in April 1973 by the Environment Coordination Board. However, it was first launched at a dinner party during UNCHE, when Shirley Temple Black, a US delegate, is reputed to have said: 'I like GEMS, let us call it GEMS.'

9 For the secretariat report on the proposed development of the programme, see doc. UNEP/GC/5. Paragraphs 37–9 deal with GEMS *per se*, while the sections on major priority areas, i.e., pollutants, oceans, atmosphere, energy, land, and water, contain additional specific recommendations related to monitoring. Some delegations were critical of the exaggerated weight given to environmental monitoring in the secretariat document.

This displeased developing countries who were not interested in pollution monitoring in particular.

In practical terms, the controversy revolved around two questions, namely, the place of GEMS in the UNEP programme, and its content and orientation. In the process of deciding on UNEP's priorities, GEMS found itself relegated to the background. It was given less emphasis and weight in the programme than had been anticipated and hoped for by its proponents. This was reflected in the reduced share of the financing that would be made available to it from the Environment Fund.[10]

As regards the content of GEMS, a compromise agreed to by the Governing Council – while calling for a system of monitoring to be 'developed first for pollutants liable to affect weather and climate, and persistent and widely distributed substances liable to accumulate in living organisms and move through ecological systems, particularly along pathways leading to man' – stated that the 'monitoring process should concern itself not only with chemical pollutants but also with the identification, by all appropriate means, of those environmental problems affecting the development process, such as vector-borne diseases'.[11]

The action of the first Governing Council, shaped by North–South differences, had important consequences for the direction and nature of GEMS. It reduced GEMS' share in the Environment Fund, and, at the same time, expanded its scope by placing among its priority concerns the environmental issues affecting the development process. This was an example of a countervailing decision, so common in the UN, where governments find it easy to add up different claims and demands but difficult to allocate resources needed for implementation. The decision meant thinning out the already limited resources and further reducing their reach. It meant increasing the burden on the secretariat in its role of devising and coordinating the system. It reduced the declared enthusiasm of the industrialized countries for GEMS and it revived their interest in regional approaches to monitoring. On the positive side, however, from the very beginning it opened the way for a more comprehensive approach to global environmental monitoring and its being linked with development concerns of the developing countries, secured their political support and increased their interest in taking part in the system.

10 It was decided that, between the first and second sessions of the Governing Council, GEMS would receive 5.5 per cent of the total allocated, or approximately $0.3 million out of $5.5 million during the first fiscal year. See Council resolution 3(I) in UN doc. A/9025. While no figures were presented formally by the developed countries, this was much less than had been anticipated for this activity. An illustration of what they thought would have been required is to be found in the US submission to the 1974 Intergovernmental Meeting on Monitoring. It was estimated that the cost of implementing GEMS would be $3.8 million in 1974, $6.3 million in 1975, $8.2 million in 1976, and $9.6 million in 1977. In its proposal the US suggested that functional tasks, such as environmental monitoring, should not be seen as competing with other priority areas for financial support, but should be considered as their integral components. For example, in 1976, $2.48 million of the estimated $8.29 million of GEMS costs were to be spent in the area of human settlements, health, habitat, and well-being. (See the document entitled 'Proposal of Initial Implementation of Global Environmental Monitoring', December 1973, prepared by the US Interagency Committee for Global Environmental Monitoring.)

11 Governing Council resolution 1(I)VII in UN doc. A/9025.

44 The quest for world environmental cooperation

What to monitor?

The next step involved choosing areas of concentration and variables to be monitored. The need to set priorities was advocated by everyone.[12] This was not easy, however. UNCHE had left a long and comprehensive list of recommended monitoring activities. Every specialized agency had its own preferred environmental monitoring proposal, and the countries had different views about priorities. UNEP's programme priorities had also been formulated in a relatively comprehensive manner by the first session of the Governing Council.

The task of deciding on the priorities was placed on the agenda of the 1974 Intergovernmental Meeting on Monitoring. The developing countries pressed successfully for the incorporation of environmental variables and/or processes other than pollutants, including desertification and the degradation and depletion of soil and living resources such as forests, grasslands, wildlife, and aquatic ecosystems. The main policy and strategy issue was resolved when it was decided that this type of monitoring should start concurrently with pollutant monitoring, and not at a later stage, as had been proposed by developed countries in yet another attempt to steer GEMS in their preferred direction.[13]

The Meeting, which was open-ended in terms of participation, was marked by the North–South debates and tensions. It became too politicized and polemical for the taste of many experts who took part. However, it ensured that greater attention would be given to the needs and views of developing countries in GEMS' orientation.

The key decision taken was to identify seven programme goals for GEMS, based on the priority subject areas agreed upon at the first session of the Governing Council. These programme goals provided the focus for subsequent action. They included:

- establishing a human health warning system;
- monitoring global atmospheric pollution and its impacts on the climate;
- monitoring the extent and distribution of contaminants in biological systems, particularly food chains;
- monitoring critical environmental problems relating to agriculture, and to land and water use;

---

12 For example, in its recommendation No. 86, UNCHE flagged the need to establish priorities in the case of pollutants by calling for an 'internationally accepted procedure for the identification of pollutants of international significance and for the definition of the degree and scope of international concern' (doc. A/CONF.48/14/Rev.1, pp. 21–2). The pre-UNCHE view expressed by SCOPE was that the monitoring system must be at first 'rigorously selective as to which variable to monitor globally'. It should choose only those variables where it is absolutely certain that acquiring global knowledge is urgent and where reliable and reproducible methods for sampling and measurement can be successfully operated on a global scale. SCOPE No.1, *op. cit.*, p.63.

13 For the report of the Group see doc. UNEP/GC/24. This approach was given policy endorsement by the UN General Assembly. In its resolution 3326(XXIX), it asked that other environmental parameters be given equal priority as pollutants in the implementation of GEMS because of their importance for the solution and prevention of those environmental problems that principally affect developing countries.

- monitoring the response of terrestrial ecosystems to pressures exerted on the environment;
- monitoring the state of ocean pollution and its impact on marine ecosystems; and
- monitoring the factors necessary for the understanding and forecasting of natural disasters.[14]

The Meeting devoted a good deal of effort to determining the criteria to be used in deciding which pollutants should be monitored.[15] It also devoted a lot of time and energy to weighting various pollutants and to deciding on priority pollutants and environmental stress indicators in different media for monitoring in the early stages of GEMS. The priority list that emerged was headed by sulphur dioxide and suspended particulates in the air, and by radionuclides (strontium 90 and caesium 137) in food. These were followed by ozone (air), DDT and organochlorine compounds (biota, humans); cadmium and compounds (food, humans, water); mercury (food, water), lead (air, food), and carbon dioxide (air); etc.[16]

The amount of effort invested in producing the priority list was not matched by the use to which it was put. It was shelved as being of little value in trying to bring the system into operation and not applicable in a global setting. As far as Africa was concerned, for example, mycotoxins and microbial contaminants in food, which were ranked last in the list, had the highest priority, while sulphur dioxide in air, which was ranked first, was of little or no concern. This did not really matter for the follow-up action; the fact that the Intergovernmental Meeting had agreed on the seven programme goals gave sufficient guidance with regard to the general direction in which to proceed.

In view of the Chernobyl events more than a decade later, it is worth noting that the Meeting assigned top priority status to monitoring of radionuclides in food. This was done at the initiative of Peru, backed by the Group of 77. Peru was concerned with the French nuclear testing in the South Pacific and the radionuclides that drifted over parts of South America.

The UNEP secretariat sought the opinion of UNSCEAR and, basing itself on

---

14 The Meeting also identified environmental factors that needed to be monitored to support a pollutant-monitoring programme and/or to help interpret the data, including: (a) water and soil quality indicators; (b) indicators of health of humans, animals, and plants; (c) selected meteorological, hydrological, and geophysical variables; (d) dietary intake and composition; and (e) indicators describing the state of the climate (doc. UNEP/GC/24).

15 These criteria included the severity of their effects on human health and well-being, on climate, and on terrestrial and aquatic ecosystems, as well as their persistence, resistance to degradation, and accumulation in humans and food chains. Other factors to be taken into account were the possibility of chemical transformation within physical and biological systems resulting in secondary substances more toxic and harmful than the parent compounds; ubiquity and mobility; concentration in the environment and/or humans; frequency and/or magnitude of exposure; feasibility of measurement at given levels in various media; suitability for uniform measurement; and the potential value of information for assessing the state of the environment.

16 For other proposed priority lists that preceded the decision of the Intergovernmental Meeting, see 1971 SCOPE list in SCOPE No.1, *op. cit.*, p.7; 1973 SCOPE list, in SCOPE No.3, *op. cit.*, pp.31–3; 1973 Interagency Group list in doc. UNEP/IG.1/2, Annex I; and 1973 US 'Proposal of Initial Implementation of Global Environmental Monitoring', *op. cit.*

46 *The quest for world environmental cooperation*

the reply that it received, the matter was quietly dropped from the GEMS work programme. In its review, UNSCEAR stated that such priority ranking of radionuclides was not justified and should not be allowed to divert the already limited resources available in GEMS for environmental monitoring from other more pressing tasks. Thanks to the atmospheric N-weapon test ban, the global concentrations of long-lived radionuclides such as strontium 90 and caesium 137 in food had levelled off and decreased, and were not likely to be increased by the ongoing tests. At any rate, UNSCEAR was already keeping global inventories of these radionuclides. As regards contamination of food by strontium 90 and caesium 137 released from nuclear energy facilities, UNSCEAR argued that the existing levels in food from these sources were too low to be detected except very near such facilities. According to its view, this did not warrant a priority status in a global system such as GEMS and was something to be determined locally. As concerned French nuclear testing, which was the target of the Peru-led initiative, the Committee was of the view that priority should be given to iodine 131 in milk and short-lived gamma emitters in geographical areas exposed to fresh radiation fallout.[17]

What system? Some institutional questions

The 1974 Intergovernmental Meeting on Monitoring also had to look into institutional questions. This matter had not been examined since before UNCHE, and many issues had been deliberately left unclear.

In deciding on the nature of the system and its institutional characteristics, it was obviously not possible to start with an ideal model and to proceed by working out its practical requirements. In the SCOPE report, scientists had argued before UNCHE that a simple and low-cost structure should be designed that at the same time would be effective in terms of regional and global coverage, accuracy, reliability, speed, and flexibility of response. The substantive goal of an effective system was self-evident. That it was possible to build such a system at little cost and in a simple manner was not so obvious. This premise, however, reflected the mood of the governments.

17 For the UNSCEAR statement see doc. UNEP/GC/31/Add.2, Annex III. The Interagency Group had already argued on similar lines when it stated that fission and activation products resulting from N-weapons testing were extensively monitored and kept under international review, and that there was no need for GEMS to become involved. In any case, caesium 137 and strontium 90 were generated by high-yield nuclear tests, which were no longer conducted. The ongoing atmospheric tests were in fact yielding iodine 131 and other short-lived nuclides. As regards radionuclides released by industrial, medical, and research activities, the level of individual nuclides, the Group argued, was so low that it was difficult to measure or even detect in the global environment. The Group felt that, in view of the expanding uses of nuclear energy, it was nevertheless important to assist governments in determining when, to what extent, and how radiation systems should be set up to measure emissions. It pointed out that WMO and IAEA had a joint programme in this respect (doc. UNEP/IG.1/2, p.9). It should be noted that countries with developed nuclear energy programmes argued that, thanks to the most stringent criteria, precautionary measures and the negligible chance of accidents, there was no possibility of global contamination from their facilities, and that consequently there was no need for global monitoring of radionuclides generated by nuclear reactors.

There were certain parameters that largely predetermined what could be done. These included the decisions of the Stockholm Conference endorsing the monitoring programmes or plans of different specialized agencies, and its *de facto* choice of a coordinated system to be built on the basis of the existing components. Also, the agencies were on the way to forming their environmental programmes, with strong backing from their domestic constituencies. As the global coordinating organization for the environment, UNEP had no capability to attempt to build an integrated global environmental monitoring system itself. The governments were far from clear about the meaning of, or the need for, an integrated monitoring system, and, indeed, no one was quite sure what it entailed and how to go about it.

In terms of starting and obtaining some results quickly, it was preferable to pursue an adaptive approach, using what was already in place and allowing the system to grow naturally, through a learning process and through interaction and conflict between different interests, perceptions, institutions, and scientific needs and insights. There were some obvious drawbacks to this strategy, however. The major one was that the planned and ongoing programmes, both national and international, were generally designed for single-medium monitoring and to fit into the existing sectoral divisions and jurisdictions of the UN system. Yet, the environmental variables and processes in many instances required an integrated monitoring and assessment approach spanning different media and disciplines. In the case of some pollutants, it was often necessary to monitor not only their levels but also their sources and pathways, as well as the roles of different media and their transport mechanisms, to establish their receptors and flux rates in the environment, and to determine the cause–effect relationships (a task complicated by the frequently synergistic effects). There was little experience and knowledge for such monitoring, even in the scientifically most advanced countries.

As was to be expected, in preparing for the 1974 Intergovernmental Meeting, a wide divergence of views about the desirable institutional traits of GEMS appeared between the specialized agencies and the UNEP secretariat.

The agencies adhered to the position that GEMS was essentially a collection of their independent monitoring activities, mutually coordinated through an interagency mechanism. In preparing for the Meeting, the Interagency Working Group spoke of the future need for 'an overall policy-making, coordinating and reviewing intergovernmental machinery for GEMS, which would be particularly concerned with the intersectorial coordination of the several components of GEMS, their interrelationships and priorities within the framework of the Earthwatch'. Such machinery would be backed by an interagency coordinating mechanism under the auspices of the Environment Coordination Board, and would be serviced by a secretariat staff from UNEP and the specialized agencies as necessary.

Throughout the lengthy report of the Interagency Group, there was scarcely any mention of UNEP. Indeed, there was no mention whatsoever of UNEP in the context of the various operational activities, and, *de facto*, none regarding matters of coordination. The agencies had argued that experience had shown that monitoring would have to be developed at the local level and on a sectoral basis before it would be possible to consider the intersectoral and global integration of such activities. Under the existing conditions, integration at the global level was only feasible and useful with respect to the exchange of relevant information among systems and with respect to harmonization and coordination of their specific monitoring activities.[18]

UNEP, on the other hand, felt that GEMS should not be simply a collection of sectoral programmes bankrolled by the Fund, but that it must have cross-sectoral connections in order to be indicative of changes occurring in the environment and of their significance. Only UNEP could ensure an overall view, coordination, and direction. In the speech he made to the Meeting, the UNEP representative stated that his organization was not prepared to parcel out pieces to specialized agencies or to fund them without having a say in the substantive design and connection among the various programmes. He added that many variables and problems were not dealt with or covered by the current work of the agencies, and only an intersectoral body such as UNEP could assume the responsibility for them.[19]

UNEP also conceived the role of a GEMS office in very ambitious terms, based more on what was required to run and coordinate such a system than on what was feasible in the context of UNEP's mandate, capabilities, and available resources. Among the functions that were proposed were: formulating the outline of a basic network of monitoring programmes or stations; negotiating agreements with those who do the monitoring to contribute data to a global system; agreeing on channels of communication for the flow of data; setting up a small central organization to receive, store, process, assess, and provide for the quality control of data; providing the means for laboratory analysis of sample material and target species, and for theoretical work on models and pathways; providing technical assistance for the establishment of new national networks; arranging for training programmes and for the establishment of new stations and instruments in developing countries.[20]

In particular, the specialized agencies were alarmed by the proposal to establish a unit for GEMS in the UNEP secretariat and the related suggestions as to how UNEP should perform its coordinating role with regard to monitoring activities. They felt that the establishment of a GEMS office could lead to operational activities on UNEP's part, which, they argued, were contrary to and exceeded UNEP's role. They were also worried that such an institutional locus would make it easier for the UNEP secretariat to encroach upon their domains and to press its ideas of how GEMS ought to be run and organized. They felt that GEMS was so complex as to make it virtually impossible for any single UN body to attempt to coordinate operationally; all that UNEP could do with its small staff and resources

---

18 For the report of the Interagency Working Group see doc. UNEP/IG.1/2. For the views of the specialized agencies on the global integration of monitoring activities, expressed at the time of UNCHE, see doc. A/CONF.48/IWGM. I/Inf.1, p.8.

19 For the UNEP secretariat position, see the report it submitted to the Intergovernmental Meeting (doc. UNEP/IG.1/3). For the statement to the Meeting, see doc. UNEP/GC/24, p.3. The agencies were upset that the paper prepared by the Interagency Group was not used by the UNEP secretariat as an action plan for GEMS. Also, they objected to UNEP recommendations implying a basic network principle. They argued that this was contrary to the Group's recommendation that reliance be placed on the association of the existing networks and stations.

20 Doc. UNEP/IG.1/3, para.16.

was to undertake the overall coordination of monitoring programmes carried out by the specialized agencies.

The governments largely favoured UNEP's proposal. They argued that without a permanent institutional home in the UNEP secretariat, there was little chance of attaining an adequate degree of coordination among the various monitoring activities in the UN system, not to mention the central vision of GEMS.

The Meeting thus approved the establishment of a GEMS office in UNEP, to initiate the implementation of the monitoring programme within the framework of GEMS, in consultation, when appropriate, with the specialized agencies concerned. However, it did not pronounce itself explicitly on the functions of this office; nor did it inquire into the feasibility of proposals made, given the resources and institutional constraints inherent in UNEP.[21]

In another decision, significant for the evolution of GEMS, the 1974 Meeting rejected the Australian proposal to institutionalize the intergovernmental guidance of the system by establishing a permanent steering committee of government experts, to be selected on a basis to be determined by the Governing Council. Australia argued that *ad hoc* meetings did not provide adequate access to the management and guidance of GEMS for those governments which, like the Australian, intended to set up monitoring installations for the system. It felt that these governments should be involved in the details of the design and development of GEMS, and that there was therefore a need for a formal mechanism through which they could offer guidance to UNEP.[22] The UNEP secretariat's proposal to create a GEMS advisory group of governmental and specialized agency representatives was rejected as impractical. It was also objected to because it did not conform to the standard UN practice, whereby international civil servants are not allowed to sit at the same table with an equal status with the representatives of governments.

This rejection of creating a permanent intergovernmental body seems to have been motivated mainly by the nature of the 1974 Intergovernmental Meeting itself. Many participants from industrialized countries became convinced that such open-ended meetings in terms of participation should not be repeated because they tend to become too politicized, are difficult to control, and deviate from the subject matter, essentially owing to the presence of the Third World representatives. An *ad hoc* approach proposed by the US was agreed to by the Meeting instead. Accordingly, the governments would nominate experts in their personal capacity to sit on *ad hoc* expert groups, convened as the need arose to deal with various aspects of GEMS. It was assumed that such small groups were less likely to become politicized and would remain strictly technical. Also, they would be easier to control, *inter alia*, through the ability to select participants. Moreover, this approach, while giving the UNEP secretariat and the specialized agencies the benefit of the governments' advice and guidance, also allowed a greater latitude and

---

21 Doc. UNEP/GC/24, pp.17–18.
22 *Ibid.*, pp.15–18.

50 The quest for world environmental cooperation

flexibility of action in GEMS' formation, making the whole process subject only to the overall policy and programmatic control of UNEP's Governing Council.

The two recommendations of the 1974 Intergovernmental Meeting – one on the establishment of a GEMS office within the UNEP secretariat, the other on the nature of intergovernmental coordination – provided the basic institutional framework. The green light to proceed was given by the second Governing Council, in March 1974.[23] The Council authorized the appointment of a director for the GEMS office with a mandate to continue to design, develop, and implement the system. GEMS was finally on its way.

WHAT IS GEMS?

What is GEMS? There is no clear-cut answer to this question. To begin with, there was no well-defined or agreed-upon idea of the system. The controversial and uncertain institutional role of UNEP, the differences of opinion within and among governments, the jurisdictional and substantive concerns of the specialized agencies, the lack of experience and the gaps in scientific and technical knowledge, and the very diversity of the phenomena, variables, and processes to be monitored all accounted for this situation.

None the less, it is possible to deduce some basic assumptions from the early discussions and documents as to what it was thought GEMS should ideally be like. Drawing heavily on the World Weather Watch model, these premises were implicitly shared by a small group of officials and scientists involved in the preparations for the Stockholm Conference. Briefly, the premises can be summed up as follows:

- The notion of a global monitoring network, or networks, was basic to GEMS. Some spoke of a single, multipurpose network, others of specialized networks. GEMS was to be global in scope, effectively covering all areas, including those beyond national jurisdiction. It was to be universal in terms of participation, with developed and developing countries alike taking part.
- The idea of a core unit, capable of providing the focus and guidance for various activities, was one of the prerequisites of the system.
- GEMS was to be a 'system' in the true sense of the word. Relying on increasingly sophisticated techniques and technologies, and based on the systems perspective required by the environmental phenomena and variables that had to be monitored, it was expected that GEMS would rise above and surmount the existing disciplinary, sectoral, and institutional cleavages.
- The monitoring data gathered in the field would flow continuously upwards, from the monitoring sites to data centres, where they would be processed, verified, stored, and evaluated, and where they would remain accessible to potential users.
- The processed monitoring data would indicate baselines and trends in the environment and effects on the targets. They would serve to conduct environmental assessments, which would primarily be usable in decision-making and

---

23 See decision 8(ii)II.1, in doc. A/9625.

in environmental management and control, and would also serve to improve scientific knowledge and understanding.

These expectations were coupled with optimism that GEMS would become operational quickly. Thus, for example, on the eve of the first session of the UNEP Governing Council in June 1973, the UNEP secretariat stated that the international capability to measure on a repetitive basis the level of pollutants that were of the greatest international concern should be developed by 1974.[24] Even the sceptics did not dispute the view that by the beginning of the 1980s the system would be fully operational. No one seemed too concerned about the disparities between the objectives and the means for their implementation, or about the complexities involved in launching and implementing global environment monitoring programmes.

Soon after UNCHE, however, these expectations had to confront and adjust to reality, including the weak institutional capability of UNEP and the nature of the response from those on whom the implementation of GEMS ultimately depended.

The UNEP secretariat, in its first attempt to define GEMS, described it as a 'coordinated effort on the part of member states, UN agencies and UNEP to ensure that data on environmental variables were collected in an orderly and adequate manner for the purpose of giving the Governing Council a quantitative picture of the state of the environment'.[25] Similar to the definition of Earthwatch discussed in the preface, this definition of GEMS dealt only with function and shunned the question of structure. This was a consequence of the institutional weakness of UNEP, of its initial experience in trying to get the system going and to assert its role, and of its having scaled down its expectations and ambitions.

GEMS had its real start immediately after the first Governing Council, as the sum of the agencies' monitoring projects that were considered for approval or approved for financing by the nascent UNEP secretariat.[26] The pressure to act immediately was great. Resolutions had been passed and decisions taken. The contributions were flowing into the Environment Fund and the allocated money had to be spent within the first fiscal year. The agencies, basing themselves on the

---

24 See doc. UNEP/GC/5, p.78.
25 UNEP/GC/32/Add.2, p.1. This document was prepared by the Director of GEMS for the third session of the Governing Council. It represented the distillation and interpretation of what had transpired in the preceding period, including the recommendations of the 1974 Intergovernmental Meeting on Monitoring, and provides the broad lines of UNEP's initial approach to GEMS and its development.
26 Among the early projects that were approved or began to be considered immediately after the first session of the Governing Council were the monitoring of food contamination – FAO and WHO; air pollution monitoring in urban areas – WHO; monitoring of background atmospheric pollution – WMO; a pilot project on monitoring marine pollution by petroleum – International Oceanographic Commission (IOC) (doc. UNEP/GC/14/Add.2, pp. 100–2). There was also the environmental monitoring-related research project with the Monitoring and Assessment Research Centre (MARC) at the University of London. The basic objective of this project was to establish a research component in GEMS and via MARC to maintain an operational link with ICSU/SCOPE, representing the international scientific community. See A. I. Sors and G. T. Goodman, 'MARC – Center for Environmental Assessment', *Environmental Science and Technology*, Vol. 11, Nov. 1977, pp. 1061–5.

Stockholm recommendations, were clamouring for financial support for their monitoring activities. The UNEP secretariat was also eager to show that something concrete was being done, and GEMS was among its priority and tangible activities.

Thus, operationally, GEMS began before much thought could be given to the strategy and tactics of its implementation, and before the 1974 Intergovernmental Meeting on Monitoring met. GEMS as an integrated system was hardly considered, and lines of action corresponding to the jurisdictions of the main specialized agencies were followed. There was little else that could be done anyway, in view of UNEP's limited institutional capability and its mode of work, as well as the need to rely on others for concrete action.

The hard-pressed UNEP secretariat faced the multiple task of quickly elaborating the programme, preparing for the first session of the Governing Council, recruiting the staff, arranging for the physical move to Nairobi, and pleasing the gallery of its eager fans. The first version of the programme document was basically prepared by clipping, classifying, and pasting together various project proposals submitted from within and outside the UN system. UNEP's position of dependence on others for the execution of activities was accentuated by the fact that it did not as yet have a GEMS office. It was reacting rather than initiating, with little choice beyond what was offered, and no capability or opportunity to mount something different. Ever since, having established this early pattern, it has been at a disadvantage in striving for a more assertive and guiding role *vis-à-vis* the agencies, with the latter having an obvious preference for the maintenance of the status quo.

GEMS, then, started in a somewhat unsystematic manner. All along, it has continued as a compromise between an ideal global system, which was more of a wish and a verbal assertion than a reality, and the actual fact of a series of discrete environmental monitoring projects carried out by agencies and partly or fully funded by UNEP. The all-encompassing notion of *Level Two* has been used by UNEP to maintain a global system vision, by including virtually all monitoring or monitoring-related activities within the scope of GEMS, even in those instances where there is no institutional linkage and where those engaged in monitoring are hardly aware of GEMS' existence.[27] This tendency is accentuated by UNEP's need for institutional self-assertion, constrained as it is by its non-operational role and its wish to prove to itself and others that it is doing something useful with respect to the real needs and its wide-ranging subject matter.

The five GEMS clusters and the Global Resource Information Database

Based on the programme goals agreed upon at the 1974 Intergovernmental Meeting on Monitoring and the environmental monitoring activities carried out within the UN system with the financial support of UNEP, five separate GEMS clusters were eventually defined. These are:

- *Health-related environmental monitoring*, with WHO taking the lead role;

---

27 An example of this is the in-depth review prepared for the ninth session of the Governing Council (doc. 'Earthwatch, an In-Depth Review', UNEP, Report No.1, 1981).

- *Climate-related environmental monitoring*, with WMO taking the lead role;
- *Marine pollution monitoring*, with IOC and UNEP taking the lead roles;
- *Natural resources and integrated monitoring*, with FAO, UNESCO, and UNEP taking lead roles; and
- *Monitoring of long-range air transport of pollutants*, with the UN Economic Commission for Europe (ECE) taking the lead role.

The clusters are basically made up of the various environmental monitoring projects executed by different agencies and supported financially by UNEP.[28] The five-cluster structure raises a number of issues. For example, it does not provide adequately for those variables or problems which span environmental sectors and cut across agency jurisdictions, such as integrated monitoring. This is why the question of long-range air transport of pollutants, which is related to health, climate, ecosystems, marine pollution, etc., was alternatively classified under health-related and climate-related monitoring. Eventually, it was assigned a cluster of its own, even though it consisted of only one project, is regional in scope, and is least important in terms of the resources that it has claimed from the GEMS budget line.

Towards the end of the period under review, in 1984, an important addition was made to the system when the pilot phase of the *Global Resource Information Database (GRID)* was launched, as an internal project of UNEP managed and executed directly by the GEMS Programme Activity Centre (PAC).

The international secretariat level

A key role in the evolution of GEMS is played by the secretariats of international organizations, principally that of UNEP but also of the specialized agencies and other bodies involved in the execution of its various components. In their task, they depend on cooperation and support from the scientific community.

As noted earlier, an important element in assuring a degree of integrated and centralized control and insight into the system and its evolution was the decision by the governments to set up a GEMS office within UNEP. Equally important was the follow-up move by the Executive Director of UNEP to establish a GEMS Programme Activity Centre. This step was taken to give the GEMS office higher status and visibility *vis-à-vis* the outside world and its users. Most of all, however, it was taken to assure for the GEMS office greater autonomy and operational flexibility within the UNEP secretariat. This was of special importance because GEMS PAC has remained very small, in line with the UNCHE prescription. With three or four professional staff members at the most, it had little spare capacity to devote to internal UNEP procedures. The PAC status relieved it from many intrasecretariat responsibilities, and gave it a significant degree of autonomy to pursue and build its programme by placing it directly under the Executive Director. Most importantly, it endowed it with its own budget line, which could not be

28 The clusters were made explicit, for the first time, at the fifth session of the Governing Council, following its request to the UNEP secretariat to show how the various programme goals of GEMS agreed in 1974 were integrated with each other (see decision 63(IV), in doc. A/31/25). The secretariat's response is to be found in the first overview of GEMS (doc. UNEP/GC/Information 2).

utilized for other programme needs in UNEP. This gave it assured resources to plan and implement its work.

GEMS PAC is supposed to act as a central, integrating core of the system, while the departments or units within the secretariats of the specialized agencies are in charge of the operational and substantive aspects of the environmental monitoring activities within their jurisdictions. These departments or units play the key role of linking up with field monitoring and of providing substantive and technical guidance and back-up for the various components of GEMS.

As regards the coordination mechanisms at the secretariat level, the GEMS subgroup of the Earthwatch Working Group, which functions within the framework of the Administrative Committee on Coordination (ACC), was set up to perform this task. In reality it has not been very successful and it fell into disuse. Rather, informal contacts and forms of interaction were preferred and have been used both bilaterally and multilaterally, and most often at the project level, where coordination seemed to work best. In addition, the overall coordination was kept up in the more general context of UNEP's System-Wide Medium-Term Environment Plan (SWMTEP) and Designated Officials on Environmental Matters (DOEM).

The intergovernmental level

The initial refusal by the 1974 Intergovernmental Meeting on Monitoring to endorse the establishment of permanent intergovernmental machinery for GEMS is probably the main practical reason why the governments, collectively, have largely remained on the sidelines with regard to the development and management of the system.

In its capacity as the intergovernmental organ of UNEP, the Governing Council has acted as the central coordinating body responsible for GEMS. However, in practice its role and influence has been somewhat marginal. With an hour or two devoted to GEMS by the Council's sessional committees, it was difficult to expect much more than a few generalized statements and rather perfunctory treatment of the whole subject matter.

At the third session of the Governing Council, the issue of the governments' role was raised once again. The UNEP secretariat proposed that a more systematic and detailed form of guidance from the member states be institutionalized. It suggested either a standing advisory group of experts on monitoring, to be appointed by the Executive Director in consultation with or on the recommendation of the governments concerned, or a steering committee of government representatives, to be selected on a basis that was to be determined by the Council. Both alternatives were turned down, and the Council opted for convening intergovernmental expert groups, as needed, to deal with given aspects of GEMS' development.[29]

These groups, however, in addition to being composed of experts only, have

---

29 See Council decision 29(III), para. 9(i), in doc. A/10025. It requested the Executive Director to convene small groups of government experts to work in close harmony with other UN bodies concerned with designing and implementing GEMS.

proved to be too sporadic and unrepresentative of the constituency, have been sheltered from the public eye, and have not provided effective access to governments in designing and managing the system. Nor have they helped to involve the governments in those instances where they did not show adequate interest – a tendency that can be countered to a degree if there is a standing body. Being specialized and dealing with given monitoring clusters, these groups were not meant or equipped to work from an integrated perspective that viewed GEMS as a whole, and this opportunity was denied to interested governments.

The counterpart intergovernmental bodies within the specialized agencies (and the ECE, in the case of transfrontier air pollution monitoring) offered governments an opportunity to have a more direct and regular say in the execution of environmental monitoring clusters. The government representatives, however, were not always aware of, or concerned with, what was taking place within the framework of UNEP and GEMS. This contributed to lack of coordination and the centrifugal pull within the system.

The basic components of GEMS

The viability of the concept of GEMS conceived in Stockholm largely depends on the money available in the Environment Fund. In other words, it is the UNEP-funded projects that have been the principal means of encouraging, coordinating, and executing various GEMS-related activities.

The Fund money has been used to pay for the staff and operation of GEMS PAC and to undertake activities inherent in its role as the centre of the system. It has also paid for the international facets of the various environmental monitoring activities coordinated and executed by the specialized agencies. In some instances, it has also supported national activities, when a government has not been able to undertake the required actions on its own.

It is the Fund money that gave weight to UNEP's claim to be the coordinator of a global environmental monitoring system. It also made it possible to forge a link, however tenuous, between the idea of a globally integrated system promoted by UNEP and the sectoral monitoring activities of the individual agencies. The availability of money in UNEP, combined with the lack of resources for monitoring in the agencies and their dependence on the Environment Fund, made this critical connection possible.

UNEP has advocated *Level Two* and *Level Three* concepts in tandem. It has stressed the need for the cooperating/supporting organizations to conduct activities on their own, or with only partial support from the Environment Fund. Within the UN system, however, GEMS has been limited mainly to *Level Three*, in the sense that virtually all environmental monitoring projects executed by various organizations have been dependent on financing from the Environment Fund. UNEP has also argued that the agencies themselves should gradually assume the financing of monitoring activities, after an initial period of its support. The agencies, however, true to their traditional view and backed by their governing bodies, have maintained that UNEP should pay for environmental monitoring on a continuing basis. This divergence of opinion continues to persist.

In the case of governments, the notion of *Level Two* has functioned much better. Their contributions in kind – ranging from funding of local monitoring projects to performing various tasks on behalf of the system (e.g., processing and publication of data) – have accounted for the lion's share of the operational costs of GEMS. These contributions are not seen among GEMS' expenditures. They could not be truly costed, as most of the time they were not budgeted for directly and were hidden among other domestic appropriations.

GEMS was conceived of primarily as a global system, although the importance of the regional approach and regional problems was acknowledged. The local monitoring programmes, undertaken by countries for their own benefit, were not originally considered as something that should be encompassed within GEMS. Gradually, however, this type of monitoring came to play an important role in the system. It is usually of immediate interest to countries and decision-makers, which are thus willing to commit the necessary resources. If carried out within a common global framework, it can provide useful and important information not only for local and national actions but also of regional and global relevance.

The nature of the projects, that is, of the different monitoring activities subsumed under GEMS, varies greatly in terms of subject matter, technical and scientific sophistication, and scope. This, combined with the institutional model that has been followed, has contributed to the loose, decentralized nature of GEMS. This tends to be frustrating for those who are used to a greater semblance of order and structure. Moreover, they would have to work hard to find out what GEMS is all about. The pertinent data and information are widely scattered, and a regular and systematic overview and insight into the system and its various components are lacking.[30]

Viewing GEMS implementation

Implementation and build-up of GEMS can be viewed as a process that consists of four phases. These can be sequential in the early stages of a given monitoring activity, eventually to become its coextensive and integrated elements. They are:

- Developing and testing monitoring methodologies, techniques, instrumentation; setting up, making operational, and maintaining monitoring programmes and

---

30 Available to an interested observer are several sources of information: (a) the UNEP programme documents, which speak in very general terms of the UNEP-funded GEMS projects, the progress made, and plans for the future; (b) the reports of the Governing Council, which summarize in a few sentences the relevant discussion and which give the decisions; (c) the occasional GEMS overviews, prepared by GEMS PAC for submission to the Governing Council; (d) the Environment Fund project documents, which include the purposes and objectives of the project, its financial aspects, etc. (these documents, however, are not usually available to the public); (e) the various specialized reports, manuals, assessments, etc., issued by the specialized agencies and by UNEP, in the context of different monitoring clusters; (f) the papers prepared for the various intergovernmental *ad hoc* expert groups, and the reports of these groups; and (g) an occasional popular brochure on GEMS, such as the one issued for the tenth anniversary of the Stockholm Conference.

networks, and carrying out intercalibration and data quality control; setting up the infrastructure in general, as well as securing the participation of as many countries as possible;
- Collecting, validating, storing, and systematizing the monitoring data in permanent reference centres, and making them available for use;
- Evaluating and interpreting the monitoring data and using them to make environmental assessment; and
- Utilizing such assessments and monitoring data as inputs in policy formulation, decision-making, and management; for improved knowledge; and for further research, evaluation, and monitoring.

The above can be a useful reference in assessing progress made in various environmental monitoring activities, which are reviewed in the six chapters that follow.

Part II
GEMS in action

Chapters three to eight focus on the constituents of the Global Environment Monitoring System (GEMS), namely the five monitoring clusters and the Global Resources Information Database (GRID), while chapter nine discusses GEMS as a whole.

For those who may wish to read only a chapter that happens to be of special interest to them, the questions asked and the content of the discussion cannot be fully understood without the background reviewed in the preceding section. For more technically oriented readers, what has been said in the introduction also bears repeating: this is primarily a study of patterns and models of international cooperation and institution-building focusing on the UN Environment Programme and does not address critically the outputs and findings emerging from GEMS.

The six mini-case studies – in the thirty pages or so allotted to each – can offer only a limited view of the fast evolving and inherently complex endeavours. They are meant to provide an insight into the content and to depict the main factors and processes that have played a role in the evolution of these activities over the years. The case studies are also intended to serve as a reference for assessment of GEMS as a global system and for suggestions on how to proceed with the further build-up of this still tentative instrument of international cooperation. A discussion along these lines is attempted in the last chapter of this section.

3 Health-related environmental monitoring in GEMS

Driven by the dynamics of economic growth and development, by rapid advances in science and technology, and by their own increasing requirements, modern societies have been introducing an ever greater number and quantity of substances and pollutants into the environment. The air people breathe, the water they drink, the food they eat, and, in general, their physical environment have thus come to contain constituents and properties harmful to their health. Often, the effects of these are insidious and hence are not directly perceived by those affected. They may take years or even generations to become manifest, as is the case with carcinogenic or mutagenic effects of low-level exposure to certain pollutants or radiation.

Although these effects tend to be more serious and complex in highly industrialized societies, they are also of serious and urgent concern to developing countries. These countries are experiencing rapid and uneven development that all too often has not been coupled with adequate sensitivity to and concern for threats to the environment and/or with the ability of the society to find the necessary resources and mount appropriate remedial and preventive actions.

While the majority of populations in developing countries continue to be exposed primarily to the classic bacteriological and viral pollutants associated with unhygienic conditions of life in poverty, increasing numbers are also beginning to suffer marked effects from chemical pollution. This has been the case, in particular, in the rapidly growing urban areas where conditions have become worse as a consequence of the acute development crisis of the 1980s. For example, as Global Environment Monitoring System (GEMS) data, cited later in this chapter, have shown, the median level of lead in the blood of inhabitants of Mexico City is almost four times higher than in Tokyo, and cities with the highest levels of sulphur dioxide and suspended particulate matter are predominantly in developing countries. It has also been realized that populations in the Third World ingest disproportionately large amounts of harmful substances because of uncontrolled use of agrochemicals, excessive reliance on a limited number of staple food items in their diet, and dependence on an untreated drinking water supply.

On the whole, then, these problems affect the entire world. To some degree, this is due to the integrated nature of the global environment and to the circulation and long-range transport of certain ubiquitous pollutants and radionuclides. Other

62 *The quest for world environmental cooperation*

contributory factors are the flows of international trade and the worldwide diffusion and homogenization of food and agricultural systems, industrial production, consumption patterns, and lifestyles, as well as the recent phenomenon of the export and dumping of toxic and radioactive waste worldwide by some industrial nations.

Health-related environmental monitoring was assigned a high priority by the Stockholm Conference, in the light of the work already undertaken or planned by the World Health Organization (WHO).[1] It was highlighted by the UN Environment Programme (UNEP) Governing Council at its first session, and was given top ranking among GEMS' programme goals, as spelled out by the 1974 Intergovernmental Meeting on Monitoring. Virtually all the pollutants that the Meeting recommended should be monitored were relevant to human health.[2]

NATURE AND CONSTITUENTS OF GEMS HEALTH

In a situation where the objective is to maintain or restore a clean environment, health-related monitoring would be comparatively simple in conceptual and operational terms and would be limited to monitoring the quality of air, water, and food. The data thus collected would make it possible for authorities to keep watch in order to maintain the quality of these elements within given norms.

Unfortunately, the objective of a truly clean environment has in most places become illusory in practice. The concept of a safe or low-risk environment has been advanced in its stead. What is safe, however, is a relative notion. Defining it is surrounded by political and scientific controversy, uncertainty, and often ignorance, as illustrated by the attempts to apply the risk-assessment analysis; it is further complicated by humans' resilience and adaptability to adverse conditions, and by the varied exposures and reactions of individuals to widely differing combinations of pollutants from different media. The time-lag between impact and effect, and the synergisms of various environmental agents in the human body make it even more difficult to ascertain what constitutes safety and risk.

Moreover, this scientific controversy and uncertainty are tied in with a series of underlying social tensions and conflicts. Most of the measures to keep the environ-

---

1 In particular, see Stockholm Conference recommendation No.76, which calls for a major effort 'to develop monitoring and both epidemiological and experimental research programmes providing data for early warning and prevention of the deleterious effects of the various environmental agents, acting singly or in combination, to which man is increasingly exposed'. Recommendation No.77 calls upon WHO to assist governments 'in undertaking co-ordinated programmes of monitoring of air and water and in establishing monitoring systems in areas where there may be a risk to health from pollution'. Similarly, recommendation No.78 calls for the establishment of 'internationally co-ordinated programmes of research and monitoring of food contamination by chemical and biological agents'. (A/CONF.48/14/Rev.1, pp.20–1.) For a comprehensive review of the early work and plans of WHO, see 'WHO Health-Related Environmental Monitoring Programme', WHO doc. EHE/74.1.

2 Sulphur dioxide and suspended particulates were to be monitored in air; radionuclides in food; ozone in air, and DDT and other organochlorine compounds in biota and humans; cadmium and its compounds in food, humans and water, and nitrates and nitrites in drinking water and food; mercury in food and water, lead in air and food, and carbon monoxide in air; fluorides in fresh water; asbestos in air and arsenic in drinking water; and mycotoxins and microbial contaminants in food. See doc. UNEP/GC/24 and page 45 in chapter two above.

ment safe for human beings can be considered as circumscribing the dynamism and freedom of action of important and powerful economic agents and forces; such measures tend to be disturbing in financial and economic terms and to entail costs that the society and/or those directly responsible may find difficult to pay or that they are not prepared to bear voluntarily.

This is where the monitoring of people comes into the picture. Personal exposure monitoring is intended to help determine the levels of exposure to environmental pollutants which cannot be deduced by monitoring environmental quality alone. Biological monitoring of body tissues and fluids provides further essential data on the levels of pollutants in the human body.

This type of exposure monitoring is needed to establish and revise environmental standards related to human health. It should yield vital data required for the study of cause–effect relationships between environmental pollution and health. In principle, such data should also alert the public and decision-makers and strengthen the case for action, which is often not taken on account of the lack of appropriate data on exposure and detrimental health effects.

The above considerations provide the basic operational outline of GEMS HEALTH. Its initial thrust was toward monitoring environmental quality, i.e., the quality of air, water, and food. The explicit link with human health was established through exposure and biological monitoring, which represents the second facet of this GEMS cluster. These components were conceptually linked when, one decade after the Stockholm Conference, the idea of a comprehensive health monitoring programme of Human Exposure Assessment Locations, or HEALs, was put forward for implementation.[3]

In the wake of the Stockholm Conference, health-related environmental monitoring was poised for a quick start.[4] The initial move was towards monitoring pollutants in air, water, and food, with UNEP approving and financing three projects executed by WHO: GEMS/Air, in cooperation with the World Meteorological Organization (WMO), GEMS/Water, in cooperation with WMO and the United Nations Educational, Scientific and Cultural Organization (UNESCO), and GEMS/Food, in cooperation with the Food and Agriculture Organization (FAO). This type of monitoring reflected the Stockholm Conference recommendations to monitor pollutant levels in the environment and to use that information for developing and implementing environmental standards to safeguard human health. An intergovernmental group on health-related monitoring, which met in 1974,

3 Referred to as integrated monitoring in the early days, the notion was mooted at the very beginning of GEMS HEALTH. See 'Integrated monitoring systems as tools to develop environmental indices, elucidate critical pathways and model human exposure', in annex 3 of the 1974 WHO document on environmental monitoring (EHE/74.1, *op. cit.*). Noting that this was a completely new area in which no experience had been gained even in the most advanced countries, it was proposed that a pilot project on integrated monitoring be initiated, using lead as a model pollutant.

4 For a detailed description and review of the initial phase of this cluster of GEMS, see 'Health-related monitoring of environmental pollutants', a background document prepared by the UNEP and WHO secretariats as a contribution to the deliberations of the intergovernmental expert group that met in 1982 (WHO doc. EHE. EFP/81.20). This document, prepared on the eve of the tenth anniversary of the Stockholm Conference, represents the first integrated overview of GEMS HEALTH. It is included as annex II to the 1982 intergovernmental expert group's report (WHO doc. EFP/82.28).

approved the lines of action charted under the leadership of WHO and formulated a more detailed framework for action than had been available up to that time.[5]

GEMS/Air consists of air-quality monitoring at residential, commercial, and industrial sites in urban areas. Its initial operational objective was limited to the monitoring of two parameters only, namely sulphur dioxide and suspended particulate matter (SPM), although there is also an expanded list of variables, which includes nitrogen oxides, carbon monoxide, ozone, and air-borne lead, to be monitored at certain urban sites only. GEMS/Air involves impact monitoring. The data are therefore location-specific and are intended primarily for local use, in contrast to the background air pollution monitoring performed by WMO's Background Air Pollution Monitoring Network (BAPMoN), which is discussed in the next chapter, whose purpose is to determine baseline air pollution levels in the global environment.[6]

The chief purposes of GEMS/Air are to develop and disseminate methodologies for monitoring and analysing air quality in urban areas, to encourage countries to establish their own domestic air-quality monitoring networks, and to achieve the harmonization of national monitoring practices, as well as to ensure the validity and comparability of air-quality measurements.

In other words, GEMS/Air is primarily intended to provide countries with a tool for their own local use and action. The size and density of the global network, then, is ultimately a function of local needs, interest, and resources. However, while this type of network produces sets of location-specific data, its applications and value are not confined solely to local needs. Uniform and standardized data and information help to create a collective awareness of conditions and to impart a significant degree of rationality to the overall effort of the international community to cope with common situations and shared problems. Data generated by the network can be used for comparative studies and research on environmental health and in devising effective responses and strategies. They can help in maintaining an overall picture of situations and trends in localities in different parts of the globe, and make possible and encourage comparisons between the effectiveness of various national policies, programmes, and measures. The network can also provide for an international flow of information, ideas, and potentially useful back-up for international health-related programmes, such as WHO's Environmental Health Criteria.

GEMS/Water is concerned with the monitoring of the quality of bodies of fresh water, such as rivers, lakes, and groundwater, that are used for municipal supplies, irrigation, industry, and livestock. A large number of water-quality variables are measured, which fall into two broad categories. The first encompasses the basic physical, chemical, and microbiological variables that are important in assessing general water quality (e.g., biochemical oxygen demand, acidity, phosphates, suspended solids, coliform bacteria) along with optional variables that are import-

5   For the report of the group, see WHO doc. EHE/75.1.
6   For a review and comparison of GEMS/Air and BAPMoN, see H. W. de Koning and A. Kohler, 'Monitoring Global Air Pollution', *Environmental Science and Technology*, Vol. 12, August 1979, pp. 884–9.

ant only at certain locations and for specific water-quality objectives; the second category, which is monitored for the purpose of determining the incidence and long-term trends of water pollution by selected hazardous substances that have global and long-term significance due to their persistence, ubiquity, and bioaccumulation in the environment, including *inter alia* cadmium, mercury, lead, polychlorinated biphenyls (PCBs), and organochlorine compounds. This division was eventually revised and simplified into two groups: basic variables and use-related variables.

Like its air counterpart, GEMS/Water is intended primarily to promote the establishment and/or strengthening of national water-quality monitoring systems and to generate data of local significance relevant to public health. By using standard and/or comparable methods and procedures, the validity and comparability of local data between countries and globally are eventually to be achieved. Unlike GEMS/Air, however, GEMS/Water also partly involves background monitoring for globally significant, persistent, and hazardous variables in areas remote from sources, in order to assess the incidence and long-term trends of freshwater pollution caused by such substances. In other words, GEMS/Water is not just limited to human health concerns, but also produces water-quality data of broader significance. Accordingly, the WHO-coordinated freshwater monitoring programme is also an attempt to perform a more generalized water monitoring function and to compensate in part for the institutional void in the UN system in which a specialized agency for water does not exist.

GEMS/Food has as its objective the monitoring of food and animal feed contamination in order to gain a knowledge of the causes, sources, and incidence of key food contaminants, to establish baseline levels and time trends for dietary intake, and to help protect consumers and implement food safety and health policies. Unlike GEMS/Air and GEMS/Water, which entail field monitoring and global networks of monitoring stations, GEMS/Food consists of the examination of food and animal feed samples in a network of national laboratories chosen on the basis of location, climate, ecological zone, willingness to take part, and laboratory capability. The pilot project on food monitoring was launched for the purpose of monitoring organochlorine compounds, lead and cadmium, and aflatoxins in selected foods and total diet; animal feed monitoring, however, was left pending for future implementation.

The Intergovernmental Expert Group on Human Health-Related Environmental Monitoring, which met in 1977, agreed that ambient monitoring alone – while useful in providing a picture of the quality of the environment and of the relative exposures of the populations involved, as well as for control purposes – was not sufficient in itself to provide an insight into the effects of pollutants on human health. The Group therefore recommended two pilot projects, one on human exposure to air pollution, the other on biological monitoring.[7] As a consequence, the scope of GEMS HEALTH was expanded to include the receptors (i.e., people)

---

7   For the report of the Expert Group, see WHO doc. CEP/77.6.

in addition to the concern with sources and levels of pollution through environmental quality monitoring.

In brief, the purpose of the pilot project on the assessment of human exposure to air pollutants, including personal and indoor monitoring, was to determine the extent to which the ambient air monitoring data alone could be used to assess the exposure of populations, and what other information was necessary. The ultimate objective was to develop an improved methodology, and a more powerful tool for analysis and evaluation of the urban air quality data produced by the monitoring networks. The pilot methodology was to be tested under different field conditions – in cities with differing climates, industrial activities, heating methods, transportation systems, lifestyles, and pollution profiles – by studying different points of exposure to pollutants in diverse surroundings: the home, workplace, street, car, etc. The long-term objective, to be achieved on the basis of the exposure assessment studies and field experience, was to identify high-risk groups in different urban settings and to estimate their exposure (statistically or with the help of models) to ambient concentrations of pollutants obtained from standard air-quality monitoring networks.[8]

The biological monitoring pilot project consisted of two sub-projects – one to monitor cadmium in the kidney cortex and cadmium and lead in blood, the other to monitor organochlorine compounds in breast milk. Its principal objective was to develop an internationally agreed methodology for determining body burdens and milk concentrations, and for assessing the rates at which humans absorb, modify, accumulate and dispose of selected pollutants. It was also intended to promote the build-up of national capabilities for this type of monitoring.

The experience with the three monitoring networks and the results of the two pilot projects on exposure to air pollutants and on biological monitoring showed the need for more and better information, for pulling the different elements and data together, and for integrating the knowledge acquired. Ten years after Stockholm, health-related environmental monitoring was thus poised for an important step forward. It came in the form of the HEALs proposal to integrate the effort by carrying out, at single locations, the monitoring of environmental components and of complete human exposure, as well as biological monitoring.

The idea was endorsed by the 1982 Intergovernmental Expert Group on Health-Related Monitoring.[9] The proposal involved the creation of a global network for research, monitoring, and exposure assessment, with 15 sites during the initial stage, mostly in industrialized countries. At first, focusing on persistent and ubiquitous pollutants that occur in more than one medium, the HEALs were intended to help develop methods for integrated monitoring and assessment of human exposure to environmental pollutants. They were also expected to lead to the formation of a permanent global network to undertake advanced work and research, and to disseminate knowledge and experience for field application.[10]

8 See 'Estimating Human Exposure to Air Pollutants', WHO Offset Publication No.69, Geneva, 1982, p.12.
9 See the Group report, doc. EFP/82.28, *op. cit.*, pp.7–8.
10 For the initial proposals on HEALs, see *ibid.*, annex II, pp.30–2 and WHO doc. EFP/83.52.

The above elements, then, are the basic constituents of GEMS HEALTH, as it developed in the decade after Stockholm. Conspicuous by its absence is the monitoring of ionizing nuclear radiation. As noted in chapter two, it was dropped from the GEMS priorities in the light of the expert opinion of the UN Scientific Committee on the Effects of Atomic Radiation (UNSCEAR), communicated to the UNEP Governing Council. For the record, the initial WHO working document on health-related monitoring contained an extensive section on environmental radiation. Among other things it expressed the view that environmental radiation monitoring activities should be expanded and that greater emphasis should be given to local monitoring of radioactive waste disposal and the operation of nuclear power reactors. It also noted that one difficulty encountered in interpreting and evaluating information obtained by UNSCEAR from national monitoring networks, mainly in developed countries, was the lack of comparability of data due to the use of different methods of measurement and instrument calibration, and varying standards of data reporting by governments that are not obliged to provide such information.[11]

While the challenge was not picked up within the framework of GEMS until after the Chernobyl accident more than a decade later, as will be discussed below, WHO managed to keep alive the idea; with an ambitious objective to assess the exposure of the world population to radioactive contamination of the environment, a limited international programme to monitor ambient radioactivity was maintained. Since 1970, data on milk radioactivity have been collected and published on a quarterly basis; since 1979, data relating to levels in ground-level air and precipitation have also been included.[12]

INDICATORS OF PROGRESS

In the case of GEMS HEALTH, the basic indicators used to appraise the status and progress of the initial phases of various GEMS clusters may be grouped as follows:

- progress in the basic design and development of the programme, primarily standardized methods and procedures for monitoring;
- formation and growth of the infrastructure for the system and the expansion of

---

11 WHO doc. EHE/74.1, *op. cit.* pp. 23 ff.
12 The Central Service for Protection against Ionizing Radiation (SCPRI) of the French Ministry of Health at Le Vésinet was designated in 1969 as the WHO International Reference Centre for Radioactivity. It has centralized and published data, and organized periodic intercomparisons of environmental samples, etc. As of 1980, France, Sweden, Canada, the USA, Japan, New Zealand, the USSR, and Australia were taking part in the programme. Caesium 137 and strontium 90 were monitored in milk, with the participating countries providing average values for gathering areas. The approaches varied between countries, ranging from climatic regions to major dairies and to large cities. In the USA, 80 per cent of the milk consumed was monitored, while the USSR provided regional samples. Ground-level radioactivity was monitored mainly for worldwide fallout. The 140 stations making up the network were unevenly distributed among the participating countries. In the case of 32 stations in France and 22 stations in the USA belonging the Environmental Protection Agency's (EPA's) Environmental Radiation Ambient Monitoring System, or ERAMS, nuclear facilities were also monitored. 'Environmental Health Monitoring, Data on Environmental Radioactivity', first quarter 1980, report No.36, WHO.

68   *The quest for world environmental cooperation*

the global monitoring networks, with the countries using GEMS procedures and methods in their national monitoring activities;
- the scope and quality of monitoring data; and
- uses of monitoring data for assessment and management, for expanding knowledge, and for further improving the system.

On most of these counts, significant progress has been achieved in the different components of GEMS HEALTH.[13]

Methodologies

Monitoring methodologies and procedures (i.e., designs for monitoring programmes, detailed measurements procedures, quality control, reporting, analysis, and interpretation of data) were developed, tested, and published by WHO in a series of manuals.[14] Interaction among scientists and institutions from different countries and their joint efforts made methodological advances possible and enriched the results. The methods were disseminated for application in the field. Participating national laboratories and institutions were provided with standardized instructions and reference methods, a prerequisite for harmonizing their monitoring activities and achieving the comparability of monitoring data. The provision of such ready-made packages to countries – reinforced by training efforts directed to the developing world – not only made it possible and easier for those wishing to take part in this GEMS activity to do so, but also acted as an inducement to action and participation.

Infrastructure

The existing WHO worldwide institutional network has proved to be of great value. It consists of six WHO regional offices and global, regional, and national collabor-

---

13 For a review of the initial period, see WHO doc. EFP/82.28, annex II, *op. cit.* More detailed reports are available on each component of GEMS HEALTH. For example, for GEMS/Air see WHO progress reports in documents CEP/77.2, CEP/78.12, ETS/79.2, ETS/80.1; for GEMS/Water, see CEP/77.8, ETS/78.3, and especially 'GEMS/Water Data Evaluation Report 1983', doc. EFP/83.55; for GEMS/Food see HCS/FCM/78.3; etc. For a popularized overview of the results of the first ten years of GEMS HEALTH, see a brochure 'Global Pollution and Health, Results of health-related environmental monitoring', UNEP and WHO, 1986. See also UNEP, *Environmental Data Report*, Blackwell, Oxford, 1987. Most of all, see three reports prepared for the September 1988 Meeting of UNEP/WHO Government Experts on Health-Related Environmental Monitoring, i.e., 'Assessment of Urban Air Quality', UNEP and WHO, 1988; 'Assessment of Freshwater Quality', Report on the Results of the WHO/UNEP Programme on health-related environmental monitoring, UNEP and WHO, 1988; and 'Assessment of Chemical Contaminants in Food', Report on the Results of the UNEP/FAO/WHO Programme on health-related environmental monitoring, UNEP, FAO and WHO, 1988. (Note that the three reports were prepared in cooperation with the Monitoring and Assessment Research Centre [MARC], London.) See also the report of the Meeting of Experts, 'The Quality of the Environment: a Health-Based Global Assessment', doc. WHO/PEP/88.15.

14 See, for example, *Selected Methods of Measuring Air Pollutants*, WHO Offset Publication No. 24, Geneva, 1976; 'GEMS/Water Operational Guide', WHO docs ETS/78.5–78.11; 'Guidelines for Strengthening National Food Contamination Monitoring Programmes', WHO doc. HCS/FCM/78.1; etc.

ating centres and reference laboratories. Moreover, WHO enjoys direct access to and the cooperation of national health authorities, institutions, and laboratories. It made it possible to plan, design, and put into operation a corresponding global infrastructure and networks for health-related monitoring projects in a relatively short time. Some of the functions performed by this infrastructure are: centralized and computerized data storage and processing; data publication; technical support; intercalibration exercises to achieve the comparability of results arrived at through different measurement methods and quality assurance programmes; and training and information exchange.[15]

Networks

As regards network development and expansion, it should be recalled that in the case of GEMS/Air and GEMS/Water, impact rather than background monitoring is carried out. This means that the data are location-specific (e.g., within a major river basin, or in a given neighbourhood within a major metropolitan area). Hence, the size, scope, and density of a network are primarily a function of local and national interests, specific needs, and the resources available for this purpose (e.g., 4000 air-quality monitoring stations were established in Japan within the framework of its national air-quality monitoring programme, unrelated to GEMS).

In the years following the Stockholm Conference, when there were virtually no internationally linked health-related monitoring stations or activities, a significant degree of global network expansion took place. Already by 1981, GEMS/Air had made considerable progress towards its goal of covering major urban centres around the world; close to 200 stations were operational in 75 major cities in 50 countries (of which 25 were developing ones), spread over the six WHO regions.[16]

15 In the case of GEMS/Air, technical support bodies were appointed to assist with these tasks in Brisbane, Lima, London, Moscow, Nagpur, and Tokyo. The Centre for Air Pollution Control of the US EPA and the UK Centre on Clinical and Epidemiological Aspects of Air Pollution were appointed to act as the WHO collaborating centres for the project. EPA also processes and publishes data on behalf of GEMS. For purposes of intercalibration, a series of monitoring sites act as comparison stations, using their own equipment as well as the comparison equipment supplied by WHO. In GEMS/Water, the WHO Collaborating Centre for Surface and Ground Water Quality, at the Canada Center for Inland Waters, was designated as the global data centre, and was also placed in charge of data analysis and publication. In the case of GEMS/Food, the WHO/FAO Unit coordinates and processes data, while certain aspects of the project, including the quality control of data, are handled by different laboratories, i.e., organochlorine compounds by the Toxicology Laboratory in Uppsala, lead and cadmium by the British Ministry of Agriculture, and aflatoxins by the International Agency for Research on Cancer (IARC). In the case of biological monitoring, coordination of the metals component of the pilot project was assigned to the Department of Environmental Hygiene of the Karolinska Institute and to the National Swedish Institute of Environmental Medicine, Stockholm; coordination of the organochlorine component was assigned to the Food Research Department of the National Food Administration in Uppsala. In addition, reference laboratories were designated at the International Atomic Energy Agency (IAEA), at the Commission of the European Communities Joint Research Centre at Ispra, and the US National Bureau of Standards, which provide standard substances and reference materials.

16 See table 1 in WHO doc. EFP/82.28, Annex II, *op. cit.*, p.9. This represents a considerable advance compared with 1973, when only 14 developed countries were taking part in the pilot project, or compared with 1977, when there were 45 stations in 15 countries, of which only two were developing nations.

In spite of important gaps in global coverage (e.g., Africa, USSR, Central America), the existing network is sufficiently representative to give a global view of urban air pollution. In evaluating this expansion, it should be kept in mind that limited and controlled growth of such networks is required in a situation where the participation of many countries is dependent upon expert advice, training, and equipment provided internationally, and where resources for this purpose are limited. Moreover, the need to carry out costly intercalibration and quality control also imposes significant limits on the speed of network expansion.

GEMS/Water also managed to build up its network considerably, mainly by 'piggy-backing' its needs on the already existing national and local water monitoring networks and stations. It was estimated that only one-third would need to be newly established stations, i.e., where it was not possible to rely on the existing ones. The network started operating in 1979 with 286 lake, river, and groundwater stations initially designated as representative of regional and typical water-quality regimes; the number had grown to 448 in 59 countries by mid-1983. In 1981, 166 stations in 33 countries were reporting data; by mid-1983, this number had increased to 344 stations submitting data on a routine basis.[17] Based on the initial experiences and network operation, which showed gaps and weaknesses in the programme, a major redesign and refinements were proposed for the global network.[18]

The GEMS/Food network consisted initially of 25 collaborating laboratories in 23 countries, which analysed retrospective data on food products from their national territories. Although biological monitoring and monitoring of exposure to air pollutants were pilot activities and were primarily research-oriented, they also represented a contribution to the build-up of global networks.[19]

Finally, the concept of HEALs, planned for 15 representative sites during the pilot stage, is an important addition and could serve as the hub for an integrated global human health-related environmental monitoring network.

Parallel with the expansion of various networks, network-capacitating actions

17 See *ibid.*, pp. 13–15, and 'GEMS/Water Data Evaluation Report 1983', *op. cit.*
18 See the section 'Proposed Modifications of the Global Network', in doc. EFP/83.55, *op. cit.*, and the 'Report of the Inter-Regional Review Meeting on Water Quality Monitoring Programmes', Burlington, Ontario, 17–21 Oct. 1983, WHO doc. EFP/83.59.
19 In the case of biological monitoring the project did in fact lead to the formation of a network of laboratories that, through collaboration, intercalibration, and practice, acquired the capability of providing high-quality data. See doc. EFP/82.28, *op. cit.*, annex II, p. 22. Note that, due to financial constraints, the pilot project avoided including developing countries that needed financial support or technical assistance. (Though designated, Iran did not participate because of religious objections to monitoring of mother's milk. Paradoxically, it was the delegate of Iran at the 1974 Intergovernmental Meeting on Monitoring who insisted with passion on the need to monitor DDT in mother's milk.) The pilot project for monitoring exposure to air pollution started with the two studies aimed at developing a methodology, which were carried out in Toronto and Zagreb. The methodology was to be tested in three categories of cities: (a) where intensive monitoring has been taking place, i.e., Los Angeles, Rotterdam, and Tokyo; (b) where air-quality monitoring networks exist, i.e., São Paulo, Mexico, and Manila (parenthetically, the network in Mexico was largely non-operational as the government was not keen to display publicly the figures on air pollution levels, while that of Manila did not function on account of broken equipment which was not replaced); (c) where, in spite of heavy pollution, no systematic monitoring has taken place (Bombay, Lagos, and Jakarta). See WHO doc. ETS/80.2.

were taking place. Extensive training was undertaken, intercalibration and data quality control exercises were carried out, and procedures for monitoring, data ordering, flow, and storage, and for data analysis and evaluation were developed and improved. One of the key aspects of the monitoring process is quality control assurance, which ideally should cover all aspects of monitoring, including sampling and measurement processes and the data, and which should be one of the principal tools available to the global centre of the system. However, it is very costly, involved, and complex, and had to be downplayed owing to lack of resources.[20]

Monitoring data

Monitoring has been carried out most notably and on a regular basis in GEMS/Air, which has generated useful data on urban air pollution in different cities of the world. On the basis of these data, initial attempts were made to analyse and interpret urban pollution data, to compare them with exposure limits suggested by WHO and to analyse trends. Comparable data are becoming available, in some cases for a number of years successively. This information makes initial assessments possible, including that of a worldwide situation in urban areas as concerns sulphur dioxide and suspended particulate matter, and provides useful back-up inputs to other health-related international programmes, such as Codex Alimentarius, the International Drinking Water Supply and Sanitation Decade, and Environmental Health Criteria.[21]

National authorities are beginning to obtain the monitoring data they need to understand, assess, and follow changes in local situations and, if they so desire, to use such data and assessments for management, control, and remedial action. This support for domestic action is one of the major intended functions and benefits of GEMS HEALTH. Moreover, fifteen years after its inception, using the data that had been collected, three reports of general interest were prepared and made public: on air quality worldwide, on global freshwater quality, and on the global situation and trends regarding chemical contaminants in food. The three documents offered for the first time an approximation of the global situation and an insight into the major problems being experienced. The two key policy messages conveyed con-

20 The degree of difficulty and complexity of these procedures varies between different segments of GEMS HEALTH. See doc. EHE/EFP/81.20, *op. cit.*, for an overview of these. The relevant manuals in the WHO Offset Publication series provide a more detailed description.
21 For an initial assessment summary and overview of the data received from the various programmes, see the report prepared for the 1982 expert group meeting, WHO doc. EFP/82.28, *op. cit.* As regards air pollution monitoring, see the WHO series entitled 'Air Quality in Selected Urban Areas', which first appeared in 1973/4 and has been published biennially ever since. See also *GEMS, Urban Air Pollution 1973–1980*, prepared in cooperation with the Monitoring and Assessment Research Centre (MARC), Chelsea College, University of London, London, Geneva, 1984. Concerning food, see 'Summary and Assessment of Data received from FAO-WHO Collaborating Centres for Food Contamination Monitoring', National Food Administration, Uppsala, 1982. With respect to the metals component of biological monitoring, see 'Assessment of Human Exposure to Lead and Cadmium through Biological Monitoring', edited by M. Vahter, National Swedish Institute of Environmental Medicine and Department of Environmental Hygiene, Karolinska Institute, Stockholm, 1982. See also the UNEP/WHO brochure 'Global Pollution and Health', *op. cit.*

72 The quest for world environmental cooperation

cerned the important role of control measures in improving the situation in the North (e.g., a downward trend in sulphur dioxide concentrations at approximately 50 per cent of the monitoring sites), and the steadily worsening situation in the Third World, precisely for lack of such measures and on account of a development crisis. The documents also presented an overview of the difficulties experienced in the build-up of various networks, including the major gaps or insufficient coverage in many geographic areas in the Third World, and also in Eastern Europe.[22]

Refining the approach

The refinement of the overall approach to health-related environmental monitoring has been of no less importance than the progress indicators listed above. It reflected the increasing understanding of practical applications and the value of monitoring data, and a desire to make these data useful and usable both in the quest for further knowledge and in decision-making and management. The initial steps in this direction were taken with the two pilot projects on the monitoring and assessment of human exposure to air pollution and on biological monitoring.[23]

The knowledge acquired through the two pilot projects and the experience gained in the air, water, and food monitoring programmes led to recognition of the need for integration and to the formulation of a proposal to undertake comprehensive monitoring and exposure assessment of air–water–food–humans at single locations and for a variety of substances. The HEALs sites are to be representative of the major classes of conditions, including levels of development, climate, geography, environmental degradation, health and nutritional conditions. The aim is to involve the institutions already working under GEMS HEALTH and therefore to benefit from their experience and the work undertaken thus far. The three GEMS HEALTH segments – Air, Water (with special emphasis on drinking water), and Food – are to be located at the HEALs sites and are to be backed up by exposure and biological monitoring.[24]

Both operationally and scientifically demanding and difficult to carry out, the

---

22 See documents on urban air quality, freshwater quality, and chemical contaminants in food, cited in footnote 13 earlier in this chapter. See also doc. WHO/PEP/88.15, *op. cit.*, pp. 2–7, for a review of the status of the three programmes. The assessments drew on a number of secondary sources, in part to compensate for the gaps in coverage and data in GEMS HEALTH.

23 Monitoring of urban air pollution has been used as an example of how needs change over the years and how the changes make it necessary to modify the approach to respond to new requirements. A six-stage programme was identified: (1) monitoring to describe ambient air quality; (2) monitoring to predict air quality; (3) checking that the prediction system remains in calibration; (4) working on the attainment of air quality criteria and standards; (5) carrying out health effects research; and (6) integrating with a broader programme of health-related monitoring involving food and water. Yu. A. Izrael and R. E. Munn, 'Monitoring the environment and renewable resources', chapter 13 in W. C. Clark and R. E. Munn, eds, *Sustainable Development of the Biosphere*, IIASA, Laxenburg, Austria, and Cambridge University Press, Cambridge, 1985.

24 For a detailed proposal on the HEALs project, see WHO doc. EFP/83.52, *op. cit.* As regards the participation of developing countries in the pilot project – based on the assumption that countries taking part will pay the internal costs – it was noted that major technical support for them was dependent on financial resources to be obtained from funding and donor organizations (*ibid.*, p. 10).

HEALs idea reflects what has been known all along, namely that environmental media are not compartmentalized and that pollutants move between them, that humans are exposed simultaneously to pollutants through media and that human exposure is strongly influenced by physical and social environment (e.g., lifestyle, diet, mobility, workplace, home, social status. Moreover, HEALS mean the coupling of research with monitoring.

In sum, then, a brief review of the chosen indicators shows considerable progress within the framework of GEMS HEALTH, in spite of unavoidable weaknesses and deficencies related to the implementation and functioning of such international activities, which will be discussed in the next section.[25] The necessary components of the global system were developed and applied, infrastructure was established, and a significant number of countries were brought into the effort. One of the tangible benefits was the provision of tools to national authorities to help them understand and assess local situations, follow changes, and carry out control, regulation, and management actions related to the improvement of environmental qualilty. Another achievement was the contribution to the improvement of the national databases through analytical quality programmes. Most of all, monitoring data became available – which is essential not only for improved knowledge and for action but also for judging the value and usefulness of the system itself and for working on its betterment.

FACTORS AT WORK

Health-related monitoring is first and foremost a tool for national action to meet local needs. It is relevant to daily life and to decision-making. Countries should be interested in monitoring for domestic purposes. Also, they should be willing to take part in health-related environmental monitoring provided that such participation results in some tangible benefit for them. While in the case of many developing countries, the initial impetus comes from the technical assistance, the ready-made methodological packages, and the training and equipment that are provided through GEMS HEALTH, the real attraction for all countries – including the most developed and self-sufficient – is primarily the scientific and technical advances that can be made through joint endeavours and the exchange of knowledge, information, and comparable data that becomes possible.

This national interest is important for the successful functioning of the key premise of *Level Two*, i.e., that the governments will assume and discharge their commitments under the global scheme. However, a number of obstacles hamper

---

25 The defects and problems become evident after the initial runs of a given programme, an organized feedback, and a critical evaluation of network structure and performance and of data quality, etc. For example, the first expert evaluation of GEMS/Water exposed a series of problems and deficiencies. Among these were gaps and imbalances in coverage, problems with the monitoring sites chosen, poor data quality, missing variables among those that were supposed to be monitored, significant delays in reporting data . These insights made it possible to propose improvements and changes to be implemented in the next phase of system build-up, e.g., a revised definition of monitoring stations, additions to the network, a shortened list of variables to be monitored to include those of special relevance to human health. See doc. EFP/83.59, *op. cit.*

progress towards this objective. These include the relatively high costs involved and the lack of resources for monitoring in most countries. The institutional obstacles that have hindered domestic efforts in responding to environmental concerns in general have also played a role – in this instance, different mixes of agencies responsible for sectoral monitoring networks in each country. For example, one of the difficulties experienced in trying to launch HEALs was to bring persons concerned with the quality of food, water, and air together so that they could talk to each other and take action jointly with the health authorities. These jurisdictional and disciplinary barriers often proved more difficult to overcome and manage than the complex technical side of HEALs.

A factor that has contributed to the relatively successful functioning of the UNEP scheme is that WHO acted as the cooperating organization in charge of the implementation of GEMS HEALTH. WHO's capability at its headquarters, its global infrastructure, its regional orientation, its experience with field work, the fact that it is sensitized to different characteristics of countries, all contributed to the launching and implementing of this GEMS cluster. This provided a shortcut through many time-and resource-consuming logistical and organizational intricacies connected with the mounting and implementing of global monitoring programmes and it ensured the continuing momentum of the build-up and the maintenance of communication within the system.

A further beneficial factor was the possibility of relying on the informal global WHO network of scientists and health officials, which functions primarily on the basis of solidarity, common interest, commitment, and friendship evolved through years of joint work and cooperation. This has helped to advance GEMS objectives and, especially, to cope with the practical difficulties of implementing loosely structured, basically voluntary worldwide environmental monitoring schemes.

Most other entries in the GEMS HEALTH balance sheet are not as positive. On the whole, they tend to attenuate and hold back programme implementation, are liable to channel it in directions that are not necessarily desired or optimal, and/or demand special efforts, resources, and time.

North–South differences

One group of obstacles is accounted for by North-South disparities. The majority of developing countries do not have adequate back-up infrastructure, experience, or, in particular, resources for this type of activity. In many, the build-up must virtually start from scratch; in most, some form of international support and encouragement is required and is usually expected. One of the principal substantive and policy goals of GEMS is the establishment of a global network and the full involvement of all countries, especially developing nations. Their needs are such that they require special attention, resources, and a corresponding adjustment of the substantive and technical approaches and pace of implementation is called for.

This development orientation has had a series of practical consequences for the nature and evolution of health-related environmental monitoring. The limited financial resources available on an international basis for the build-up of GEMS

HEALTH were not really meant to be used to respond to, much less satisfy, the needs of developing countries. However, in the absence of other financing, these resources have to be used for what amounts to technical assistance. On balance, the consequence is that money made available and allocated in order to fulfil other, financially less demanding, requirements of the global system build-up are reduced accordingly, whereas the real needs of developing countries are met only superficially and unevenly.

This dilemma was recognized by the 1982 intergovernmental expert group when it noted the need to tap alternative sources of financing to support developing countries' involvement and participation. For this purpose, it singled out the UN Development Programme (UNDP).[26] Failing to obtain these resources, the group felt that the objectives of GEMS HEALTH should be scaled down and the programme should be stretched over a longer period of time and adjusted to the resources that were available internationally.[27]

In order to make access and participation possible for all countries, at least in principle, the system must be kept within the limits of the capabilities and resources of the majority. This limitation obviously exerts a simplifying and reductionist influence. For example, the number of variables monitored in a given programme may have to be reduced, the methods used may be simpler and sometimes less reliable than is theoretically and practically possible, and the standard instrumentation used will have to be relatively inexpensive (and hence, by definition, less sophisticated than it would otherwise be).[28]

This inevitable pull in the direction of the lowest common denominator is accentuated by some countries that have consistently failed to organize their efforts, or were not adequately motivated (e.g., Egypt in the case of monitoring the Nile). The resulting lag in progress can tax the patience of industrialized countries. They

26 Support from UNDP had been envisaged in the early outline of the health-related environmental monitoring programme prepared by WHO. It speaks of limited technical assistance and small amounts of equipment within the regular WHO programme, as distinct from UNDP-financed projects, which would pay for laboratory facilities, demonstration monitoring networks, etc. See doc. EHE/74.1, p.7. As experience has demonstrated, however, even the limited technical assistance entailed in the regular programme of WHO is costly and requires considerable resources. Globally, the resources needed could come only from a development assistance organization, such as UNDP. Its support has failed to materialize, in part because developing countries do not attach enough importance to GEMS and health-related monitoring to include it among priority claims in their country programmes. These activities and needs are too small and scattered to figure under the regional programmes of UNDP. The attempts to offset the resulting multilateral financing gap via bilateral aid agencies have not shown notable results.
27 See doc. EFP/82.28, *op. cit.*, p. 3.
28 The methodology proposed by the International Council of Scientific Unions (ICSU) in its project on the transport of carbon minerals in major world rivers is in contrast to the modest approach applied in GEMS/Water. The project design envisaged advanced scientific objectives and the use of sophisticated techniques and instrumentation. Samples were to be collected worldwide and shipped to a central laboratory in Hamburg, where analytical work was to be carried out by a top team of scientists. This was supposed to ensure homogeneity and reliability of the data – a key prerequisite in global monitoring programmes of this kind that is difficult to attain when data are provided nationally. Moreover, annual workshops were to be held, to be attended by the Hamburg group and the scientists in charge of the programme in the field. The purpose of this was to ensure a continuing exchange of views and experience, a joint review of data and procedures, and, in general, a team effort. See 'GEMS/Water Data Evaluation Report', *op. cit.*, section 4.

can, and in fact do, undertake this type of monitoring both nationally and in their own regional groupings, and could sidestep GEMS HEALTH. Their frustration, however, is counteracted by their policy commitment to GEMS and their wish to see a global system build-up. Moreover, they are quite interested in a relatively sophisticated stream of activities, exemplified by projects for monitoring food and air pollution exposure, by biological monitoring, and in particular by the HEALs experiment.

In fact, one of the objects of formulating the HEALs proposal was to revive the sagging interest of industrialized countries by addressing their current concern about how to use and integrate the ambient monitoring data that had been accumulating. It was also reasoned that developing countries would be interested in HEALs because of the existence of 'hot spots' in many Third World urban areas, where pollution-related conditions affecting health are similar to and often more serious than those in industrialized countries.

As a consequence of these different situations and needs, the GEMS HEALTH effort branched into two distinct streams of activities. The first consists of setting up functioning and representative monitoring networks, along with a corresponding build-up of national capabilities for domestic use in places where these are weak or do not exist. Conceptually, this is relatively simple, but it is operationally demanding, costly, lengthy, and often frustrating. The second stream of action is sophisticated and advanced, involves research and monitoring, and addresses complex problems about which little experience or knowledge exists. However, the fact that it has been limited to a relatively small number of advanced institutions from developed and some developing countries, which have been producing monitoring data of standard required quality, makes it less complex to put into operation and manage.

This diversification within the framework of GEMS HEALTH was buttressed by the recognition of the need for a flexible system to respond to problems of regional concern.[29] It helped to balance the effects of the lowest common denominator drift, promoted reliance on institutions of proven excellence to lead and propel the system, often as a voluntary contribution, and made it possible to link up GEMS HEALTH with the advanced work being undertaken mainly in the developed countries.

Nature of the problem

Another set of factors that has influenced the development of GEMS HEALTH arises out of the very nature of the phenomena and variables that are monitored and assessed. Starting from relatively simple expectations that focused primarily on monitoring environmental quality, it has gradually diversified and moved into a much more complex sphere of research and monitoring, involving humans and linking different components. This represents a whole new field in which experience is lacking, knowledge is limited, and methods and instrumentation are still in

29 See the report of the 1982 intergovernmental expert group, doc. EFP/82.28, *op. cit.*, p.7.

the developmental stage. For example, the notion of HEALs is a genuine offspring of GEMS, where all health-related environmental monitoring is under one roof, so to speak. It should not be surprising that an international programme has shown the way to national agencies, which labour within traditional sectoral boundaries.

In practical terms, this has meant that target dates have had to be pushed back, the implementation of the global programme has become more complex, and attention and resources have had to be devoted to research and the quest for new knowledge. In addition, the achievement of some of the concrete benefits and operational goals that were charted somewhat optimistically at the start has suffered a delay.

International funding

The question of funding is pivotal to GEMS HEALTH. Many of its problems and characteristic features hinge on funding. UNEP has been the principal source of international finance. During the period 1975–81, the Environment Fund contributed $3.5 million to GEMS HEALTH, equivalent to 55 per cent of the total $6.23 million that was spent, the other 45 per cent being supplied by the contributions of specialized agencies, principally WHO, as well as by Sweden for the biological monitoring pilot project.[30]

The agencies' contributions are essentially in kind (e.g., personnel, time) and are allocated out of their regular budgets, which do not specifically provide for health-related environmental monitoring. Consequently, the only money available for the international promotion and execution of this global programme during the six-year period were those coming from UNEP.

As noted above, a considerable portion of the resources available for various GEMS HEALTH projects has been used to secure the participation of developing countries; in particular, these resources are used for training and for the provision of some monitoring equipment.[31] The fact that some money was channelled in the direction of developing countries may have given the impression that their needs were being met, though this was far from true. Also, it strengthened their belief that they would be obtaining additional international support, which, it turned out, was very slow in coming. This gap between what was needed in terms of international support and encouragement for broader and full-fledged participation by developing countries, and what was actually available, explains in part the relatively slow

---

30 According to the UNEP project documents and their summaries in the periodic 'Report to the Governments', UNEP paid out $815,000 for GEMS/Air (out of a total of $1,366,000) for the period Nov. 1975 – Dec. 1981; $742,800 for GEMS/Water (out of a total of $1,329,000) for the period Oct. 1976 – Dec. 1981; $664,100 for GEMS/Food (out of a total of $884,900) for the period Feb. 1976 – Dec. 1981; $870,000 for biological monitoring (out of a total of $2,070,000, of which $1,000,000 was provided by two supporting national institutions from Sweden) for the period July 1978 – Dec. 1981; and $411,000 for the pilot project on the assessment of human exposure to air pollutants (out of a total of $677,000) for the period 1978–81.

31 For example, in the case of the pilot project on biological monitoring, of the total UNEP contribution of $870,000, $218,000 was used for training purposes and $195,000 for the provision of equipment, amounting to $413,000, or close to 50 per cent of the total.

pace of expansion and the gaps in the networks and activities of this monitoring cluster.

The usually high costs of health-related environmental monitoring undoubtedly have had a discouraging effect on many developing countries. To place the issue in perspective by way of one example, the total contribution of UNEP to GEMS/Air over a six-year period ($815,000) would be barely sufficient to pay the external costs of an urban air pollution monitoring programme for a medium-sized city over a four-year period.[32]

On the whole, the cluster build-up would have benefited had more resources been available for paying consultants to assist in system build-up and management, travel to and organization of meetings, subcontracting technical support centres to carry out data quality assurance, data handling and analysis, publication, and, in general, the costs incurred in regular and direct interaction within the networks/programmes and in maintaining scientific exchange.

None the less, the small sum provided by UNEP in support of GEMS HEALTH clearly had a catalytic value in making the whole enterprise viable and in keeping the various global networks going. For example, paying for the trips made by consultants to visit countries, discuss their participation, assess their capabilities, and consider possible monitoring sites was the essential first step in the network build-up. Fifty countries were thus visited in the case of GEMS/Air. Resources for this kind of travel and contact were not available in WHO. Indeed, considerable progress can be made with a minimum of resources, when these are used wisely and are supplemented by counterpart contributions from countries and accompanied by an interested and committed global peer-group network.

The point to be noted, however, is that opportunities were missed, and that the pace and scope of progress and the ability to cope with constraints were limited by lack of international financing reasonably proportionate to the needs.

Voluntary nature

Yet another important factor in the implementation of GEMS HEALTH is its basically voluntary nature. The global programme rests on the assumption that once the governing bodies of UNEP and WHO have adopted the relevant resolutions and these have been further developed and given programmatic and substantive content by the intergovernmental expert groups, the governments will proceed to implement whatever actions are required of them. This is but a tenuous link.

Without intermediate and more binding instruments, and given the varying attitudes and responsiveness of governments to the general recommendations and objectives, the single most important element in getting national authorities interested and involved has been the ability of the international officials, in this instance

---

32 It was estimated by WHO, using 1976 prices, that an urban air pollution monitoring programme of four years' duration, based primarily on automated equipment, would have external costs of between $650,000 and $1.3 million, in addition to the costs of local personnel (54 man-years) and facilities. See 'Air Monitoring Programmes Design for Urban and Industrialized Areas', WHO Offset Publication No. 33, *op. cit.*, 1977, pp. 38–42.

from WHO, to make the system attractive and to promote it, as well as to persuade, mobilize, and involve the peer group of scientists and health and environmental officials.

This approach has some practical advantages. The cumbersomeness and formalities of intergovernmental instruments as well as administrative build-up at the centre are avoided, thus relieving the few officers of tasks that detract from their substantive work. Communications within the peer group and its common concerns, commitment, and solidarity play a key role in binding the programme. Substantive issues and objectives tend to be dominant, as opposed to the more politicized concerns of intergovernmental action. Also, many of the existing institutional/resource constraints are overcome thanks to the in-built flexibility. Moreover, because the voluntary scheme has to be made attractive to governments in order to secure their participation efforts are made to improve its operation and to upgrade its output.

On the other hand, there are also some obvious drawbacks. The lack of formal instruments and mechanisms and of a more binding commitment on the part of governments to provide data on a regular basis accentuates problems of spotty implementation and response.[33]

In practical terms, the absence of a well-defined link between intergovernmental decision-making and governmental action has been manifested in a marked gap between the detailed blueprints generated at secretariat and expert group level and the follow-up undertaken by governments. The reason is that in putting together the programme packages, the secretariats and the experts are guided primarily by what they consider would be substantively important. They do not give adequate consideration to the usually high costs of implementation or to the willingness of governments to foot the bill and meet the requirements. Nor does what they say and agree to at the expert group level commit anyone to act.[34]

An additional and significant element had to do with the unwillingness of many governments to take part for fear of losing face internationally upon the publication of monitoring data illustrating the bad state of their environment. For example, they did not allow air monitoring stations in urban areas to be properly located, agreed only to selective releases of data, or simply did not sumbit data.

Of course, were all governments to exhibit the level of interest and commitment and to provide the resources specified by the model, as some of them have done,

---

33 An example of this is the performance evaluation (PE) study carried out by EPA. A total of 290 PE samples were sent out, and results were expected in three to four months; after eight months only 33 of 290 samples had been returned. In contrast, a similar study in the Western Pacific Region had a very high response rate, due to prodding and follow-up work by the Western Pacific Regional Reference Laboratory at the Institute of Public Health in Japan. Rather than just sending samples out and waiting for responses, the Regional Reference Laboratory chose a more structured approach. See doc. EFP/83.59, *op. cit.*, p. 12.

34 This point is illustrated by the 1982 intergovernmental expert group. Its report noted that the plan of action developed by the group entailed significant personnel and financial commitments. However, it was not within the group's terms of reference to discuss this matter, nor did its report and recommendations commit anyone to action. See EFP/82.28, *op. cit.*, p.2. Experts from fourteen countries took part: Argentina, Australia, Brazil, Bulgaria, China, Egypt, Federal Republic of Germany, Hungary, India, Israel, Kenya, Sweden, USA and Yugoslavia.

the voluntary and informal approach could work quite successfully in practice. As it is, such responses are the exception rather than the rule. In view of this situation, then, a more formal approach to secure the compliance of countries with the agreed blueprints could result in a higher degree of responsiveness, including the allocation of more domestic resources for the purposes of health-related monitoring. Similarly, the explicit recognition of the difficulties experienced by the developing countries in fulfilling their tasks and commensurate international support to help them overcome these, could assist greatly in network build-up and its improved performance.

The coordination issue

In the case of GEMS HEALTH, the coordination and institutional tensions within the UN system were not prominent. Indeed, with some slight variations, the symbiotic relationship between UNEP and WHO largely corresponded to the original expectations of the model.

WHO, which had started the air segment of health-related environmental monitoring before the UN Conference on the Human Environment (UNCHE) and which had conceived and nurtured the very concept all along, considered it as part of its regular programme. However, it had to subsume the programme under the comprehensive aegis of GEMS and to concede both access and a share of the credit to UNEP. The reason was not only that UNEP through its financial support provided the key missing ingredient for the process of implementation, but also that it played a role in the substantive and organizational build-up of the system, especially of those components that were developed after UNCHE.

UNEP argued that its financial inputs were temporary and would continue until such time as GEMS HEALTH became self-sustaining and financed on a regular basis out of the budget of WHO. Both the secretariat and the domestic constituency of WHO saw things differently; for them, the Environment Fund was to finance health-related environmental monitoring on a continuing basis. This was in line with standard WHO practice of relying on extrabudgetary multilateral and bilateral funding for a large part of its operational budget. In the absence of UNDP financial inputs, WHO was mostly dependent on UNEP for financial support, however insufficient. It was in fact institutionally favourable to UNEP; were such financing to cease, UNEP would deprive itself of an effective way to secure access to and gain influence in the direction and management of health-related environmental monitoring.

UNEP could not secure such access through general resolutions and recommendations adopted at the intergovernmental level. Rather, the points of leverage and programme influence are to be found in the project cycle. The project conception and approval stage thus offers an important opportunity for access and influence, as does project revision, evaluation, and extension. Also, the preparation and guidance of periodic intergovernmental expert groups, done jointly by WHO and UNEP, represents one of the key points for assessing, influencing, and steering GEMS HEALTH.

It should be noted that UNEP was relatively successful in using these opportunities primarily thanks to the existence within the GEMS Programme Activity Centre (PAC) of relevant expertise and experience, at least during the initial stages of programme development. This carried weight in WHO and in the scientific community at large, and made possible the establishment of a solid working relationship. For example, GEMS PAC played a role in conceiving and elaborating the idea of HEALs. In this particular instance, the UNCHE assumption regarding the potential of a small, high-calibre, influential secretariat seems to have been borne out. As if to underscore this point, when changes took place in GEMS PAC staff, its interest in health-related monitoring waned and UNEP's role in the operational and substantive guidance of this cluster was visibly reduced.

UNEP has also demonstrated some of its potential in securing interagency cooperation in regard to GEMS HEALTH. Once again, it was its ability to fund projects that made it possible to bring agencies together. UNEP deserves some credit for encouraging WMO to take part in GEMS/Air, specifically to advise on the selection of sites for air-quality monitoring stations. Another example relates to GEMS/Water. After lengthy negotiations and some tugs-of-war between the major specialized agencies, which were responsible for the operational delay in the start of the programme until 1979, a cooperative project was begun: with WHO as the lead agency, UNESCO was placed in charge of training and of scientific aspects, while WMO was made responsible for uniformity and standardization in network design and in the collection and transmission of hydrological data. FAO, however, did not participate because it refused to recognize the lead role that was assigned to WHO.[35]

Logistics

Last but not least, the logistics and requirements of a global system build-up need to be mentioned. The amount of organizational effort, energy, and time demanded by logistical considerations is seldom appreciated by those who have not tried their hand at executing some, at first sight, simple operations. It is these details that often determine the speed and efficiency of progress and that occasionally make the difference between failure and success for a given activity. Trying to carry out on a global basis a very simple task, for example, sending quality-control samples and obtaining confirmation of their receipt by participating laboratories, can prove to be tantalizing. The difficulties are magnified by the differences in countries' capabilities, geographical distances, costs of communications and transport, break-

---

35 The interagency tensions are a reflection of what goes on at the national level. Since water has multiple roles and uses, as a resource, as a hydrological subject, as a waste recipient, for drinking, for irrigation, for shipping, etc., it is subject to different sectoral approaches and jurisdictional responsibilities. Hence, each UN agency has some jurisdictional claim over water. In the context of environmental monitoring, UNESCO had conducted scientific studies on water quality and was responsible for the programme on River Inputs into Ocean Systems (RIOS) and for the World Register of Rivers Flowing into the Oceans (WORRI). WMO had worked on hydrological network designs and was in charge of operational hydrology.

downs and maintainenance of equipment, etc.[36] Obviously, logistics tend to overburden the centre of the system and to divert resources, time, energy, and talent from substantive matters.

ASSESSMENT AND PROSPECTS OF GEMS HEALTH

It is possible to visualize the model of GEMS HEALTH being implemented successfully in, for example, the countries of the European Economic Community. They have the infrastructure, resources, expertise, and experience that are required. These countries are relatively homogeneous, share certain concerns and conditions, and exhibit an adequate degree of commitment and, above all, the capability to respond to the prescriptions of the institutional model. The same criteria and expectations do not apply to the global programme, and the evaluation of progress has to be adjusted to the circumstances and conditions under which implementation is taking place. The preceding section has outlined some of the conditions and factors that have had a strong if not decisive influence on the pace, nature, and quality of GEMS HEALTH development.

What matters crucially are the differences in resources, capabilities, and interest that countries bring to health-related monitoring, as well as in the very scope and diversity of such a programme on a global scale. This accounts for the difficulties in obtaining the full participation of countries, and for the slow pace of technical and substantive progress. Also, this heterogeneity weakens some of the key premises of the UNEP model that guide the whole undertaking. For example, if some countries are not in a position or are unwilling to supply the type of counterpart inputs that are expected of them, parts of the system are deprived of the major building blocks called for by the model. Moreover, if development finance organizations, such as UNDP, do not take part in the system build-up in order to fill the resulting gaps, yet another of the system's major initial assumptions ceases to be valid. This, in turn, undermines another premise, i.e., that the Environment Fund is not intended for development assistance.

As noted, the fact that GEMS HEALTH was being introduced into an unequal and heterogeneous global setting resulted in imbalances in programme implementation that required special efforts to correct. This fact was formally acknowledged in 1982, a recognition that was intended to introduce greater realism into what is planned, done, and expected with regard to GEMS HEALTH. This hesitancy in stating the facts openly is partly due to reluctance to tamper with policy or institutional and substantive packages devised and agreed to earlier, especially when this may cause institutional commotion or may require additional financial outlays on the part of the governments. It is also a consequence of the experience that it makes little practical difference whether or not an appeal is made for more funds. The tendency, then, is to carry on, improvising and adjusting to the institu-

---

36 A case in point is organizing intercomparison and checking the analytical capability of participating laboratories. In the case of sulphur dioxide monitoring, this exercise could not be carried out worldwide and had to be confined to European laboratories owing to sample deterioration over long distances. See doc. EFP/82.28, Annex II, *op. cit.*, p. 11.

tional and financial status quo, a tendency reinforced by the reluctance to criticize publicly the shortcomings in the implementation of a given programme.

The voluntary and largely informal approach used may have been responsible for overlooking some of the institutional possibilities which could have helped to buttress the system. For no other reason than those having to do with the institutional and bureaucratic formalities of domestic structures, the majority of governments are geared to respond better to formal instruments requiring them to undertake actions aimed at agreed objectives.

The foregoing statements should not be construed as meaning that the UNEP model has failed. Indeed, the story of health-related environmental monitoring corroborates its basic validity and shows that progress was made towards meeting the substantive objectives. Trying to establish rudiments of a global system as the first objective and then, as time went by, improving upon it and pursuing substantive and quality objectives was a sensible strategy to follow. Whether the programme and the model could have functioned better is a moot point. There was, however, considerable scope for more systematic attention to be paid to the conditions in which the system was being organized, the forces at work, and the various constraints, as well as opportunities for progress.

The first fifteen years of GEMS HEALTH provide some clues to its prospects. The system build-up will remain an important concern. Improved global coverage and involvement of additional countries and institutions, especially in the Third World, will continue as a priority. Upgrading the quality and standardizing of the data base, which had to be downplayed in the early stages owing to the high costs of comprehensive quality-assurance programmes, will figure as a major preoccupation.[37] Quality assurance, in particular, will require greater attention and resources, because the availability of data of certified accuracy is vital not only for assessing exposure and risks but also for setting standards and control strategies. More effective and adequate monitoring and assessment techniques and instrumentation, such as multi-elemental analysis tandem mass spectroscopy, open new possibilities and provide shortcuts to many objectives of the system.

The experience gained in the past and the infrastructure that has been built will prove of great help in speeding up the task and attaining greater efficiency and quality. It is a commonplace that human ventures have to go through a learning process. Thus, as noted above, the assessment of the first sets of data and network performance resulted in considerable modifications to the initial network design of GEMS/Water. This was made possible only through practical experience. Similarly, conceptual advances and learning in the field played a role in formulating the HEALs proposal. In this context, national and regional efforts and the experience of the industrialized countries could be usefully applied to the global programme.[38]

---

37 For example, it is estimated that, for routine ambient monitoring of several gaseous pollutants and suspended particulate matter, instrument checks and calibration, cross-reference calibration with certified standards, instrument maintenance, data validation, etc. can exceed 80 per cent of operational costs. See 'Estimating Human Exposure to Air Pollutants', *op. cit.*, p.38.

38 An example of such cross-fertilization is to be found in drinking water monitoring, as experience with municipal water surveillance exists in a number of European countries and is being used in devising the GEMS/Water drinking water component. Doc. EFP/83.59, *op. cit.*, pp. 14–16.

With the expansion of monitoring networks and capacity, monitoring of pollutant levels and trends on a routine basis will continue as an important element of GEMS HEALTH. In this context, the special nature of GEMS/Water, which involves background monitoring and goes beyond health concerns needs to be recognized. Its relevance to other GEMS clusters calls for appropriate linkages with these. A network composed of relatively few, strategically located trend stations which monitor a number of variables on a long-term basis would provide the data needed for global assessments and trends of the quality of freshwater resources, and thus fill an important gap in the current knowledge.

Parallel with network operation and build-up, management, evaluation, interpretation and use of data, and the integration of research, monitoring, and assessment, as embodied in the concept of HEALs, are bound to become a central activity in the development of the cluster. HEALs will be of particular importance in dealing with the complex issues of cause-and-effect relations between environmental pollution and human health, and in contributing to the overall objective of better public and environmental health through appropriate social responses and policies.[39]

It is to be expected, therefore, that HEALs will become the centrepiece in GEMS HEALTH evolution, with Air, Water, and Food playing a secondary and supporting role. Indeed, HEALs has gained prominence because of the high policy and scientific interest shown by many countries in the concept and its application. Some developed countries have extended direct financial support to WHO work, thus supplementing the limited money contributed by UNEP for the global effort and the participation of developing nations. Still, the programme has been slow in taking off; more than five years elapsed from the time when the proposal was first made to the start of the pilot monitoring programme in 1987.[40]

The need to balance the objectives and the available resources, primarily by augmenting the latter, should play a role in the further build-up of GEMS HEALTH. This is implied in some of the conclusions reached by the secretariats and governmental experts involved. Obtaining a firm and regular commitment from governments to participate in and contribute to system build-up could bring greater rationality and effectiveness to programme development and help to clear some of the existing bottlenecks. Additional and adequate financial resources, from governments, international institutions such as UNDP, and bilateral donors, would help to reduce the improvisation that is manifest and to place the system build-up on a more solid footing.

---

39 As regards specific recommendations for programme development of each segment of GEMS HEALTH, see the report of the 1982 Intergovernmental Expert Group, doc. EFP/82.28, *op. cit.*, pp. 5–7, and subsequently the report of the 1988 Government-designated experts, doc. WHO/PEP/88.15, *op. cit.*, pp.7–11.
40 The foci of initial work were Yokohama in Japan, Stockholm in Sweden, Chattanooga in the USA, and Zagreb in Yugoslavia, with plans to include Pôrto Alegre in Brazil, Beijing in China, and Bombay in India. The first set of pollutants to be monitored included selected metals (lead and cadmium), pesticides (DDT and PCBs), and air pollutants (oxides of nitrogen and benzo[a]pyrene). UNEP's contribution for 1987 was $70,000, with the remainder accounted for by Japan ($123,000) and the United States ($50,000).

A growing challenge for GEMS HEALTH is the use of the monitoring data. One important application concerns assessments carried out for scientific and research purposes in a quest for knowledge. Such uses of the monitoring data have become relatively common, as indicated in the above discussion. Other uses of data and assessments are related to control, regulatory, and preventive actions to safeguard human health from pollutants and other environmental agents and, of course, in the broader policy debate concerning these questions. It is here that GEMS HEALTH could fulfil one of its most important functions, nationally and thus internationally. The mere collection of data on air pollution in a city, for example, is of little use to decision-makers and for action unless it is part of a broader framework, where emissions and sources of pollution are also known and corrective measures undertaken. Measurement for measurement's sake is of little use.

GEMS HEALTH has been directly and indirectly responsible for providing many countries and localities with the monitoring tools, standardized methods, and procedures for control and action. The provision of this capability has been one of its major goals and achievements. Yet, little or no feedback has been received at the centre of the system about the uses of monitoring data in the field. This notable information gap as regards system functioning and build-up is, in some degree, a reflection of the fact that it is still in the early stages of its development.

It is in part the consequence of the staff and resource constraints under which GEMS HEALTH operates and of the difficulties of organizing and assimilating such feedback. It has also to do with an underlying orientation, common in the specialized agencies, that it is for the countries to decide what they will do with the monitoring capability and the data that are generated. This attitude reflects a technical bent and mandate, and the desire to avoid becoming involved in social conflicts and undercurrents at the national level, especially when such politically volatile issues are involved as pollution effects on human health, for many governments are not eager to have their environmental situation publicly exposed. However, GEMS HEALTH is unavoidably entering the phase where it will trigger scientific, social, and political controversies and responses and where greater commitment and involvement will be necessary in scientific and social controversies, nationally and internationally, related to monitoring data assessment and uses. This stage was foreshadowed in the proposal that any assessment carried out by specialized institutions or groups should be subject to evaluation by intergovernmental expert groups. It was a like consideration that was implicit in the refusal of some countries to submit regularly the data on air quality in some of their urban areas, or that explains why they provide data that do not reflect the true situation.

This means, of course, that a stage intrinsically suited to UNEP's mandate and greater direct role and initiative is gradually being entered upon. Having been in the shadow of WHO in a specialized area where it has relatively little substantive weight, UNEP finds itself in a favourable position to promote national and international uses of GEMS HEALTH outputs. This is a role that a specialized agency cannot easily fulfil, both because of its sectoral mandate and because of its traditionally more technical stance.

Time will tell whether and how UNEP will face up to challenges of this kind. In part, it will depend on the vision, inspiration, and strength the organization and its staff can muster in the years to come, its rethinking of its own role, and its search for innovative uses for the resources at its disposal, in general and as regards health-related monitoring and assessment. The opportunity is there and it represents a return on the funds and effort contributed to the build-up of the global instrument and capability embodied in GEMS HEALTH.

This potential of GEMS HEALTH was demonstrated on the occasion of the September 1988 meeting of the UNEP/WHO Government-designated Experts on Health-Related Monitoring. The three background documents cited in footnote 13 in this chapter, on urban air quality, freshwater quality, and chemical contaminants in food, were noted by the world media. The extrapolation from GEMS/Air data, covering the period 1973–84, showed that of the global urban population of 1.8 billion, only one-third, or about 625 million, reside in urban areas where sulphur dioxide pollution is below the WHO recommended guidelines, while two-thirds, or 1.2 billion, reside in areas with annual sulphur dioxide levels that are unacceptable (625 million) or marginal (550 million). As concerns exposure to particulates or smoke, less than 20 per cent, or 350 million, live in conditions that are acceptable, while 80 per cent live in unacceptable (1250 million) or marginal (200 million) air-quality conditions.[41] Thus, the world public at large heard and learned of GEMS, possibly for the first time after its establishment fifteen years earlier. In turn, those in charge of GEMS may have realized that they had an important policy instrument in their hands and should have recognized the need to bring the findings to the attention of decision-makers and public opinion on a regular basis.

The story of monitoring radionuclides is also of some relevance in this context and is a good way to round up this discussion of the formative stages of global health-related environmental monitoring. As mentioned above in chapter two and in this chapter, global monitoring of radionuclides was sidelined as an activity within the framework of GEMS at the very start. This happened in spite of its potential importance, the top priority assigned to it by the 1974 Intergovernmental Meeting on Monitoring, and the concern shown by many governments and other actors. There was thus some poetic justice, during and after the Chernobyl accident, in the confusion, panic, and unpreparedness that prevailed in some of the industrially most advanced countries in the world. They realized that they did not have the regional monitoring capability required for early warning and timely exchange of information. They were not able to assess fully and on time the dispersal and deposition of radioactive substances that were released into the atmosphere and carried across national borders. Nor were they prepared or certain how to present and interpret the facts to their panicking populations.

There was also a touch of irony in the fact that some people looked to GEMS for assistance, as well as in UNEP's offering GEMS' services. Had the opportunity been seized in 1974 – when the proposal was made initially and the mandate in fact given to UNEP – GEMS, the specialized agencies, and most of all the countries

---

41 'Assessment of Urban Air Quality', *op. cit.*, pp.20, 34.

concerned would have been in a better position to monitor, understand, and deal with the massive release of radioactive isotopes into the environment by the Chernobyl accident. However, it seems that in international cooperation things seldom happen in a most logical and simple manner. This time one could hardly place the blame on developing countries or even on the inefficiencies of the UN system, as is common in current debates on multilateralism.

It took fifteen years and what could have been a major catastrophe in a densely populated region to demonstrate the obvious need for such knowledge and capability, and to lay bare the unpreparedness to deal with nuclear emergencies, especially when many countries are concentrated within a region affected by radioactive release and deposition. Only then did it become possible to secure the political acceptance of the requirement to set up an appropriate, internationally coordinated mechanism to monitor radioactivity, as a key component of a comprehensive system of national and international preparedness to deal with the impact of accidents in nuclear facilities and installations.

This system was defined in two international conventions that were negotiated rapidly after Chernobyl, adopted and brought into force in 1986, one on Assistance in the Case of Nuclear Accident or Radiological Emergency, the other on Early Notification of a Nuclear Accident. The first Convention provides, among other things, for assistance to member states in developing appropriate radiation-monitoring programmes, procedures, and standards, and in establishing radiation-monitoring systems. Article 5 of the Convention on Early Notification calls for parties to submit information on general characteristics of the radioactive release (i.e., nature, physical and chemical form, quantity, composition, and effective height), meteorological and hydrological conditions for forecasting transboundary release of such materials, and the results of environmental monitoring relevant to transboundary release of radioactive materials. Quite obviously, appropriate and standardized radiation monitoring systems were required for the achievement of the objectives of these two Conventions.

The attaché of the Embassy of Peru in Cairo, who at the 1974 UNEP Intergovernmental Meeting on Monitoring argued passionately against the French nuclear tests in the Pacific and pressed for the monitoring of radionuclides, is sure to have felt vindicated in 1986. In 1974, many of the experts at the Meeting wondered whether the Peruvian delegate, and the other representatives of developing countries who supported him, really understood the technical side of what they were talking about. Whether they did or not hardly matters. Had they been listened to, the decade preceding Chernobyl could have been used to gain skills and experience, set up a monitoring system, institutionalize the effort, and achieve an adequate level of preparedness.

In 1974 several factors were responsible for assigning to the monitoring of radiation a secondary position within GEMS. In part, such monitoring was seen, both by many governments and by the nuclear establishment led by IAEA, as being anti-nuclear, an interference with and policing of vital national activities. After the ban on nuclear arms tests in the atmosphere, there was widespread complacency about the safety of nuclear installations and much less concern for radioactive

88 The quest for world environmental cooperation

contamination of the environment – at least in the northern hemisphere. There was also a considerable degree of tension and conflict, nationally and internationally. IAEA was not favourable to such an undertaking under the aegis of GEMS, nor were UNEP and GEMS PAC convinced of the need to devote resources to this type of monitoring. The positive case for action could make little headway against the blocking power of various actors, which amounted to a veto in a situation where general consensus was required. One is thus led to conclude that if it takes a major disaster to shake loose the blocking coalition and dissipate the institutional inertia, to change perceptions and mobilize decision-makers and public opinion, to open up possibilities and upgrade the definition of common interest, then accidents such as Chernobyl can play a strategic role and can be beneficial in the longer run.

Of course, the Chernobyl accident did not just expose the inadequacy of national and international capabilities, the weakness in information exchange, and the lack of a base for a joint response. It was above all a new experience, which taught significant lessons. Among other things, it showed that an international monitoring system was essential for an effective response in such emergencies if the exposure of populations to dangerous doses of radiation, especially through air and food, was to be avoided. It was also essential for forecasting the movement of the radioactive cloud, taking rehabilitation measures and controlling the impact of the situation on the national and international economy. It showed the total insufficiency of the existing WHO network for routine monitoring of radiation: only a few countries were taking part and the network was not equipped to make the immediate response or to arrange for the information exchange required by this type of accident. As a consequence of the accident the procedures for sampling and monitoring under emergency conditions were revised.

The effects of nuclear radiation on human health have been the main reason for public concern and alarm. But, as it turned out during the Chernobyl episode, there is much more to accidental releases of massive doses of radiation into the environment, both in terms of effects and of the ways to understand and manage these. Indeed, such releases also have serious impacts on ecosystems, agriculture, fisheries, national economic activities, international trade, and so on. The measures to protect the public's health and psychological well-being can have serious disruptive effects on economic activities and inflict high costs on certain sectors of the economy and particular economic interests and social groups in different countries. It is these effects that seemed to concern politicians and decision-makers as much as, if not more than, the figures about radiation doses received by the population through the radioactive contamination of the environment, or the estimates of increase in cancer deaths per 100,000 people thirty years later.

Chernobyl provided conclusive evidence of the well-known requirement for an integrated approach and an international cooperative effort on a global scale as concerns monitoring of radiation in the progressively nuclear world. The UN agencies and organizations were accordingly invited to sit around the table and the Interagency Committee for the Coordinated Planning and Implementation of Response to Accidental Releases of Radiation Substances was set up. With the Economic Commission for Europe (ECE), WHO, UNEP, FAO, WMO, IAEA, as well as UNSCEAR taking part, GEMS was to provide the global framework within which to place the internationally coordinated monitoring of key radionuclides under different environmental headings (air, water, soil, and vegetation) and basic foodstuffs.

Based on this, a proposal for a WHO/UNEP Global Environmental Radiation Monitoring Network (GERMON) was elaborated. Taking as a starting base the WHO health-related radiation monitoring network mentioned earlier in this chapter, GERMON is supposed to carry out monitoring of global radiation data on a routine basis. In addition, it is to transmit and evaluate rapidly environmental radiation data during emergencies, when the flow of information between states is regulated by the Convention on Early Notification of a Nuclear Accident, with the objective of enabling countries to respond in a timely and effective manner in protecting public health.[42] In view of the lack of radiation monitoring facilities in the majority of developing countries and in order to fill gaps in global coverage, it was also proposed to utilize the existing GEMS networks on which to piggy-back radiation monitoring needs whenever possible.[43]

In planning the concept, launch and implementation of GERMON, the underlying themes of health-related environmental monitoring, and indeed of other GEMS clusters, as will be seen below, surfaced again. Among the questions raised were how to secure standardized and good quality data, how to achieve global coverage and develop national capabilities where none existed, how to secure technical guidance and advice from the scientific community, how to coordinate various institutions, and ultimately, how to use radiation monitoring data nationally and internationally. Moreover, and similar to GEMS as a whole, a start was being made from the existing pieces, in an attempt to weld them together into a global system. The traditional boundaries between sectors, institutions, and scientific disciplines, and indeed between the existing clusters of GEMS, had to be bridged in order to conceive and implement GERMON.

Thus, thanks to the impetus generated by the Chernobyl accident, some two decades after UNCHE the international community seemed poised to set up internationally coordinated worldwide monitoring of environmental radioactivity and data exchange. This represented an essential capability for responding to and managing increasing environmental and human health risks generated by the

---

42 For the initial concept of GERMON see 'Basic Principles of the WHO/UNEP Global Environmental Radiation Network', WHO, Geneva 1988 (doc. PEP/88.8). According to the proposal, the minimum requirement was to measure external radiation dose rate at ground level at all times, airborne activity at least weekly, and precipitation and milk at least quarterly. Prepared by a group of experts from seventeen countries invited in their personal capacity, the proposal was intended to guide governments and the secretariats of UNEP and WHO in developing a global network of national institutions for environmental radiation monitoring. In parallel, IAEA was preparing a guide book on determination of radionuclides in food and the environment, and a publication on monitoring for the radiation protection of the public following a nuclear accident.

43 For a global survey of radiation monitoring see 'Radiation Monitoring in Countries: a Global Survey for the United Nations Inter-Agency Committee for the Co-ordinated Planning and Implementation of Response to Accidental Releases of Radioactive Substances', GEMS Information Series No. 6, Feb. 1988, Nairobi. Those countries with developed radiation monitoring have two types of programmes, one to monitor the environment for radioactive contamination regardless of its origin and the other which is site-specific to facilities such as nuclear power plants or reactors.

proliferation and diversification of peaceful uses of nuclear energy and for informing the public adequately. Also, it was a significant though somewhat belated step in building up GEMS.

In conclusion, the future of GEMS HEALTH could be rather dynamic. Impelled by the requirements of political change and by the precarious environmental health situation in Eastern Europe, it could come to play an important role in domestic and regional environmental policies of this region. The growing need for health-related monitoring and assessment services in the Third World, reeling under environmental stress and economic hardships, presents yet another gap in the global coverage that needs to be filled in years and decades to come, and is a challenge for the international community as a whole. Finally, the advanced industrialized countries who have their own national and regional institutions to deal with these issues and have a considerable experience, could share their knowledge with others and thus contribute to the build-up of a greatly improved global system of health-related environmental monitoring which could respond adequately to real needs and demand for its services.

4 Climate-related environmental monitoring in GEMS

One of the concerns that gave rise to the Stockholm Conference (UNCHE) had to do with the possibility that human activities were making a significant contribution to the natural variability of the global climate.[1]

For thousands of years humankind has been contributing its share to local, regional, and even global climatic change and variations. It has done so by modifying land-surface properties through the conversion to agricultural uses of the ground covered by forests and other natural vegetation. This process continues on a massive scale today, with the principal changes occurring in regions that were less affected in the past. Non-agricultural uses also modify the surface of the land substantially, as an ever-increasing proportion of land is covered with settlements, roads, buildings, and other types of infrastructure. Modern urbanized and industrialized society is introducing large amounts of heat, gases, and materials into the environment, especially the atmosphere, which can contribute to regional and global climate change. These impacts have been classified as 'inadvertent climate modification'. Some countries, however, are also developing capabilities for deliberate weather and climate modification, which represents an added factor of importance for climate change.

As its contribution to UNCHE, the scientific community had prepared a comprehensive overview of humanity's possible impacts on climate and proposed a programme of study and research.[2] The attention focused on the rising concentra-

---

[1] Global climate is determined by the following basic actors: the intensity and spectral distribution of the sun's radiation; the geometry of the Earth's orbit and the axis of its spin relative to the sun; the speed of the Earth's rotation; the mass and composition of the Earth's atmosphere; and the features of the Earth's surface (physical relief, distribution of land and water masses, surface properties including albedo, etc.). The external variability of climate is caused by changes in one or more of the basic factors noted above. However, even were these to remain unchanged, there would still be the internal or inherent variability of the climatic system, which is caused by highly non-linear dynamic and thermodynamic interactions between atmosphere, oceans, cryosphere, land surfaces, and biota. See 'Climate-related monitoring', the background paper prepared for the UNEP/WMO Government Expert Group Meeting on Climate-Related Monitoring, Geneva, 10–14 April, 1978, pp.5–6.

[2] The initial US-based formulation of the problem came from the Study of Critical Environmental Problems (SCEP), *Man's Impact on the Global Environment*, MIT Press, Cambridge, Mass., 1970, pp.39–112. It was further developed in the Study of Man's Impact on Climate (SMIC). The study was prepared at a three-week meeting of 30 scientists from 14 countries, held in Stockholm in June/July 1971. See *Inadvertent Climate Modification*, MIT Press, Cambridge, Mass., 1971.

tions of carbon dioxide in the atmosphere, caused mainly by fossil fuel combustion. Higher levels of carbon dioxide in the atmosphere, which would act as a 'heat trap', were expected to cause a global warming trend, especially pronounced in the higher latitudes, leading to the melting of polar ice, a rise in the sea level, and the flooding of low-lying coastal areas. The warming was also expected to result in significant modifications in global circulation and precipitation patterns, altering water and energy supply, shifting major ecological zones and affecting in particular climate-sensitive latitudinal and altitudinal margins of crop areas, and leading to changes in national and international economies. The 'greenhouse effect' was thus placed on the international agenda, becoming eventually one of the key environmental themes of concern both to the scientific community and to the public at large.

A fuller understanding and knowledge were required in order to assess society's impacts on climate and to predict the effects on the planetary heat budget. In this task, climate-related monitoring of the environment had an important role to play.[3] A uniform global database generated through such monitoring was essential in trying to assess humanity's impact on climate and its extent relative to natural fluctuations and causes. It was an important addition to the more traditional global observation of climate components, which had been organized and coordinated for more than 100 years, first by the International Meteorological Organization, and then by its successor World Meteorological Organization (WMO). It was needed for studying variability of climate and trends, understanding the global climatic system, and modelling and predicting climate behaviour.

GEMS AND HUMAN-INDUCED CLIMATE CHANGE: THE FIRST STEPS

In its recommendation No. 76, the Stockholm Conference gave prominence to the 'appreciable risk of effects on climate' of human activities. It specifically recommended that 10 baseline stations be set up in remote areas, far from sources of pollution, to monitor long-term global trends in atmospheric constituents and properties that might cause climatic changes. It also recommended that, in parallel with these 10 stations, a larger network of 100 stations be established for monitoring properties and constituents of the atmosphere on a regional basis. Subsequently, the Governing Council of the UN Environment Programme (UNEP) and the 1974 Intergovernmental Meeting on Monitoring singled out climate-related environ-

---

3 At the end of the nineteenth century, a hypothesis about possible changes in world climate due to fluctuations of carbon dioxide levels was advanced by the Swedish scientist S. Arrhenius. The issue was studied and sporadic measurements of carbon dioxide were undertaken by individual scientists in the period that followed. The first systematic effort to monitor carbon dioxide started in 1954, with the establishment of a monitoring network of 15 stations in Scandinavia. On the global scale, monitoring of carbon dioxide started during the International Geophysical Year in 1957. Two monitoring and research stations were established by the US, one at Mauna Loa and one at the geographic South Pole. The Mauna Loa station has been recording carbon dioxide levels ever since; the South Pole station recorded data sporadically until 1975, and regularly thereafter. For an overview of the history of monitoring of the atmosphere, see R. F. Pueschel, 'Man and the Composition of the Atmosphere, the Background Air Pollution Monitoring Network (BAPMoN)', WMO, 1986, pp.16–24.

mental monitoring as one of the priority activities of the Global Environment Monitoring System (GEMS).

In broaching this task, the initial focus was on monitoring background levels of pollution in the atmosphere, principally carbon dioxide concentrations. By the time of the Stockholm Conference, an incipient Background Air Pollution Monitoring Network (BAPMoN) coordinated by WMO was beginning to take shape.[4] After the policy endorsement of climate-related environmental monitoring, WMO turned to UNEP for financial support needed to help it in the build-up of BAPMoN. This was one of the first monitoring projects to be funded by UNEP and marked the beginning of the climate-related environmental monitoring cluster in the framework of GEMS.[5]

In the case of climate, the variables, trends, and processes monitored are of global relevance. The monitoring data show global trends, the monitoring networks are based on a global rationale, and national monitoring activities are valuable primarily in so far as they form an integral part of the worldwide network and contribute to the global data base.

Climate-related environmental monitoring was oriented primarily to scientific research and study, the objective being to build the database and a permanent capability essential for a better understanding of both the global climate system and how it is affected by and interacts with contemporary society. Some, though not all, anthropogenic impacts on climate have been identified. The exact nature and extent of these effects – many of which are very long term – is far from certain, as is their weight and role in total climate change and variability, which itself is far from being fully understood.

In this context, it is useful to make a comparison with health-related environmental monitoring, discussed in the preceding chapter. In most instances, health-related monitoring is concerned with delimited situations that, at least in principle, lend themselves to corrective approaches and management in local, national, and regional settings. For example, air pollution in a metropolitan area, a

---

4 Some of the variables related to human impacts on climate began to be monitored at a few sites as early as 1956 and by 1961 it was accepted that WMO should centralize the collation of certain data on atmospheric chemistry. The proposal to establish a global background air pollution monitoring network was raised for the first time in 1965, in WMO's Working Group on Atmospheric Pollution and Atmospheric Chemistry of the Commission for Atmospheric Sciences. In 1969, the WMO Executive Committee endorsed this proposal, recommended the establishment of a global network, and set up a panel of experts for this purpose (which prepared the first manual on sampling and analysis procedures). The WMO Congress in 1971 called for the establishment of a global network of 10 baseline and 100 regional stations. For a discussion of atmospheric chemistry, air pollution monitoring, and BAPMoN origins, see R. F. Pueschel, *op. cit.*, and G. W. Kronebach and R. E. Munn, 'Air Pollution Monitoring and the WMO Background Air Pollution Monitoring Network', WMO doc. POL/DOC.5, 1977. Also see H. W. de Koning and A. Kohler, 'Monitoring Global Air Pollution', *Environmental Science and Technology*, Vol. 12, August 1979, pp. 884–9, B. Martin and F. Sella, 'Earthwatching on a Macroscale', *Environmental Science and Technology*, Vol. 10, March 1976, pp. 230–3, and C. C. Wallen, 'Global Atmospheric Monitoring', *Environmental Science and Technology*, Vol. 9, January 1975, pp. 30–4.

5 For an overview of the first ten years of climate-related monitoring in GEMS, see WMO document 'Review of GEMS climate-related monitoring: achievements and future activities', pp. 5–59, prepared for the UNEP/WMO Government Expert Group on Climate-Related Monitoring that met in March 1982.

carcinogen in a food product, or seepage of toxic waste into a groundwater aquifer are occurrences where it is possible to identify culprits and victims, often to see and feel the effects, to generate public concern and pressure, and to mount corresponding action.

The effects of climatic change on the biosphere, society, economy, and the well-being of individuals, although potentially enormous, are more diffuse, with a multitude of contributing sources and with response times on much longer time-scales than those traditionally used by humans. With the exception of a rise in sea level and its effects on low-lying islands, coastal areas and settlements, their exact nature and impacts are very uncertain. To make matters even more ambivalent in terms of societal responses, some of the projected climatic changes may be beneficial to some countries and geographical regions. Indeed, some scientists have argued that global warming is desirable, would increase world agricultural production and should be welcomed.[6]

It follows that, unlike the majority of health-related issues, the climate transcends the national level; national responses make sense only within the framework of a global programme and global management. The first goal of GEMS was therefore to promote the establishment of a global monitoring capability and a global data base concerning those human-induced environmental changes that have an impact on climate. By helping to generate new knowledge and insights, this monitoring capability would make possible and give rise to the required collective responses. It was also recognized that in many instances the monitoring data may not be useful or subject to proper interpretation until some time in the future, at which point the time series of data would prove invaluable to science and decision-making.

As for BAPMoN, its main objective was to determine, on a global and regional basis, background concentration levels of atmospheric constituents, their variability and possible long-term changes. BAPMoN consists of a network of baseline (global) stations in remote locations. These stations monitor background levels, variability and trends of atmospheric carbon dioxide and precipitation chemistry. Also, they monitor turbidity (extinction of solar radiation by particles in the atmosphere, i.e. aerosols), and concentrations of suspended particulate matter (SPM), and, as an option and where suitable, trace gases including surface ozone, methane, sulphur dioxide and nitrogen oxides. BAPMoN comprises, as well, regional stations which identify changes in regional atmospheric conditions due to

---

6 One of those arguing along these lines is M. Budyko, a Soviet climatologist who was among the first to warn that the greenhouse effect could warm the earth. Unlike the Western climatologists who base their predictions on numerical models of climate, Budyko and others base theirs on reconstruction of past climates, i.e. paleoclimatology. While his colleagues in the West predict aridization of continents, Budyko argues that these will experience improved moisture conditions. Initial drying will be replaced by wetter conditions by the year 2050, as the warming trend continues. By the year 2050, when the levels of carbon dioxide will be double those in the pre-industrial era, temperatures will be as high as those in the Pliocene era of 4 million years ago, the so-called 'Pliocene optimum'. A wetter and warmer world, combined with the increasing fertilization effect of carbon dioxide, would raise agricultural yields. See 'Soviet climatologist predicts greenhouse paradise', *New Scientist*, 6 Aug. 1989. For a review of the greenhouse effects see a GEMS popular brochure 'The Greenhouse Gases', UNEP/GEMS Environment Library No. 1, pp.24–32, Nairobi, 1987.

human activities, in particular land-use practices that can affect evaporation, surface albedo, turbidity, SPM, water runoff rate, etc.

In the context of BAPMoN build-up, UNEP has also partly supported the operation of a central carbon dioxide laboratory, which carries out the calibration of measurement instruments against an absolute system, and undertakes continuous precision testing and adjustment of instrumentation. Both of these functions are essential for obtaining consistently accurate global carbon dioxide data.[7]

In promoting network expansion, especially in those areas of the globe not adequately covered, UNEP supported a feasibility study concerning the establishment of a BAPMoN baseline station on Mt Kenya. The specific purpose of the project was to determine, on an experimental basis, the suitability of the area for the establishment of a baseline station, to select a site, and to work out the logistical requirements. Mt Kenya was chosen because of its high altitude and remoteness, its location on the Equator and at a boundary between a continent and a maritime region, and because it is in an area where a substantial part of interhemispheric exchange takes place. This proposed baseline station did not come into being as hoped, for reasons that will be discussed below.

The project for supporting a World Glacier Inventory (WGI), coordinated by UNESCO, was also included in the initial phase. Glaciers are one of the most sensitive indicators of current climatic change, and historical glacier variations provide data on past climatic fluctuations. The purpose of the project was to strengthen and broaden the system for monitoring the dimensions of glaciers in different parts of the world (length, area, volume, thickness, and median and lowest elevations) and to develop methods for using the data in the diagnosis of climatic fluctuations, for general climatology, and for designing ice core programmes for climatic and pollution monitoring. Existing data were too heterogeneous, with priority being given to regional over global analyses of climatic change. Thus, the initial effort was aimed at two principal objectives: filling the gaps in national activities, and standardizing data and making them computer-compatible.[8]

---

7 The central laboratory maintains primary reference gases and provides secondary reference gases that are used by national networks for calibration purposes. As early as 1956, during the International Geophysical Year, under the guidance of C. D. Keeling, the Scripps Institution of Oceanography began to support the two US monitoring laboratories (Mauna Loa and the South Pole). It developed a concentration scale for carbon dioxide, on the basis of which an informal international system has been built up for consistent calibration. Following the recommendation of the 1975 expert group meeting that Scripps should be appointed as the central carbon dioxide laboratory for BAPMoN, a GEMS project supporting its work was approved.

8 At the start of the International Hydrological Decade (1965–74), it was recommended that an inventory of permanent snow and ice masses be made. In 1973, the International Commission of Snow and Ice established a temporary technical secretariat at the Swiss Federal Institute of Technology to compile national glacier inventories. Nine countries with glacierized regions, but without inventories, were brought into the programme to join countries with advanced inventory work. The idea of incorporating UNESCO's WGI into GEMS met with objections within WMO. Some were not happy about the prospect of the already very limited financial resources for climate-related monitoring going to UNESCO and WGI. They argued that this was not a funding priority. According to this view, WGI was useful primarily for studies of the world water balance and for local hydrological purposes, and was not of much use in studies of climatic change, as glaciers were only a small part of the global cryosphere. They admitted that glaciers could be indicators of climatic variations, but argued that it was difficult to interpret such data on a global scale because

96 The quest for world environmental cooperation

The significance of WGI was that it signalled a more comprehensive approach to monitoring of climate variability and change. It was related to an improved understanding of what needed to be done and to the beginnings of an integrated approach to climate change within the UN system, and WMO in particular. The task of outlining a more comprehensive approach within GEMS was undertaken by an intergovernmental expert meeting on climate-related monitoring, which met in 1978. It sounded a note of adaptiveness and flexibility when it said that 'climate-related monitoring is tailored in scope, detail and quality according to the need to know, continues in a given form only as long as necessary, and can be revised, changed or terminated as new knowledge is gained'.[9]

The basic conclusion was that monitoring and data collection should provide a picture of the past, present, and future climate, the aim being to understand it and predict its trends and variability. Data must be collected on the variable qualities of the climate and on the variable qualities of the environment that are, or could be, causally related to climatic change. The human impacts have to be placed in perspective, as being only one set of elements in a complex climate system. If they are to be understood and managed, they need to be monitored and studied concurrently with other variables. Moreover, it was recognized that the atmosphere is not the sole medium to be monitored and that attention should also be given to other media, as well as to the interaction between media, which is important in shaping the global climatic system. For example, oceans play a crucial role in the carbon cycle, both as a source and a sink for carbon dioxide, which requires measurement of dissolved carbon dioxide in ocean water.

With a number of climate-related monitoring activities in place, it was proposed that they should be integrated and executed in a mutually supportive and interactive manner.[10] A broad distinction was made between monitoring characteristics and status of climate and climate variability, on the one hand, and monitoring and understanding conditions that cause climatic variability and change, on the other hand. The first group of activities is concerned with the present climate and with past climates (e.g., studies of proxy and fossil records in bottom sediments in lakes, ice sheets, pollen, tree rings); the second group subsumed monitoring of variable qualities of the environment that cause or could cause climatic change, both natural (e.g., solar activity, volcanism, cryosphere) and anthropogenic (e.g., atmosphere composition changes, and land-use changes affecting albedo, surface roughness, and evapotranspiration).

glaciers were influenced by local conditions. UNESCO, on the other hand, argued that the glacier variations provided important data on past climatic fluctuations and thus were a basis for forecasting future trends.

9 See WMO document 'Report of the UNEP/WMO Meeting of Government Expert Group on Climate-Related Monitoring', 10–14 April 1978, p.8.

10 The background paper prepared for the 1978 expert group gave an overview of such activities and made it possible for the expert group to recommend linkages and integration, as well as to note gaps and to propose additional activities (*ibid.*, pp. 8–28). In the UNEP scheme of things, this represented *Level One* of the programmatic process and, at least in theory, it should have been done at the very beginning, before the Environment Fund resources were committed to climate-related monitoring projects.

Climate-related environmental monitoring 97

This comprehensive conceptual approach was reflected in an omnibus project whose purpose was to prepare a base for operational climate-related monitoring activities.[11] These included:

- Continuation of BAPMoN and of related activities focused on atmospheric compositional changes;
- Monitoring the heat budget of the Earth's atmosphere system, to consist initially of expert meetings on cloudiness and radiation, aerosols, land-surface processes, oceanic transport and storage of heat, etc., all of which were intended to provide guidance and assistance in preparing a monitoring plan;
- Monitoring of the cryosphere, to include WGI, a programme for monitoring snow and ice, standardizing data on sea-ice that were available from current monitoring, etc. ; and
- Monitoring volcanic activity, including the vertical distribution and concentration of volcanically derived particles in the global stratosphere, and expanding the BAPMoN network to cover observations of aerosols and dust load in the atmosphere from volcanic activity.

Reflecting the recognition by the 1978 expert group of the importance of advanced monitoring and data management technologies, it was also decided to initiate a project on the use of satellites for climate-related monitoring. It included plans for long-term use of satellite systems for monitoring such climate-related variables as sea-surface temperature, wind stress, ozone distribution, stratospheric aerosols, the Earth radiation budget including solar ultraviolet flux and surface albedo, precipitation over oceans, soil moisture, snow, and ice fields. One of its objectives was to develop a standardized format and facilities for international processing of climate-related data. Additional activities were also incorporated within the omnibus project, including historical and proxy data for the study of past climates and climatic cycles (e.g., the analysis of unicellular micro-organisms of the sea in ocean deposits), a climatological data inventory, a data referral system, data bank centres.

The recommendations of the 1978 expert group and the basic objectives of the omnibus project were reflected in the framework of the fledgling World Climate Programme (WCP). Actually, the omnibus project, including several expert groups held in its framework, was used to prepare the ground for WCP. UNEP's financial and catalytic support was thus of importance in the planning stages of WCP and in preparing the ground for its operational launch.[12]

WCP has four components:

- World Climate Data Programme (WCDP), with WMO as the responsible agency;

---

11 The various sub-projects followed the scheme adopted by the 1978 intergovernmental expert group. See the report of the expert group meeting, *op. cit.*, pp.18–41. For a description of the omnibus project, see *UNEP Report to Governments*, No. 27.
12 For the first proposals for a world climate programme formulated by the Secretary-General of WMO, see WMO document EC-XXX/Doc.8 (16 Feb. 1978). Also see 'Outline Plan and Basis for World Climate Programme, 1980–1982', WMO Publication No. 540.

98 *The quest for world environmental cooperation*

- World Climate Research Programme (WCRP), with WMO, in cooperation with the International Council of Scientific Unions (ICSU), as the responsible agency;
- World Climate Applications Programme (WCAP), with WMO as the responsible agency; and
- World Climate Impacts Studies Programme (WCIP), with UNEP as the responsible agency.

WCP is broadly patterned on the UNCHE scheme of monitoring, research, assessment, and management. Monitoring is but one component of the programme, which combines the need for sustained and longer-term scientific observation and the study of the climate system and climate change, with the current requirements of management and with the growing necessity to take into account and understand the role of climate variations and changes in national economies and society, to predict such changes, and to mount appropriate responses. Anthropogenic impacts on world climate are viewed in the broader context of the world climate system.

Seen in the context of UNEP's catalytic approach, WCP represented a significant *Level Two* achievement. The initial work undertaken on climate-related environmental monitoring in the framework of GEMS and the direct financial assistance provided in the preparatory stages of WCP both played a catalytic role in the process of shaping and launching WCP itself.

After the start of WCP, climate-related environmental monitoring under the auspices of GEMS continued with the traditional support for BAPMoN and WGI, both of which had remained operationally outside WCP. An additional tranche of financing was allocated to WCP itself, in support of monitoring and publication activities dealing with the behaviour of the world climate system within the framework of the World Climate Data Programme.[13]

In addition to BAPMoN, and other climate-related monitoring activities, it is also important to mention the existence of another global monitoring network, namely WMO's Global Ozone Observing System (GO_3OS). Because of its many and varied implications, the ozone problem could not be easily accommodated within the GEMS five-cluster breakdown and rationale. The thinning of the ozone layer and the resulting increase in UV-B radiation at the surface of the planet have a bearing on human health and biological systems; increases of tropospheric ozone have harmful effects on human health as well as on vegetation; increases in UV-B radiation are likely to affect plankton in the oceans which act as one of the major

---

13 UNEP supported publications in a series called 'Climate System Monitoring'. These include monthly bulletins, advisories on climatic events, and biennial summaries. The biennial publication, which is aimed at decision-makers and the public, had its origins in the 1982–3 El Niño/South Oscillation (ENSO) episode which is thought to have been responsible for weather anomalies, including abnormal droughts or prolonged wet periods and floods, and for consequential dislocations in economic, especially agricultural, activities in various regions of the world that were affected. The publication was initiated in order to review at regular intervals the global climate system. For the first such biennial summary, see 'The Global Climate System, a Critical Review of the Climate System during 1982–1984', World Climate Data Programme, WMO. For the second summary, see 'The Global Climate System, Autumn 1984-Spring 1986', document CSM R84/86, WMO 1987. It was more ambitious than the first one, and included such topics as greenhouse gases, carbon dioxide and the annual cycle in photosynthetic production, the ozone hole, Earth rotation changes, changes in solar radiation, geomagnetic jerk.

sinks for excess carbon in the atmosphere and thus indirectly contribute to the greenhouse effect; and, CFCs implicated in the destruction of stratospheric ozone, as well as the tropospheric ozone, are also active greenhouse gases in the atmosphere, supplementing the effects of carbon dioxide (e.g. while the latter accounts for about 51 per cent of the warming effect, ozone is responsible for 9 per cent, CFCs for 20 per cent and methane for 16 per cent).

These classification dilemmas largely explain the absence of a GEMS ozone monitoring project. Indeed, UNEP's support for WMO's ozone monitoring activities came via its 'outer limits' budget line. Operationally, this meant that GEMS PAC was not in charge of this project and that ozone monitoring did not figure formally among GEMS-related activities.

Ozone monitoring network embodies the essence of the model inherent to GEMS. GO_3OS was formally initiated during IGY. It was decided then that WMO should act as the data centre and, building on measurements done by individual scientists since the 1930s, should set up a long-term ground-based network for monitoring ozone using Dobson spectrophotometers. International procedures were set up by WMO for standardized and coordinated observation, publication of data and related research. In 1960, Canada began to operate in Toronto a World Ozone Data Centre for the network, which has since published a bimonthly bulletin entitled 'Ozone Data for the World'. The network grew over the years; three decades after it was started, it consisted of 140 operational ground stations, with 60 countries, mostly in the northern hemisphere, taking part. The ground measurements are supplemented by atmospheric and stratospheric monitoring and by remote sensing from satellites.

It is the monitoring data generated by this network that have demonstrated the long-term decline in the total ozone column over the northern hemisphere, i.e. an average decline of 3 per cent over a 20-year period, with an increase in the tropospheric ozone and a more rapid decline above the altitude of 35 km.[14] The resulting research and major international assessments of the ozone layer undertaken in 1975, 1980, 1985 and 1988 gave rise to mounting worldwide concern. They provided the basis for the adoption of the legal instruments to reduce production and emission of CFCs, i.e. the Vienna Convention for the Protection of the Ozone Layer, and the Montreal Protocol on Substances that Deplete the Ozone Layer. After thirty years of build-up, an essential capability was thus in the hands of the international community to monitor and assess ozone changes in the atmosphere and the stratosphere.

EARLY INDICATORS OF PROGRESS

The indicators of progress in climate-related environmental monitoring are broadly similar to those in GEMS HEALTH. They are related to the initial build-up of a global monitoring capability embodied mainly in BAPMoN, which has been the

---

14 See 'The WMO Global Ozone Observing System', Fact Sheet No.1, WMO, 1989. For a popular review of the ozone problem see GEMS publication 'The Ozone Layer', UNEP/GEMS Environment Library, No. 2, Nairobi, 1987.

principal *Level Three* activity in this cluster and to which most of these comments pertain. Also, progress is reflected in the gradual adoption of a systems approach to the question of climate, in refining the approach to climate-related monitoring, as reflected in the World Climate Programme, and in reorganizing the monitoring efforts more comprehensively, starting from the medium, i.e. the change in global atmospheric composition.

Methodologies

Standardized methodologies and guidelines concerning the siting, establishment, and operation of monitoring stations were prepared and published in a manual that has been continuously updated to reflect the advances made and the experience and new knowledge acquired. The methodologies have been diversified, expanded upon, supplemented, and continuously revised, essentially by means of periodic specialized meetings held to deal with given aspects of background air pollution monitoring.[15]

Infrastructure

An infrastructure in support of the global effort to monitor background levels of air pollution has been set up to perform the following functions:

- provision of globally exchangeable standards and the interlaboratory intercalibration of instruments;[16]
- processing, storage, and publication of data;[17]

15 See an updated version of 'WMO Operations Handbook for Measurement of Background Atmospheric Pollution,' WMO publication No. 491. The manual contains, among other things, a discussion of types of stations, siting criteria, parameters to be monitored, sampling and methods of analysis, standardization and intercalibration, data handling, and procedures for logging data on standard forms. Among the many reports resulting from expert consultations, see 'Report of the WMO/WHO Technical Conference on the Observation and Measurement of Atmospheric Pollution', Helsinki, July 1973; 'Report of the Expert Meeting on Wet and Dry Deposition', Toronto, November 1975; 'Report of the WMO Air Pollution Measurements Techniques Conference', Gothenburg, 1976; 'Report of the UNEP/WMO Expert Meeting on Siting Criteria for Background Air Pollution Stations', Mainz, October 1976; 'Report of the Expert Meeting on Dry Deposition Monitoring', Gothenburg, April 1977; 'Report of the WMO/UNEP/ICSU Meeting on Instruments, Standardization and Measurement Techniques for Atmospheric Carbon Dioxide', Geneva, September 1981; 'Report of the Expert Meeting on Quality Assurance in BAPMoN', Research Triangle Park, North Carolina, USA, January 1983; Environmental Pollution Monitoring and Research Programme (EPMRP) Report, No. 16, June 1983; etc.

16 In 1980, after the initial trial period, the arrangement between the Scripps Oceanographic Institution and BAPMoN was extended, with the US Government assuming the obligation to finance this work. At the same time, efforts were initiated by the National Oceanic and Atmospheric Administration (NOAA) to set up a similar facility at its own Geophysical Monitoring for Climatic Change (GMCC) in Boulder, Colorado, which eventually assumed the role of the central laboratory providing standard gases. Similar facilities for the calibration of sun photometers used in monitoring turbidity are under consideration. Suggested sites are in the US and Switzerland.

17 The WMO Collaborating Center on Background Air Pollution Data of the US Environmental Protection Agency (EPA) receives data on a monthly basis from the monitoring stations belonging to the network. The data have been published by WMO, in a publication sponsored jointly with EPA, NOAA, and UNEP. See, for example, 'Global Atmospheric Background Monitoring for

- evaluation and quality control of monitoring data;[18]
- organization of regular meetings and continuous scientific exchange on various aspects of system operation and build-up;[19] and
- training activities directed mainly at experts and technicians from developing countries.[20]

Networks

From its very modest beginning on the eve of UNCHE, when there were three partly operational baseline stations and 15 regional ones, BAPMoN grew steadily until it met and then passed the target of 10 baseline and 100 regional stations that had been set by the Conference. At the end of 1987, the global network consisted of 197 stations that were in operation. Of these 17 were global or 'baseline' stations, 10 continental or 'regional with extended programme', and 170 regional stations, with 57 countries taking part.[21] While the network had reached a sufficient density in Europe and North America, this was not the case globally; major gaps and imbalances still remained, especially in the Third World.[22]

Selected Environmental Parameters, BAPMoN Data for 1982', Volume II, Precipitation chemistry, continuous atmospheric carbon dioxide and suspended particulate matter, EPMRP Series, No. 41. (This publication was started in 1971, under the title 'Atmospheric Turbidity Data for the World'. When carbon dioxide measurements were added in 1975, the title was changed.) Note that, since 1989, the WMO World Data Centre for Greenhouse Gases has been hosted by the Japan Meteorological Agency in Tokyo. A generous offer of staff and resources to process gaseous chemistry data made this possible. The US continued to process precipitation chemistry, turbidity and SPM data.

18 The BAPMoN Precipitation Reference Laboratory, located at the Quality Assurance Division of the EPA Environmental Monitoring Laboratory, carries out such tasks as quality assurance and data handling, interlaboratory comparison of precipitation chemistry sampling and analysis of the techniques used, and establishment of confidence limits for the data reported. These services, which are provided at no cost, are of crucial importance for the system and its operational status, specifically for improving the performance of the participating laboratories. See R. L. Lampe and J. C. Puzak, 'Fourth Analysis on Reference Precipitation Samples by the Participating World Meteorological Organization Laboratories', EPMRP Series, No. 7, December 1981, WMO.

19 Typical examples of such exchange and interaction are the 'Report of the Meeting of Experts on BAPMoN Station Operation', 23–6 Nov. 1981, EPMRP Series No. 6, and within the framework of the WMO regular machinery, 'Report of the Executive Committee Panel of Experts on Environmental Pollution', fourth session, Geneva, 27 Sept. – 1 Oct. 1982, EPMRP Series, No. 13.

20 Training seminars have been held in which more than 100 trainees from 35 developing countries took part during the period 1978–89. Moreover, a special training facility has been established at the Central Institute for Atmospheric Physics in Budapest, using the non-convertible currencies from the Environment Fund. Training activities are an efficient way to promote unified and standardized approaches to monitoring. They have been strengthened and supplemented through consultant services to countries to assist with various aspects of organizing and carrying out climate-related monitoring.

21 See 'Summary report on the Status of the WMO Background Air Pollution Monitoring Network as at 31 December 1987', EPMRP Series, No. 55. For an earlier report of the status as at 31 December 1985, see EPMRP Series, No.38. These surveys of BAPMoN implementation were somewhat passive and mechanical as they were based on the questionnaires mailed to and answered by permanent representatives of member countries of WMO.

22 For example, only four regional stations with a partial programme were in operation in South America, ten in Africa, and sixteen in Asia, as compared with 69 in Europe, 42 in North and Central America, and 22 in South West Pacific. Note also that, with the exception of the Cosmos observatory

102 The quest for world environmental cooperation

Monitoring data

Progress has been achieved as regards the number of countries reporting monitoring data, though regularity of reporting leaves a lot to be desired. According to WMO reports, nine countries were reporting in 1973, 24 in 1977, 34 in 1980, and 47 in 1982. In 1986, of 87 stations measuring turbidity, 32 reported data; of 155 stations measuring precipitation chemistry, 33 reported data; of 85 stations measuring SPM, 25 reported data.[23] Data have been subjected to quality evaluation through interlaboratory comparisons. Data have also been used in making initial assessments of trends in background air pollution.[24] In the case of carbon dioxide, the monitoring data have confirmed the long-term upward global trend, which is roughly parallel to the rate of increase in carbon dioxide released into the atmosphere by human activities.[25] The initial data reviews and assessments played a role in further development of BAPMoN and resulted, among other things, in modification in network procedures and data handling.

Promoting international scientific dialogue

The very existence of BAPMoN – its problems, needs, and outputs – has served as

in the Peruvian Andes, there were no other baseline stations in the tropical belt, and that only one baseline station was performing the full programme, i.e. Cape Grim established by Australia as its contribution to GEMS (Table II, EPMRP Series, No.55, *op. cit.*). It is also worth noting that, among the developing countries that were either taking part in BAPMoN or planning to do so, virtually all were operating regional stations, mostly with the equipment being provided through UNEP's financial support, through the WMO Voluntary Cooperation Programme(VCP), or directly from bilateral sources. For details about developing countries' participation, see *ibid.* In some developing countries where the equipment supplied by GEMS was not used, it has been suggested that it be transferred to other countries that might be better prepared and interested in carrying out monitoring.

23 See *ibid.*, tables IV–VI. For the situation during the earlier period see 1982 and 1983 reports on status of BAPMoN and R. F. Pueschel, *op. cit.*, pp.59–62. In the case of carbon dioxide, monitoring data is submitted by the baseline stations. However, considerable delays have occurred in the reporting of data due to laboratories' desire to ascertain data quality. In order to overcome this time-lag, provisional data are being published with a delay of less than one year. See 'Provisional Daily Atmospheric Carbon Dioxide Concentrations, as measured at BAPMoN sites for the years 1986 and 1987', EPMRP Series, No. 58, May 1989.

24 For a discussion of carbon dioxide data quality, problems and delays in reporting and archiving data, etc., see M. R. Manning, 'An Assessment of BAPMoN Data Currently Available on the Concentration of Carbon Dioxide in the Atmosphere', EPMRP Series, No. 9, February 1982. Also see C. Hakkarinen, 'General Considerations and Examples of Data Evaluation and Quality Assurance Procedures Applicable to BAPMoN Chemistry Observations', EPMRP Series, No. 17, July 1983. For a summary of BAPMoN data assessments, see the 1982 background paper, *op. cit.* See also C. C. Wallen, 'A Preliminary Evaluation of WMO/UNEP Precipitation Chemistry Data', MARC report No. 22, Chelsea College, London, 1980; H. W. Georgii, 'Review of the Chemical Composition of Precipitation as Measured by the WMO BAPMoN', EPMRP Series, No. 8, February 1982, and C. C. Wallen, 'Sulphur and Nitrogen in Precipitation: an Attempt to Use BAPMoN and Other Data to Show Regional and Global Distribution', EPMRP Series, No.26, April 1986.

25 See the 1982 background paper, *op. cit.*, p.45. For the results of the BAPMoN performance survey, see 'BAPMoN Newsletter', No.2, March 1987. Also see 'Report of the Joint WMO/ICSU/UNEP Meeting of Experts on the Assessment of the Role of Carbon Dioxide in Climate Variations and Their Impact', Villach, Austria, November 1980, WCP document, Geneva, 1981, and 'Report of the International Conference on the Assessment of the Role of Carbon Dioxide and Other Greenhouse Gases in Climate Variations and Associated Impacts', Villach, Austria, October 1985, WMO publication No. 661, Geneva, 1986.

a catalyst for a dynamic scientific dialogue and exchange related to the monitoring of humanity's inadvertent impacts on climate and on climatic change. This dialogue, in addition to supporting BAPMoN and climate-related environmental monitoring and promoting their development, has also been of value for the study of climate and climatic change in general.[26]

Learning by doing

As in the case of GEMS HEALTH, practical experience in siting and setting up monitoring stations, carrying out observations, processing and analysing data, and so on, has proved very valuable in refining and modifying the approaches and in shaping the objectives of climate-related environmental monitoring.[27]

Refining the approach to climate-related monitoring

Parallel to the build-up of the global climate-related monitoring infrastructure and capability embodied in BAPMoN, the truly important step forward was made in casting the foundations of a comprehensive and long-term programme of monitoring and assessment of the global climate system, which was eventually embodied in the World Climate Programme. An outcome of a series of converging develop-

---

26 See 'Papers Presented at the WMO Technical Conference on Regional and Global Observation of Atmospheric Pollution Relative to Climate', Boulder, August 1979, WMO Special Environmental Report No. 14, WMO publication No. 549, Geneva, 1980. The conference, like the preceding ones in Helsinki and Gothenburg, dealt with the expansion of BAPMoN. While the first two meetings were concerned with measurements, observation, networks, and standards, the Boulder meeting also examined data analysis. The Boulder report contains a series of papers on the state of the art and reviewing the situation. Among the topics covered were: instruments, techniques, and standardization; interpretation of measurements; status of networks and observing programmes; network design and data selection; and global and regional observations. The process was continued at the WMO Technical Conference on Observation and Measurement of Atmospheric Contaminants (TECOMAC), held in Vienna in 1983. See 'Extended Abstracts of Papers Presented at the WMO Technical Conference on Observation and Measurement of Atmospheric Contaminants (TECOMAC)', Vienna, 17–21 October, 1983, EPMRP Series, No. 20. Among the topics covered were chemical climatology; significance of air chemistry observations for understanding and predicting the state of the environment and climate variations; observation and measurement of long-lived and reactive gases, aerosols, dry deposition, chemical constituents in precipitation, and organic contaminants; instruments and analytical techniques, standardization, quality assurance; and interaction of atmospheric constituents with the adjacent environmental media.

27 For example, the Mt Kenya pilot exercise of choosing a site for a baseline station provided experience and methods valuable for siting baseline stations. Moreover, the same project contributed to the development of BAPMoN as a whole by demonstrating the need for a larger number of baseline stations than had originally been forecast; it also advanced the knowledge of carbon dioxide and its behaviour and led to a relaxation of original siting criteria, which were difficult to fulfil for continental baseline stations (i.e., no change in land-use practices for 50 years in a radius of 100 kilometres, infrequent experience of volcanism, forest fires, and sandstorm effects; being remote from population centres and transport routes; etc.). Another example is turbidity. In principle, turbidity measurements ought to be easily obtainable owing to the simplicity of procedures. However, because of problems with the stability of instruments and with the location and performance of stations, the data obtained over the years have not been very useful in establishing global trends. A redefined approach is therefore called for. 'Report of the Executive Committee Panel of Experts on Environmental Pollution', fourth session, Geneva, 27 Sept. – 1 Oct. 1982, EPMRP Series, No. 13, p. 8.

ments, WCP was ultimately an attempt at a systemic response to the need to understand and deal with climate and climate change.

Even though its basic outline and components were clear already at the time of UNCHE and were reflected in the Study of Man's Impact on Climate, reaching the point of putting together such a package was a significant achievement. It reflected the practical lessons and the accumulated experience over the years since the Stockholm Conference, and an improved understanding of the climate system and its relationship to human existence and civilization. It implied rising above some of the traditional institutional and disciplinary constraints. And it meant placing climate change among the priority concerns of WMO, a status it did not have earlier in this primarily meteorological institution. Indeed, up to 1975, climate in WMO was largely a matter of statistics. Things began to change under the impact of the increasing concern about climate change aroused by UNCHE, but primarily because of the growing awareness that the characteristics of the atmosphere were changing rapidly and significantly. Also, it came to be understood that climate change and environment was a subject which offered to WMO significant opportunities for task expansion.

This approach involved linking up firmly scientific research, monitoring, and assessment. It also involved the acceptance of long-term scientific study and monitoring as a legitimate objective in itself, equal in priority and value to the requirement to cater to the immediate needs of decision-making and management. In order to strengthen the case for such study and global observation, an effort was made to argue and demonstrate the many effects of climate and climate change on economic and social well-being of society and of the international community. The initial impulse for climate-related monitoring was given by observations of increasing levels of carbon dioxide in the atmosphere; the evidence that concentrations of carbon dioxide were steadily on the rise was accompanied by the realization that other greenhouse gases and variables also figure in processes related to climate change. Thus, climate-related monitoring grew in scope and linked different media. As for BAPMoN, it was moving incrementally towards multipurpose uses, which was a logical outcome both of the phenomena under study and of the underutilized potential of this slowly evolving global network.

In a sense, then, what transpired in the years following UNCHE was that the systems nature and character of the world climate and its interrelationships with society and the environment were imposing a greater degree of comprehensiveness and rationality on the whole approach. While the physically tangible indicators of progress were concentrated mostly in the build-up of BAPMoN, the real breakthrough was made in the framing of a long-term programme of monitoring and assessment of global climate system and climate change. BAPMoN and the early work on climate-related monitoring were important in preparing the ground for the comprehensive framework articulated in 1978 and broadly reflected in WCP.

In sum, then, the principal advances made were:

- A global monitoring network, embodied in BAPMoN and WGI, was set up and represented an important capability;

- Climate-related monitoring was firmly linked to research and assessment, and embedded in an integrated approach to climate and climate change within the fold of WCP, in particular its data and research components;
- A bridge was maintained between long-term scientific needs and the more immediate requirements of decision-making and management, through WCP publications dealing with matters of current concern;
- Gradually yet markedly task expansion had taken place. The concern about carbon dioxide spread to encompass other greenhouse gases. Attention was devoted to change in surface properties of the Earth and the impact of these changes on climate. And the concern with humanity's impacts on the atmosphere went beyond climate to include some of the more immediate threats to the Earth's environment, human health, and society, such as the depletion of the ozone layer and the long-range transport of pollution. In other words, bridges between climate-related monitoring and other GEMS clusters, including natural resources monitoring and health-related monitoring, were becoming possible, thanks to a variety of effects of atmospheric pollution.

FACTORS AT PLAY

Compared with health-related environmental monitoring, BAPMoN was initially in a less favourable position to attract the participation and interest of a large number of countries. Being primarily oriented to scientific and global needs, it was perceived as being of little relevance to daily concerns and management. This perception was partly offset by the fact that BAPMoN facilities are proving adaptable for multiple uses, potentially making them relevant to some current needs. For example, monitoring via BAPMoN of the long-range transport and deposition of air pollutants, and especially transfrontier pollution, responds to one of the growing concerns of many governments. However, as political and public concern with climate change began to mount, climate-related environmental monitoring began to gain in relevance.

In terms of implementation, however, several characteristics of climate-related monitoring placed it in a favourable position from the very start. Unlike GEMS HEALTH networks, which ultimately should reflect local problems and are thus subject to continuing expansion to meet these needs, climate-related monitoring networks are subject to global design and planned implementation. The system is therefore easier to manage and control, and attention can be directed more readily to issues critical to its functioning and build-up. Furthermore, in contrast to a new path that had to be charted in a relatively unknown field of health-related monitoring, climate-related monitoring benefited from the initial advantage that the very idea of monitoring had its origins in meteorology and that the established tradition and experience of WMO-coordinated international monitoring served as a model and could be relied upon.[28] Indeed, WMO's World Weather Watch was one of the

28 An example of this was the monitoring of solar radiation at ground level. Measurements made nationally are sent to Leningrad observatory in the USSR for collation, editing, and publication as

organizational models that inspired the idea of GEMS. Also, it offered a functioning global infrastructure that could be of assistance to climate-related environmental monitoring, for example, in data transmission.

Ultimately, the real strength of climate-related environmental monitoring comes from the fact that it embodies an evolving global capability necessary for understanding and predicting the climate. Thus it is of importance for devising appropriate strategies and for planning national, regional, and world development, as well as for controlling and modifying humanity's inadvertent and deliberate impacts on climate. The leadership of a few countries, the scientific community, and the international organizations provided the necessary drive and motivation for the start and sustained development of climate-related environmental monitoring.

As concerns the specific factors that have played a role in shaping the direction, nature, and pace of development of this GEMS cluster, principally of BAPMoN, they are broadly similar to those discussed in the context of health-related environmental monitoring.

North–South differences

The objective of getting the developing countries fully involved in various monitoring programmes takes on additional importance in the case of climate-related monitoring because of the truly global scope of this cluster. The great majority of developing countries suffer from a lack of financial resources and have a weak domestic infrastructure. They tend to be less interested in climate than in health-related environmental monitoring because they do not see the very act of monitoring as having much direct or immediate usefulness for their domestic needs.[29] The fact that they perceive this as an activity of international concern, and global climate change and atmospheric pollution as matters within the responsibility of developed countries, strengthens the expectations of many that they should receive financial and technical support as an incentive to participate, and discourages independent action.

The developing countries' capabilities and position have had several consequences for climate-related monitoring in GEMS. Most notably, the need to make it accessible to them has largely defined the scope and nature of the initial approach. Thus, in devising the first guidelines, the scientists involved in the design of the system set aside the ambitious and sophisticated schemes that would have suited

part of an arrangement with WMO to act as its World Radiation Data Centre. See SCEP, *op. cit.*, p. 171.

29 Scientists from developing countries were hardly involved in the design and direction of climate-related environmental monitoring. For example, in the Study of Man's Impact on Climate prior to UNCHE, of the 30 scientists taking part only one was from a developing country (India). In one of the intergovernmental expert groups organized by UNEP and WMO, only one of the 15 experts attending was from a developing country (El Salvador). At the TECOMAC Conference in Vienna, of the approximately 60 papers submitted, six came from developing countries, of which three were from India. This situation seems to have changed little over the years; at the 1985 Villach Conference on the Role of Greenhouse Gases in Climate Variations, of 77 participants only nine were from the developing countries. Yet by then it was already clear not only that climate change mattered but also that developing countries were likely to be affected in particular.

the abilities, experience, resources, and interests of industrialized countries. Instead, they opted for universal participation and global coverage as the first objective on the road to the future goal of a complex and advanced global climate monitoring system. The design of the monitoring programme and in particular of the minimum programme carried out by the regional stations was 'a compromise between scientific objectives and technical and practical feasibility. Global comparability, simplicity and compatibility have been given preference over a complete and complicated programme, quite desirable for many purposes'.[30]

Attempts to ensure that developing countries would take part were hampered by minor but often troublesome logistical problems occasioned by their lack of financial resources, appropriate infrastructure, and adequately trained and motivated personnel. For example, unavailability of chemicals and of spare parts for instruments, difficulties with the servicing and maintenance of instrumentation, problems with keeping trained personnel working on a continuing basis, and lack of cooperation among national institutions tended to hinder the implementation of the scheme and the more effective involvement of developing countries.[31]

As usual, the greatest difficulties were experienced in funding monitoring activities in developing countries. The lack of interest and motivation on the part of some to pay for those endeavours placed a special onus on international funding; it was the only means to fill the gaps in the global system and to start up, and often supervise and help manage, monitoring activities that would not have been carried out if one had had to rely solely on locally available means and motivation.

The amount available from international sources for this purpose was, however, inadequate. The shortfall was not felt so much in the case of BAPMoN regional stations. The basic equipment required for the minimum programme, being relatively inexpensive ($10,000 – $40,000 per station), was supplied to a number of developing countries short of hard currency.[32] In the case of baseline stations and the more complex regional stations, however, the absence of international funding produced gaps in the global network. Baseline monitoring stations are *de facto* observatories which in most instances also carry out research activities and are

---

30 This quote, which is taken from a draft version of the monograph by R. F. Pueschel, *op. cit.*, could not be found in the published text.
31 One example relates to the batteries used in sun photometers. The lack of these batteries locally caused interruption of turbidity measurement programmes in Ethiopia, Sudan, and Jordan, with the operators writing to WMO to ask for replacements. Another example relates to a frequent lack of horizontal cooperation among meteorological services, environmental institutions and universities in developing countries. Thus, technical questions that could be sorted out locally are referred to WMO instead.
32 A total of 35 developing countries had received the equipment by 1983. Among the few developing countries that operated regional stations but that were not supplied with equipment under this programme were India, Indonesia, and Malaysia. The equipment was mainly paid for by the Environment Fund and, in some instances, through the Voluntary Cooperation Programme (VCP) of WMO and by bilateral sources. See the 1983 BAPMoN Status Report, *op. cit.*, pp. 1–11. (Note, however, that in support of the establishment of a network of regional stations in Romania, UNEP contributed more than $200,000 mostly for the equipment. This was an exceptional project, which was approved as part of the general support extended by the UN system to Romania after the country had been hit by a major earthquake. The figure is illustrative of the costs involved in properly equipping regional stations.)

108 *The quest for world environmental cooperation*

costly and complex undertakings. Developed countries that established baseline stations in response to the needs of BAPMoN were prepared to invest the necessary resources and sufficiently interested to conduct baseline monitoring. Without international encouragement and support, developing countries generally were not prepared to undertake this kind of activity.

This situation is exemplified in the fate of the proposal, mentioned earlier, to set up a baseline monitoring station on Mt Kenya. It was a serious, sustained, and costly attempt to establish a major station in the developing world. The original proposal was floated at the time of UNCHE by a scientist from Sweden. Detailed pre-feasibility and feasibility studies were carried out at a cost of more than $350,000, mostly UNEP and some WMO funds, on the assumption that eventually the combination of national and international funding would make it possible to establish an operational baseline station. The inability of the interested parties to put together the final financing package, in particular the refusal by Kenya to provide the counterpart contribution, prevented the establishment of this baseline station that would have been of considerable importance to BAPMoN objectives.[33] The idea of an international station operated by WMO could not be entertained. Such a possibility was ruled out by its terms of reference, and also because countries normally wish to be fully in charge of any monitoring activity on their territory.

In brief, then, because of the situation in the developing countries and their response, important components of the global climate monitoring system were not likely to become operational by relying on national efforts alone. Special and regular international support was needed to fill gaps and strengthen the weak links in the global network, in particular as concerns baseline stations. International funding and technical support have a key role to play in this effort. As if to highlight this point, as well as to absolve developing countries of the reputation of being the only weak links in the global system, delays and frustrations that have accompanied the attempts to set up and operate a baseline station on the Canary Islands, with the technical and financial support of the Federal Republic of Germany, illustrate some of the human, institutional, and attitudinal problems that often need to be dealt with in trying to secure such global coverage.[34]

33 As regards the costs of this station, the initial estimate made in 1975 by the Swedish government was that $250,000 would be needed for the feasibility study, $1.25 million for buildings and equipment, and $300,000 annually for its operation. After the feasibility study was completed, the Kenyan government estimated that in addition to its own costs (buying the site, providing offices at the base of the mountain, assuring access to laboratories in Nairobi, etc.), estimated at approximately $143,000, it needed outside help for road construction ($300,000), for bringing in electric power ($50,000), for the construction of buildings ($90,000), for scientific equipment ($223,000), for external scientific staff ($100,000 per year), for support staff ($32,000 per year), etc.

34 The Federal Republic of Germany approached Spain and offered to equip and finance a baseline station on the Canary Islands. It did so because establishing a baseline station on German soil was not possible on account of polluted air, and because the Canary Islands are well located in relation to the European continent and offered ideal background conditions. A long delay occurred initially because the Spanish Parliament was not prepared to grant duty-free status for the importation of the equipment, which was valued at about DM 2 million. Once the station was turned over to local management, breakdowns of the sophisticated equipment occurred, which made it necessary to fly in German technicians to keep it operational. Criticism has also been voiced that little scientific work was carried out at the site, in spite of the possibilities offered by the state-of-the-art equipment.

International funding

On the eve of UNCHE, the SMIC study estimated that approximately $17.5 million would need to be spent annually to translate into practice its recommendations concerning global climate-related environmental monitoring. This included 10 baseline stations, each one projected to require $0.5 million in capital investment and $0.25 million in annual operating costs.[35] During the period 1974–82, $4.4 million was spent internationally on climate-related environmental monitoring projects in the framework of GEMS. Of this, $2.42 million, or more than 50 per cent, came from UNEP, with the remainder being essentially contributions in kind from cooperating agencies.[36]

The difference between the SMIC estimate of annual costs and the amount available internationally for promoting the build-up of climate-related environmental monitoring over an eight-year period can be in part accounted for by the distinction between *Level Two* and *Level Three* of the UNEP model: the operational activities are the responsibility of governments and it is they who incur the major costs. The international funding available via UNEP, i.e., *Level Three*, is used to promote the functioning of the monitoring cluster rather than to pay for operational monitoring. In other words, the true costs of monitoring activities are much higher than shown by GEMS figures. This having been said, the SMIC estimate broadly illustrates the magnitudes involved and the relative paucity of international funding.

This matter, however, is simply skimmed over in the implementation of GEMS because of the assumptions of the model used. While this may be justified as regards developed countries, in the case of developing ones it means avoiding the real issues. It leaves unanswered the question of how to supplement what is available locally, or how to pay fully for activities that are not likely to be funded nationally. An attempt is made to fill gaps with the money available from the Environment Fund. The support given to developing countries consists of equipment, training courses, visits by consultants to discuss their capabilities and interest in participating and to advise them on siting, establishing, and operating monitoring stations.

> Germany has expressed interest in funding another baseline station in the tropical belt through a similar bilateral arrangement with another country. Undoubtedly, the Canary Islands experience will be useful in planning and implementing future arrangements of this kind.

35 See SMIC, *op. cit.*, pp. 6–7.

36 The total figure and the project figures are derived from UNEP project documents and its 'Report to Governments'. They are a broad approximation, in view of continuing changes in allocations and expenditures. However, for the purposes of this discussion, exact figures are not required. For BAPMoN, UNEP's contribution for the period Oct. 1974 – Mar. 1980 amounted to $771,900 (out of a total of $1,326,900); for the BAPMoN carbon dioxide laboratory at Scripps, Jan. 1976 – Dec. 1977, $156,100 (of the total $169,000); Mt Kenya baseline station feasibility study, 1976–7, $245,300 (of the total $291,800), plus $71,000 in 1978 for a bridging operation; support for the BAPMoN network in Romania, Jan. 1977 – Sept. 1980, an unusual 'UNDP' type of project prompted by the wish to assist Romania after a major earthquake that it suffered, $213,090 (of the total $215,590); for the World Glacier Inventory, Feb. 1976 – Mar. 1980, $199,739 (of the total $396,139); and for the climate monitoring omnibus project, Apr. 1978 – Dec. 1981, $433,900 (of the total $1,445,000). Phase II of the climate-related monitoring omnibus project (Jan. 1983 – Dec. 1985) was costed at $1,474,500, of which UNEP provided $672,400, including $292,000 to continue the build-up of BAPMoN.

Also, their participation in various meetings is paid for. As with GEMS HEALTH, these forms of support absorb a significant portion of the available funds. Yet this assistance is far from meeting the true needs. Furthermore, disbursements on these items reduce the amount of money available for energizing the cluster as a whole. Support from the development assistance institutions was even less likely to materialize than in the case of health-related monitoring; it was not easy to demonstrate the direct relevance of climate-related monitoring to the national development process, and the recipient countries were not prepared to use scarce hard currency obtained from multilateral or bilateral donors for this purpose.

Once more, an activity was in a dilemma because of the minimal international funding available. Money is needed to support the establishment and functioning of the stations (e.g., $500,000 needed to equip a baseline station may appear insignificant, yet even in the case of such an important country as the USSR its baseline stations were delayed by lack of hard currency to pay for the necessary equipment). It is also needed to coordinate and supervise the network and its performance, a function that assumes added importance when technical back-up, including that from the centre, is needed to support the weak links in the field.

Voluntary nature

Notwithstanding decisions and recommendations of UNEP and WMO governing bodies, BAPMoN is essentially a voluntary effort. It is predicated on a high level of responsiveness of governments, their willingness to carry out prescribed activities, to allocate adequate resources and staff, to report data on a regular basis, and ultimately to make a long-term commitment to take part. Their participation is sought and encouraged by the secretariat of WMO, through visits to countries and other forms of interaction.

The work of those in the field who undertake the observations, recording, processing, and assessing of the data is as important as the position of their governments. An effort to promote greater responsibility has been made by encouraging interaction and exchange of practical experience among those taking part in BAPMoN-related activities, and by fostering a feeling of belonging to a global network and peer group.[37]

The voluntary approach is broadly modelled on similar endeavours carried out in developed countries.[38] Moreover, there is a long-standing tradition within WMO

37 Occasional technical meetings of the kind organized to discuss BAPMoN station operation are an opportunity to achieve this kind of exchange. (See, for example, 'Report of Meeting of Experts on BAPMoN Station Operation', Geneva, 23–6 Nov. 1981, EPMRP Series, No. 6.) However, there are limits to what can be done along these lines in view of financial constraints: only 11 experts took part in the meeting just cited.

38 An example of this approach within a single country is the US National Atmospheric Deposition Program (NADP). It represents a voluntary effort by interested institutions and scientists. Using standard instrumentation and procedures, they operate the deposition monitoring sites and collect weekly samples, which are sent to a central laboratory that reviews and publishes the data. See P. L. Finkelstein and J. M. Miller, 'Quality Assurance for the Acid Precipitation Measurement Programme in the United States', in the TECOMAC Proceedings, *op. cit*, p. 59. The problem of 'responsibility' is encountered also in the developed countries. BAPMoN duties are often grafted

of 'responsible members', whereby interested countries take upon themselves specific responsibilities within given international programmes, a practice also resorted to in BAPMoN.[39] The problem with global programmes such as BAPMoN that require participation of many countries is that, in the absence of a clear-cut obligation of governments, they are implemented in a highly uneven manner. This has proved to be one of the main problems in the development of the network and in the efforts to ensure the necessary quality and reporting of the monitoring data.

The coordination issue

The institutional set-up for the launch of this cluster of GEMS was favourable. WMO brought in the experience, capability, infrastructure, and leadership of a major specialized agency. UNEP provided crucial seed money needed to start and support monitoring activities. Also, it projected the environmental dimension that was helpful to WMO in its efforts to broaden and update its approach to climate matters.

Coordination between the two bodies was effective, as the direct working-level contact between GEMS PAC and WMO functioned well. The human factor was of importance here. The Deputy Director of GEMS PAC, who had been recruited from WMO, was in fact posted at the WMO secretariat in Geneva. This assured a close and direct working relationship. Details of this nature and human relations are usually overlooked in the analysis of processes within international organizations, though in terms of everyday cooperation and the development of interagency coordination, they often happen to hold the key to success and are more important than formal instruments and arrangements.

However, with the operational launch of WCP, the previously close working ties between GEMS PAC and the pilot programme weakened, exposing the vulnerability of such relationships to personnel changes and subjective factors.[40] Similarly, personnel changes in GEMS PAC deprived it of relevant expertise and sharply reduced its interest and role in climate-related environmental monitoring.

---

on to other duties of government officials, who do not have the adequate training and cannot generate the necessary enthusiasm and interest for the undertaking.

39 An example of the WMO 'responsible member' concept is found in the preparation of marine climatological summaries. World oceans and seas are divided into eight areas, each allotted to one WMO 'responsible member'. These members prepare and publish summaries for their respective areas on an annual basis. Each member also provides a databank for storing the basic marine data on the area within its responsibility. In the case of BAPMoN, for example, the US EPA has assumed responsibility for processing and publishing the data. (The same centre also handles the WHO data, as noted in the preceding chapter.) This is of great assistance to the system, as it provides a capability and resources that are not available at WMO and performs such time-consuming tasks as correspondence with countries, transferring data on to computers, providing camera-ready tables, etc.

40 GEMS is not mentioned in the 'Outline Plan and Basis for World Climate Programme, 1980–1982', WMO publication No. 540. Yet the Climate Data Programme, which was one of its proposed four sub-programmes, subsumed all monitoring activities that were classified as falling under the GEMS framework in the UNEP scheme of things.

Monitoring tools and capabilities

Monitoring instruments and techniques are important elements in the evolution of BAPMoN. Their effectiveness and quality not only are important for obtaining good monitoring data but also determine the feasibility of many monitoring operations, both in the technical sense and in terms of staff and resources needed.[41] Similarly, the degree of technical difficulty involved in undertaking given types of monitoring influences the choice of activities. Aerosols have not been given prominence in spite of their importance for the study of climate. In part this was due to the inherent difficulties of monitoring and studying their effects.[42] Moreover, as also happened with turbidity measurements, instrument shortcomings resulted in unsatisfactory monitoring data being recorded over a period of time. Hence the instrument had to be redesigned and a new series of monitoring data started.

New, space-age technologies and monitoring techniques hold the key to the successful development of climate-related monitoring and the fulfilment of many of its ambitious tasks. For example, airborne and ground-based Light Detection and Ranging (LIDAR) systems, developed by the National Aeronautics and Space Administration (NASA) for global air monitoring, represent a very important new monitoring capability. In the case of airborne LIDAR, the short pulse of the laser beam is bounced off the ground, recaptured through a telescope, and analysed, showing instantaneously the distribution and concentration of particles and gases in the atmosphere below. A three-dimensional 'slice' of the atmosphere thus obtained can then be analysed by computers. In the past, this type of information could be obtained only by direct sampling and complex laboratory analysis to determine the constituents of the sample.[43] This new capability represents literally a jump from the horse-and-buggy era to the space age in monitoring. A network of 12 to 14 LIDARs (at $250,000 per unit), placed strategically around the globe, would be able to provide data on vertical aerosol distribution. And, in fact, data are already being collected in this manner by the World Ozone Data Centre because of its importance for vertical ozone distribution calculations.

Remote sensing, in general, promises to make it possible to dispense with a series of costly, time-consuming, and personnel-intensive chores carried out on the ground or in the atmosphere. It will also permit the monitoring of a number of

---

41 Monitoring of carbon dioxide offers a good example. While scientific interest in monitoring carbon dioxide concentrations was already considerable as long ago as the 1940s and 1950s, such monitoring was not undertaken on a systematic basis because the sampling technique then in existence was very difficult to use and was not very reliable in terms of producing exact data. The development of the infra-red carbon dioxide analyser in the mid-1950s made it feasible to monitor carbon dioxide on a continuing basis, to take unattended measurements, to obtain a fair degree of accuracy, and to avoid sampling biases caused by the time of day and biological activity. The technological change that made this monitoring practicable was accompanied by increasing interest in this kind of work. R. F. Pueschel, *op. cit.*, p.16.

42 In contrast to carbon dioxide, a stable and homogeneous gas that mixes well in the atmosphere and is easy to monitor, aerosols are heterogeneous in chemical composition, size, shape, and optical properties. They do not mix well and are unevenly distributed, have to be collected at high altitudes, and have a wide variety of origins, including soil erosion, sea spray, volcanoes, vegetation, etc. (*ibid.*, p. 81).

43 D. T. Wruble, 'Lasers Help Unravel Air Pollution Mysteries', *EPA Journal*, October 1985.

variables and/or trends that would have been prohibitively expensive, very difficult, or impossible to carry out.[44] Furthermore, remote sensing will help to resolve some problems related to standardization, quality, and reporting of data for a number of variables, which in turn would make it easier to use such data for the assessment, modelling, and further study of climate.[45]

TOWARDS AN INTEGRATED APPROACH TO GLOBAL ATMOSPHERE MONITORING

When the Stockholm Conference placed on the international agenda the issue of climate change caused by anthropogenic activities, this was a rather remote and unfamiliar concern for politicians, decision-makers, and the public at large. A mere fifteen years later, at the US–USSR summit in 1987, one of the few areas chosen for long-term bilateral cooperation concerned global climate and environmental change.

As if to illustrate the relevance of this choice, the year that followed witnessed a series of natural disasters, including major droughts, floods, and hurricanes, that disrupted national and regional economies.

Many saw in these disasters, and in weather anomalies in different parts of the world, the first manifestations of a rapid global warming trend. The 'greenhouse effect' provided the plausible explanation, thus becoming not only a favoured media topic but also a subject of everyday conversation, feeding people's innate interest in matters having to do with weather and the elements of nature. More or less simultaneously with the mounting public concern for the depletion of the ozone layer, global atmospheric pollution and human-induced climate change had finally reached the point of becoming an issue of international and national politics. The many functions of the atmosphere as a vital global resource and the many atmosphere-related environmental phenomena were beginning to be better appreciated, including their implications for international relations and global security.[46]

---

44 For example, a picture of global carbon monoxide distribution has been obtained thanks to the gas-filter radiometer flown on the US shuttle as part of the experiment called Measurement of Air Pollution from Satellites (MAPS). To the surprise of scientists, monitoring data showed that burning of tropical forests and savannas yields at least as much carbon monoxide as the burning of fossil fuels, and that some of the highest concentrations are to be found in the Southern hemisphere. It is suspected, moreover, that the carbon monoxide encourages accumulation of methane, which adds to the greenhouse effect. See R. E. Newell, H. G. Reichle, Jr, and W. Seiler, 'Carbon Monoxide and the Burning of Earth', *Scientific American*, Oct. 1989.

45 For an early review of satellite uses for climate-related monitoring and the list of variables that can be monitored from satellites, see the 1978 background paper, *op. cit.*, pp. 44–7. See also the 1982 background paper, *op. cit.*, pp. 70–1, and the 1980 ICSU/COSPAR report, 'Space Based Observations in the 1980s and the 1990s for Climate Research', prepared in the framework of the World Climate Programme. Further, see F. W. Taylor, 'Satellite Measurements of Minor Constituents in the Middle Atmosphere', presented to the TECOMAC Conference, *op. cit.*, pp.73–4, and J. T. Houghton, F. W. Taylor and C. D. Rodgers, *Remote Sounding of Atmosphere*, Cambridge University Press, Cambridge, 1983. For additional discussion of remote sensing of environment, see chapter six below.

46 As an example, see 'The Changing Atmosphere: Implications for Global Security', Conference Statement, Toronto, Canada, 27–30 June, 1988.

Climate-related environmental monitoring represents an essential tool for the assessment and study of the phenomena involved, of changes and trends, and for preparing for action as called for by the international community. Though GEMS had played a role in the evolving process, and in the maturation of perceptions and of practical approaches to monitoring of anthropogenic influences on climate, it found itself on the margin of events and was not in a favourable position to respond to this opportunity and challenge.

As discussed above, via GEMS, UNEP helped to start and place on the right track the process of building up a global capability for monitoring climate, climatic change, and those variables reflecting humanity's impacts on the composition of the atmosphere and the environment that may affect the climate. But, with the rise of WCP, the climate-related monitoring in GEMS seemed to lose its initial drive and inspiration.

A number of factors account for this situation. It was in part the consequence of the very nature and inherent tenuousness of UNEP's catalytic approach, including the difficulty in relating meaningfully to a *Level Two* activity, i.e., an activity that it does not support financially and on which therefore it has no claim. It was also linked to the systems nature of the climate issue, which cannot be readily accommodated within the existing scheme of organizational responsibilities and of functional and disciplinary divides, including those established by the five clusters of GEMS. In addition, it was the consequence of the lack of clear structuring and greater coordination within and among governments, and of the unsettled relationship between the traditional meteorology establishment and the evolving environmental problématique and institutitions.

The initial policy impulse from UNCHE and GEMS-supported activities within WMO was essential to the maturation and launching of WCP. Once WCP took off, however, its relationship with GEMS weakened. Among the contributory reasons, as mentioned earlier, were the following:

- WCP is a comprehensive effort. It links research, monitoring, and assessment, while GEMS focuses on monitoring only. Also, WCP takes an integrated view of the climate system, while GEMS, in principle, is supposed to concentrate on human-induced environmental impacts that lead to climate change.
- The earlier cooperation did not continue owing to an apparent loss of interest in GEMS among those in charge of WCP once their programme had become operational. Moreover, UNEP's increasing activism regarding the greenhouse effect was viewed as interference by some of those involved in shaping and directing WCP, who consequently took an unfavourable view of GEMS as well.
- Owing to staff changes in GEMS PAC, and the lack of relevant in-house professional expertise, the interest in climate-related environmental monitoring weakened, and the close working level ties and direct involvement in project follow-up virtually disappeared.
- BAPMoN, which remained as the centrepiece of climate-related environmental monitoring in GEMS, was not made an integral part of WCP. It continued as the

responsibility of WMO's Research and Development Department, while WCP was placed within a newly created department.

The discontinuity and uncertainty in the evolution of climate-related environmental monitoring within GEMS was, to a degree, also a consequence of UNEP's project-funding mode of operation and the resulting ambivalence from the point of view of the cooperating agency as to the intrinsic nature of GEMS. It should not be suprising that the 1988–97 Long-Term Plan of WCP makes virtually no substantive reference to GEMS, except as a potential source of funding.[47]

As concerns BAPMoN, which UNEP continued to support financially, inertia was allowed to prevail. While significant progress was made in building up and improving BAPMoN, many problems and shortcomings continued to be experienced. All of these called for review and action with the objective of improving the network. However, as in many other international undertakings, they were not addressed openly and were allowed to drag on by inertia. BAPMoN thus remained largely set in its initial conceptual and institutional straitjacket, wedged between institutions, jurisdictions, and scientific disciplines, and most of all without the necessary resources to respond to expanding requirements and needs and to fulfil its basic functions.

Criticism about BAPMoN implementation surfaced in the report of the WMO Working Group on Environmental Pollution and Atmospheric Chemistry, which met on the occasion of the thirtieth anniversary celebration of the Mauna Loa observatory in March 1988. This was almost twenty years after the WMO Executive Committee had formally recommended the establishment of BAPMoN.[48]

Among other things, the group recommended that the whole concept and implementation of BAPMoN be subjected to thorough re-evaluation, including a component-by-component review of its operations, data quality, and data assimilation. Such an evaluation should be carried out by specialized groups drawn from the scientific community.

Several important points were signalled by the Working Group:

- The growing knowledge of atmospheric chemistry and pollution, and of atmospheric aspects of climate change, which was also assisted by BAPMoN data, called for re-examination of the functions and priorities of BAPMoN.
- On account of spotty performance and implementation, the very credibility of BAPMoN was at stake. Among the factors mentioned were the uneven quality of data, delays in exchange of data, stations that were not operational or that failed to submit data on a regular basis, gaps in global coverage, inadequate training relative to needs, financial difficulties of the central reference laboratory for carbon dioxide, failure to follow up and implement various recommendations made by previous expert group meetings.
- BAPMoN was too narrowly conceived and implemented in isolation from other

47 See 'The World Climate Programme, 1988–1997', Second WMO Long-Term Plan, Part II, Vol. 2, July 1986, WMO doc. CG-X/Doc. 8.
48 See 'Report of the First Session of the EC Panel of Experts/CAS Working Group on Environmental Pollution and Atmospheric Chemistry' (Hilo, Hawaii, 27–31 March 1988), EPMRP Series, No. 56.

relevant programmes and activities. It should be expanded to monitor biospheric sources and sinks of carbon dioxide, and the impacts on biomass, all of which contribute to uncertainty in projections of carbon dioxide content of the atmosphere. And efforts should be mounted to monitor other greenhouse trace gases whose combined effect on climate may be as important as that of carbon dioxide (e.g., methane, nitrous oxides, chlorofluorocarbons). It should be also used to fulfil additional functions. It was thus suggested that it be placed at the service of integrated monitoring, of monitoring long-range transport of pollutants, and of monitoring of radioactive species in air and precipitation in the context of the increasing role of WMO in this domain following the Chernobyl disaster.
- There was an increasing gap in the capabilities and knowledge of different countries as concerns atmospheric chemistry. This had to be taken explicitly into account in further implementation of the network and in assigning national, regional, and global responsibilities to countries.

In a way, the report also projected a feeling of frustration on the part of scientists – most of whom came from scientifically advanced societies and were familiar with state-of-the-art monitoring – in face of the usual difficulties experienced in implementing global programmes of this kind and the mismatch between the needs and the responses. It was reflected in their recommendation to concentrate resources on improving the quality of what existed rather than to increase the number of stations to fill the important remaining gaps in global coverage.

Among the factors responsible for this situation in BAPMoN were the familiar dilemmas experienced in implementing various GEMS projects, including the lack of clear-cut responsibility of individual governments to follow up a given commitment and to deliver according to a workplan and a timetable. It was also evident that the assumptions of the UNEP model were not entirely realistic. While this was due mainly to disparities in terms of development among countries and their differing capabilities and responses, another reason was the lack of adequate international funding, both for fostering efforts in those countries that were not willing or able to undertake monitoring activities on their own, and for promoting more vigorously the international aspects of the global system build-up and operation.

The basic trade-off, namely working to obtain universal participation and a global network at the initial cost of variable implementation, gaps in coverage, and the lowest common denominator approach, was unavoidable in the initial phase of the global system build-up. In the short term, it detracted from the speed, quality, and effectiveness of the endeavour. In the longer run, however, it was intended to secure the involvement of more countries and to contribute to the gradual development of their national capabilities, which is one of the important preconditions for the functioning of a fully fledged global climate and climate-related environmental monitoring system. However, for lack of resources and commitment, the overall response was not as expected. The project was allowed to carry on without a more determined international effort to prop up the weak links and fill the gaps. This resulted in disparities in the network; some of the baseline stations were sophisticated research laboratories (e.g., Cape Grim in Australia), while others were no more than a few instruments, often in a storage room.

Furthermore, a global monitoring network could not be fully and effectively operational without a well-staffed unit at WMO and adequate resources to back it up. In the absence of such support, it becomes difficult to perform a series of supervisory, technical, and scientific functions that are increasingly called for as the network evolves, as well as the many institutional and promotional roles inherent in such an international undertaking. This gap became more pronounced as the network grew and the management load increased. At the same time, internal WMO resources devoted to BAPMoN were reduced owing to pressures on its budget. The UNEP contribution also shrank (to $50,000 per year in 1988–9), causing unhappiness among those engaged in BAPMoN work, who felt that they were being let down by UNEP.

The institutional dimension also played a role, both in WMO and domestically. BAPMoN did not secure a strong foothold for itself in WMO. In part, this was because of the expectation of continuing support from UNEP, and in part owing to the tight-fisted attitude of government representatives and the hard choices of how to allocate the limited resources among the many priority concerns. In a meteorological organization, environmental concerns and BAPMoN were not a top priority. The result was that governments' representatives in WMO approved the objectives of BAPMoN, but failed to allocate the necessary funding.

A somewhat similar situation was replicated in many countries where BAPMoN was placed under the jurisdiction of hydro-meteorological services that showed limited or no interest in the subject and did not allocate sufficient resources for this purpose. Often, they were not equipped to deal with the broader environmental issues, and the staff assigned to the task did not have appropriate qualifications. Nor did many of these departments see much point in pursuing an activity that they felt was not of immediate value and was not likely to yield direct benefits for their countries. The fact that meteorological services were in charge was not always helpful to the efforts to mount an integrated approach to monitoring at chosen sites. The situation was made worse in some places by the smallness of these services, and/or by the jurisdictional and administrative obstacles that prevented the involvement of other national agencies with a broader interest in environmental issues.

It was high time to ventilate these issues. The politics of global warming, largely fuelled by environmental monitoring, research and assessment, were becoming a serious matter. In the framework of WCP, UNEP and WMO were among the principal movers of the climate change issue in the international arena. The objective was to build up the awareness and the feeling of urgency, on the way to adoption of an international agreement for limiting emissions of greenhouse gases and their sources, patterned on the approach used in the case of ozone, and to working out long-term strategies of adaptation to climate change.[49]

49 See, for example, 'Report of the International Conference on the Assessment of the Role of Carbon Dioxide and of Other Greenhouse Gases in Climate Variation and Associated Impacts', Villach, Austria, 9–15 October 1985, WMO document No. 661, 1986. It is estimated that, if current trends continue, the combined concentrations of carbon dioxide and other greenhouse gases would be

In this context of the rising demand for monitoring and assessment, GEMS and WCP were invited to consider what new climate-observing activities were needed, what emission data on greenhouse gases should be monitored and archived globally, and how to monitor consequences of climate change. Moreover, the need was stressed for simultaneous, integrated, multimedia monitoring of climate and of its impacts on biota and ecosystems at selected interdisciplinary sites.[50]

Climate change as a prime global systems issue had become a conceptual and political lever which was forcing an integrated and sustained view of the planet Earth as a system, including the all-important interrelationship between biosphere, geosphere and humankind.

Once more, and as in the pre-Stockholm days, it was the scientific community that responded first and led the way, both in formulating and documenting the challenge and thus stirring up the political and public concern, and in proposing that a comprehensive, longer-term view be taken of the limits of habitability of the Earth, namely its ability to support life. Such view was to subsume the global climate, atmosphere, lithosphere, oceans and the biogeochemical cycles of the major nutrients (carbon, nitrogen, phosphorus and sulphur).

The initial suggestion was floated by the US at the 1982 UN Conference on Peaceful Uses of Outer Space (UNISPACE) in Vienna, when it presented a NASA blueprint for a 'Global Habitability' programme. The initiative was not endorsed, however, in part owing to inadequate preparatory work and consultations, and in part because of the usual fear that it would interfere with and duplicate some of the ongoing programmes. The basic ideas were kept alive, and recycled soon thereafter, both internationally and within the United States.[51]

At the international level, the initiative inspired the International Geosphere-Biosphere Programme (IGBP). The concept of IGBP was first endorsed by the 1984 ICSU General Assembly and linked with the World Climate Research Programme of WCP. In the US, the 'Global Habitability' programme spawned a proposal for an ambitious Earth System Science research programme, which has as its linchpin an Earth Observing System (EOS), also known as 'Mission to Planet Earth'. EOS, to be housed initially on board polar-orbiting platforms of the US Space Station complex scheduled to be launched in the mid–1990s, represents the most ambitious

---

equivalent to the doubling of carbon dioxide levels by the year 2030 as compared to pre-industrial levels. According to the current climate models, the consequential rise in the global mean equilibrium surface temperature would amount to between 1.5 and 4.5 degrees Celsius, which in turn could lead to a sea-level rise of between 20 and 140 centimetres (*ibid.*, pp. 20–1). See also 'Developing Policies for Responding to Climatic Change', a summary of discussions and recommendations of the workshops held in Villach (28 Sept. – 2 Oct. 1987) and Bellagio (9–13 Nov. 1987), doc. WMO/TD-No.225, April 1988. Further see 'The Greenhouse Gases', UNEP/GEMS Environment Library No. 1, *op. cit.* and M. K. Tolba, 'For a World Campaign to Limit Climate Change', *International Herald Tribune*, 15 Mar. 1988. The concern led to the creation of WMO/UNEP Intergovernmental Panel on Climate Change (IPCC), which held its first session in 1988. For an overview of its work, see 'IPCC Gazette', No.1, June 1989.

50 'Developing Policies for Responding to Climatic Change', *op. cit.*, p.4.
51 See M. M. Waldrop, 'An Inquiry into the State of the Earth', *Science*, 5 Oct. 1984. Also see M. McElroy, 'The Challenge of Global Change', *New Scientist*, 28 July 1988.

plan so far in the domain of peaceful uses of remote sensing technologies for study and observation of the planet.

IGBP is modelled on its much more modest antecedent, namely the International Geophysical Year (IGY) of 1957–8. It focuses on factors and processes affecting the Earth's ability to support life. It builds and draws on a series of international programmes that, over the years, have yielded important new knowledge about the atmosphere, oceans, biosphere, etc. It focuses, however, on what occurs at the interface between these and on the role of humankind as an increasingly important factor in the processes and change. It reflects the recognition of the limitations of traditional scientific disciplines and of the need to mount a systemic study and observation of the earth and of its climate system. Moreover, and unlike IGY, IGBP is a long-term programme, projected to last well into the twenty-first century.[52]

The US Earth System Science research and observation programme has as its two principal objectives to understand the Earth as a system within the context of the solar environment, and to carry out an integrated research programme for the study of global change. The whole concept is based on the need for a systematic approach to the scientific investigation and understanding of the entire Earth system, how its component parts and their interactions have evolved, how they function, and how they may be expected to evolve in the future. One of the main objectives is to develop a capability to predict changes that will occur both naturally and in response to human activity.[53]

Both IGBP and the Earth System Science programme were designed in anticipation of advanced and sophisticated space-based monitoring technologies backed by powerful information systems that scientists will have at their disposal and that will make possible quantum jumps in their ability to observe, study and understand the Earth as a system.[54]

New tools and capabilities, backed by *in situ* monitoring networks, will dispense eventually with many operational and institutional problems that have hampered global observation and assessment so far. Also, they will help to overcome the limitations of sectoralized knowledge and bridge gaps between scientific disciplines, e.g. earth and biological sciences, oceanography and atmospheric sciences.

From the foregoing account, it is obvious that the key to the secrets of the Earth

---

52 See 'A Plan for Action', a report prepared by the Special Committee for the IGBP for Discussion at the First Meeting of the Scientific Advisory Council for the IGBP, Stockholm, Sweden, 24–8 Oct. 1988, *Global Change*, report No.4, 1988. For a US proposal, see *Global Change in the Geosphere–Biosphere, Initial Priorities for an IGBP*, US Committee for an International Geosphere–Biosphere Program, National Research Council, National Academy Press, Washington, D. C., 1986.

53 See *Earth System Science, a Closer View*, Report of the Earth System Science Committee, NASA Advisory Council, NASA, Washington, D. C., Jan. 1988. See also 'Bringing NASA down to Earth', *Science*, 16 June 1989, and 'An Expedition to Earth', *New Scientist*, 29 July 1989.

54 See *Earth System Science, op. cit.*, in particular chapter 7 on trends in instrumentation and technology, chapter 8 on observing and information systems for Earth System Science, and table 9.1 on sustained long-term measurements of global variables required for global analysis to be continued or repeated indefinitely. An advanced information system, relying on supercomputing facilities, is at the very foundation of the programme and is needed in order to handle the immense quantities of data that will be generated and to convert these into meaningful geophysical information.

system and how it functions, and the role of humankind, lies in advanced organization, big science, big technology and, of course, big money. It is estimated that an investment of $15–$30 billion will be required for the NASA satellite-based research programme over the period of two decades starting in 1991, and that EOS will need about $1.5 billion per year as of 1994. This estimate would have sounded like science fiction at the time when the idea of GEMS and Earthwatch was hatched in Stockholm in 1972. However, it is a reflection both of the increasing power and capability of humankind to understand the global environment and the Earth system, and of the increasing worry that the human species might be undermining the very ground on which it depends for its survival and prosperity.

GEMS seemed largely bypassed by these events, though, according to the model conceived at UNCHE, UNEP should have provided the institutional home for similar developments and initiatives at the global level. Undoubtedly, this bypassing was a consequence of the weak institutional capability of UNEP and of the marginal role that it had carved for itself in the ongoing processes. It was also symptomatic of the flux of international arrangements in this complex domain, the seemingly unavoidable lack of coordination and the multitude of initiatives and actors able to work in the same direction on their own and with the necessary resources at their disposal.

The IGBP, accordingly, envisages the establishment of a network of globally distributed marine- and land-based observatories (Geosphere–Biosphere Observatories or GBO). The proposed network for the time being remains an idea, with ICSU lowering its objectives on account of the complexity of the task. GBO, inspired by similar concepts and needs that gave rise to the idea of GEMS, is to measure simultaneously various earth system parameters, and to observe and document global changes, test hypotheses relative to interactions among Earth system components, serve as ground-truth location for remote sensing measurements, and provide inputs and serve as validation sites for regional and global simulation models.[55]

It is this integrating thrust that has also exerted an impact on WMO. More than two decades after the operational start of BAPMoN, its Executive Council took a decision to set up a Global Atmosphere Watch (GAW).[56]

Starting from the recommendations of the Hilo Report of the Working Group on Environmental Pollution and Atmospheric Chemistry, cited above, the GAW scheme represents an attempt to achieve an expanded and stronger monitoring and

---

55 See section 6.4, 'Geosphere-Biosphere Observatories', in 'IGBP Plan of Action', *op. cit.*, pp.103–8. GBO is based on the need for an interdisciplinary, integrated observatory network. The network is conceived as consisting of '5–10 regional research and training centres, multinationally supported, interdisciplinary in operation and multi-purpose in function'. These centres would be backed up by 100 cooperating national facilities, which would be used for experiments and as key monitoring sites. The next tier would consist of another 1,000 stations in affiliated networks, acting primarily as monitoring sites. Finally, there would be transient research and monitoring sites, both in time and space, the so-called 'tents and rowboats', aimed at capturing transient phenomena and studying specific processes.

56 For the decision adopted at the forty-first session of the Executive Council in June 1989, see WMO document EC-XLI/PINK 12. For the summary description of GAW, see WMO Fact Sheet No.3.

research programme on changing global atmospheric composition, and to rationalize and integrate the existing efforts within and outside the framework of WMO.

What this decision meant was the setting up of an integrated global atmosphere monitoring network, building on the BAPMoN, GO_3OS and the Monitoring and Evaluation of the Long-range Transmission of Air Pollutants in Europe (or EMEP, Evaluation and Monitoring of Environmental Pollution) networks and experience. GAW was to be guided by standard functions embodied in GEMS, namely, routine measurement and quality assurance, data storage and retrieval, research and development, and data assessment and interpretation.

In a significant policy departure for WMO, and in recognition of the fact that environmental issues were a key national and international concern, it was recommended that atmospheric chemical observations should be assigned very high priority by WMO member states, and should be carried out with the same attention as measurements of meteorological parameters. Were this recommendation to be taken seriously and implemented by member states, it would become possible to set up an *in situ* global network for monitoring atmospheric pollution.

Starting from the existing baseline stations of BAPMoN, it has been recommended that 20 global atmospheric observatory type stations be established, to monitor variables related to climate change and ozone change, as well as long-range transport of pollutants and radioactive substances.[57]

WMO's role in the long-term monitoring of the global atmospheric composition and preparation of scientific assessments was conceived quite ambitiously. Among the many functions envisaged were: to provide network design, standards, data collection systems for global monitoring and to ensure participation by member states; to serve as a forum to plan, coordinate and evaluate relevant monitoring activities; to act as a publication and information centre; to arrange for national centres with specific responsibilities in the global network; to promote 'twinning' of developed and developing countries to ensure more effective participation of the latter and better coverage of tropical and Southern hemisphere regions; to carry out training; to perform analysis and interpretation of data; to be responsible for harmonization of methods, development of standards, and quality assurance; and to coordinate all international monitoring programmes on atmospheric composition.

GEMS and two decades of halting and difficult progress and learning after UNCHE were at the basis of these developments. Yet, UNEP and GEMS were

---

57 The list of variables to be monitored is comprehensive and includes all greenhouse gases, ozone, radiation and transparency of the atmosphere (turbidity, solar radiation, UV-B, total aerosol load, etc.), chemical composition of precipitation, sulphur dioxide, nitrogen oxides, carbon monoxide, radionuclides (e.g. Krypton–85, Radon, Tritium), water in soil and plants, classical meteorological elements, cloud condensation nuclei, integrated air samples, etc. The work of the observatories is to be backed up by regional stations with flexible programmes, geared also to meeting regional and national needs. Finally, a mechanism was to be set up for effective coordination of WMO and non-WMO atmospheric composition monitoring networks. See annex to paragraph 5.4.4 of WMO document EC-XLI/PINK 12. For a discussion of advanced instrumentation for investigating atmospheric processes, as opposed to the more simple monitoring of trends, see D. L. Albritton, F. C. Fehsenfeld and A. F. Tuck, 'Instrumental Requirements for Global Atmospheric Chemistry', *Science*, 5 Oct. 1990.

hardly involved in this significant step in the framework of WMO. The relationship seemed to be more of the same. The decision of the WMO Executive Committee thus noted that this expanded and strengthened programme will continue to be a contribution to GEMS. And UNEP's continuing support was sought in developing and building of GAW.

The fast-moving events and the challenge of improving the knowledge of climate, atmosphere, biosphere, geosphere, biosphere and human civilization as an interactive system, and of using such knowledge for evolving and implementing political, economic, social and technological responses, raised forcefully the question as to the very nature and role of GEMS, and of the original structure and assumptions adopted in 1974.

GEMS' inherent potential to serve as the overall international framework for monitoring and study of the earth as a system remains to be tapped. It is certain that the political, economic, scientific, and technological developments taking place do call for a reassessment of the role and direction of GEMS in this broader context.

The UN Conference on Environment and Development, and the International Space Year, both scheduled for 1992, should provide an opportunity for a fresh look at the notion of a global environment monitoring system. There is a strongly felt need for a new start, taking into account the lessons learned, the increasing pace and scope of human-induced changes of the Earth's surface, oceans, and the atmosphere – because the impact of human-induced changes is beginning to approach that of the natural processes which control the planet's life-support system – and the rapidly expanding scientific and technological frontiers that will make efficient global monitoring possible.

This entails, among other things, rationalizing and defining the role of GEMS in the dual challenge of improving the knowledge of climate, atmosphere, biosphere, and contemporary civilization as an integrated system, and of using such knowledge for evolving and implementing a new agenda of political, economic, social, and technological responses. These responses would involve protecting and managing the planet's integrity, anticipating changes and adapting to them, and directing global climate change for the benefit of humanity, a feat that is no longer in the realm of science fiction only.

There is no doubt that this will require the setting up of an integrated global environmental monitoring network, a task from which UNEP and GEMS shied away in the early stages, as will be discussed further in the chapters that follow. Such a network will be an essential, supporting tool for confronting the complexity and instability inherent to these processes, and for providing the data to back up policy-making and public-opinion formation. In setting up this network, one will have to take note of and build on the early and halting experiences of GEMS. This includes taking into account the existence of BAPMoN, which to its credit has not only produced incontrovertible data on the steady rise of carbon dioxide concentrations in the atmosphere, but has also contributed to the acceptance of the idea of routine worldwide monitoring of environment-related variables that are of global interest. In addition, account will have to be taken of the broader efforts to undertake climate-related monitoring, which, rather than being a well-defined

GEMS cluster, has turned out to be primarily a process of learning and trying to set up a meaningful global approach under unfavourable conditions.

5 Marine pollution monitoring in GEMS

Protection of the marine environment was one of the major concerns at the Stockholm Conference. Marine pollution by oil from tanker spills such as occurred in the case of the *Torrey Canyon*, the health effects of polluted seafood as dramatized by the Minamata Bay disease in Japan, and a notable degradation of seawater quality in many coastal areas, caught attention and gave rise to growing alarm among scientists, politicians, and the public.[1]

The Conference emphasized monitoring as a tool for generating the scientific database needed to keep track of the health of the oceans and to assess the status and trends of marine pollution. Monitoring was to contribute both to greater knowledge and understanding of the processes involved and to the formulation and implementation of public policies for the management and exploitation of the marine environment. In line with the significance assigned to global approaches to environmental problems, the need for undertaking baseline studies and monitoring of selected sections or spots of open ocean to determine trends in global ocean pollution was given prominence. The importance of national and regional monitoring schemes was recognized, as well.[2]

The 1974 Intergovernmental Meeting on Monitoring listed monitoring of marine pollution and assessment of the health of the oceans among the priority objectives of the new Global Environment Monitoring System (GEMS). The

---

1   An Intergovernmental Working Group on Marine Pollution met in preparation for the UN Conference on the Human Environment (UNCHE). For its reports see docs A/CONF.48/IWGMP. I/5 and II/5. At the expert group level, the International Oceanographic Commission (IOC) convened a working party at Castellabate in 1971 which prepared an outline of a plan for a Global Investigation of Pollution in the Marine Environment (GIPME). For its report, see document IOC/GIPME-II/7. For the action by the Stockholm Conference, see its recommendations 86–94 in doc. A/CONF.48/14/Rev.1. Also see the general principles for assessment and control of marine pollution recommended by the Intergovernmental Working Group and endorsed by the Conference (*ibid.*, annex III).

2   For one of the early proposals on an ocean baseline sampling programme, as a precursor of ocean monitoring, see the Study of Critical Environmental Problems (SCEP), *Man's Impact on the Global Environment*, MIT Press, Cambridge, Mass., 1970, pp. 176–84. Setting up a global system of pollution monitoring was recommended by several meetings of experts. These proposals were elaborated in the 'Comprehensive Outline of the Scope of the Long-term and Expanded Programme of Oceanic Exploration and Research (LEPOR)', IOC Technical Series No.7, UNESCO, 1970. Many of these proposals drew on and were inspired by the pioneering work undertaken within the framework of the International Council for the Exploration of the Sea (ICES) in the North Atlantic.

purpose of marine pollution monitoring was formulated ambitiously and in global terms, i.e., to determine sources and levels of pollutants in the marine environment, to measure and understand pathways of pollutants to and in the sea, and pollutants' fate once in the sea, as well as the exchange of pollutants between the sea and air, sediments, land, and biota.[3] The focus on the open ocean and a global approach was in line with the Earthwatch orientation towards global environmental assessment.

As its initial response to recommendation No. 90 of the Stockholm Conference, which called for the promotion of the monitoring of marine pollution as well as for 'the development of methods for monitoring high-priority marine pollutants in the water, sediments and organisms', and the recommendation of the 1974 Intergovernmental Meeting, the UN Environment Programme (UNEP) began to provide support for the IOC pilot project on global monitoring of marine pollution caused by petroleum.

The marine pollution monitoring component of GEMS was thus off to an early start. However, someone leafing through GEMS documentation of a later date would be puzzled to see practically no reference to marine pollution monitoring and no relevant projects funded from the GEMS budget line.[4] Only upon close reading will one find that the marine pollution component of GEMS was being implemented under the auspices of UNEP's Regional Seas Programme. And, indeed, this cluster of GEMS had a different trajectory from climate-and health-related monitoring.

THE FIRST STEPS AND CONTROVERSIES

It may rightly be said that the common view taken of marine pollution monitoring at the time of the Stockholm Conference was something akin to World Weather Watch (WWW) or Background Air Pollution Monitoring Network (BAPMoN) models, i.e., a global network of baseline stations (coastal and island stations, buoys, ships of opportunity, weather ships, research vessels, etc.) engaged in systematic and regular monitoring of given variables. This was the view conveyed by its recommendation No. 90, which called for the promotion of marine pollution monitoring, preferably within the framework of the fledgling IOC/WMO Integrated Global Ocean Station System (IGOSS). Launched in 1969, IGOSS was meant to be a permanent system for monitoring the physical properties of the oceans

---

3 Doc. UNEP/GC/24, para. 24.
4 A series of references to ocean monitoring, as a component of GEMS, are made in the early documentation. The GEMS Programme Activity Centre (PAC) Director, in his statement to the 1976 World Meteorological Organization(WMO)/IOC/UNEP Preparatory Meeting of Experts on a Monitoring System for the Assessment of Background Levels of Selected Pollutants in Open Ocean Waters, noted that the pollution of oceans was one of the main concerns of UNEP and that, if approved, this would be one of the most significant activities under the aegis of GEMS. Following this expert group meeting, however, references became fewer, and the 'Earthwatch/GEMS In-Depth Review', prepared for the 1981 session of the UNEP Governing Council, does not even contain a section on the monitoring of marine pollution. The only reference to the subject is to be found in the context of climate-related monitoring, and the interface and pollution exchange between the surface layer of ocean and the atmosphere.

126 The quest for world environmental cooperation

(e.g., temperature, salinity, currents, tides), for exchange and analysis of oceanographic data and for the dissemination of oceanographic services.[5]

As noted, one of the first monitoring projects financed by UNEP was the IOC pilot project on the monitoring of marine pollution caused by petroleum (MAPMOPP). It was executed in the framework of IGOSS. Petroleum was chosen as a model pollutant, because it was considered to pose a major threat to the global marine environment and was of the greatest concern to the public and decision-makers. Moreover, it was a pollutant most likely to attract the participation of both the developed and the developing countries in the pilot project – an important policy consideration.[6]

More or less parallel with the start of the pilot project on petroleum monitoring, the UNEP Governing Council recommended that the ocean monitoring programme be launched and that it should include 'ocean baseline stations, analogous to the atmospheric baseline stations'.[7] It is on the basis of this decision that IOC (which already had its plan for monitoring and baseline studies in GIPME), WMO, and UNEP started to work on a joint plan of action for monitoring background levels of selected pollutants in open ocean. The stated objective was to extend the scope of GEMS to include open ocean waters in a planned and comprehensive manner.[8] As a first step, and relying on the facilities and experience gained within the framework of ICES, a pilot project was to be launched in the Atlantic Ocean, with eight monitoring sites initially. Among the categories of pollutants that could be studied in the open ocean the following were mentioned: heavy metals, chlorinated hydrocarbons, petroleum hydrocarbons, surface-active compounds, radionuclides, and solid waste.

During these consultations at the expert group level, however, there were

---

5 For an overview of the IGOSS, see H. U. Roll, 'A Focus for Ocean Research, Intergovernmental Oceanographic Commission, History, Functions, Achievements', IOC Technical Series Monograph No. 20, UNESCO 1979, section 14.2.4. Operationally, IGOSS is analogous to World Weather Watch, and provides it with inputs. In so far as possible, it relies on WWW facilities for observations, data processing and data transmission. The backbone of IGOSS is its observing system (IOS), consisting of weather ships and oceanographic research vessels, moored or shifting buoys and satellites, as well as voluntary observers on ships of opportunity. The IGOSS Data Processing and Servicing System consists of specialized oceanographic centres for the processing of monitoring data and the preparation of oceanographic products.

6 The purpose of the pilot project was primarily: (1) to develop and test techniques and methodologies for monitoring petroleum pollution (e.g. oil slicks, tar balls, dissolved/dispersed hydrocarbons, oil and tar on the beaches); (2) to try out the capability of the incipient IGOSS network to serve for ocean pollution monitoring; and (3) in general to review and test organizational and logistical issues related to the establishment and functioning of such a global system, including the readiness of countries to cooperate. The methods used were simple in order to make it possible for countries at different levels of technical sophistication to participate. As the IGOSS pilot project took off, thought was also given to the possible inclusion of additional pollutants, e.g. trace metals and chlorinated hydrocarbons. For the original plan of the pilot project, see document IOC-WMO/MPMSW-I. (During the period 1975–9, UNEP paid $484,120 out of the total project cost of $587,120. It also provided additional funds for related projects.)

7 Decision 32(III), in A/10025.

8 For a review of related events and decisions, see IOC/UNEP/WMO document 'A Programme for Monitoring Background Levels of Selected Pollutants in Open Ocean Waters', Open Ocean Pollution Monitoring, Geneva, 1976.

differences of opinion as to the immediate need and value of pursuing open ocean monitoring activities. A UNEP representative stated that 'only a sound assessment of the substantive value of the programme will permit us to decide whether the investments that will be required are justified by the expected results and their bearing on the long-term problems of the health of the oceans'. This signalled a change of heart and was confirmed when UNEP soon after declined to provide financial support for the activities related to monitoring background levels of selected pollutants in the open ocean. Both IOC and WMO felt slighted by this, especially because their governing bodies had adopted appropriate decisions on follow-up, based in part on the recommendations emanating from UNEP's Governing Council.[9]

In the meantime, and while these discussions went on, various activities of the IGOSS pilot project were carried out as planned and the project was completed. Methodologies for monitoring oil in the open ocean were tested; scientific and organizational aspects of such monitoring were studied; the inclusion of other pollutants into the scheme was considered (e.g., trace metals and chlorinated hydrocarbons); intercalibration exercises were carried out; manuals were published and training offered to developing countries; some equipment was supplied to coastal stations; remote sensing applications were reviewed; national oceanographic centres (Washington and Tokyo) were used for data storage and exchange; etc.[10]

On the basis of the results of the petroleum monitoring pilot project and pending its detailed assessment and specific recommendations for follow-up actions to be made by a workshop, the IOC Assembly decided to establish a Marine Pollution Monitoring System (MARPOLMON) on a permanent basis, as a data gathering arm of GIPME. The assessment made by the workshop was a positive one, and the pilot project was considered to have fulfilled its objectives. The workshop also concluded that the continuation of petroleum monitoring was both scientifically justifiable and useful for management purposes as the pilot project had already generated a large amount of data used by governments in their policy and decision-making processes. Therefore, it was recommended that the monitoring of petroleum should continue on an operational basis and should not be interrupted, so that the results already achieved would not go to waste.[11]

---

9 See, for example, WMO resolution 16(EC-XXVIII), which calls upon the Executive Director of UNEP to examine favourably the WMO application for support of the Programme for Monitoring Background Levels of Selected Pollutants in Open Ocean Waters, prepared at the request of UNEP itself, as a follow-up to decision 32(III) of the Governing Council.

10 For a presentation of the results of the pilot project, see E. M. Levy, M. Ehrhardt, D. Kohnke, E. Sobtchenko, T. Suzuoki and A. Tokuhiro, 'Global Oil Pollution', Results of MAPMOPP, the IGOSS Pilot Project on Marine Pollution (Petroleum) Monitoring, IOC, Paris, 1981.

11 See 'Third IOC/WMO Workshop on Marine Pollution Monitoring', Feb. 1980, New Delhi, IOC Workshop Report No. 22, and its various recommendations including that on converting MAPMOPP into MARPOLMON. The visual observation of oil pollution, while useful in obtaining an approximate global picture of the distribution and extent of petroleum pollution, largely produced commonsense data (oil along tanker routes and downwind). It was recommended that a similar overview exercise might be repeated, after a period of time, provided that more accurate observation methods were developed and the existing patterns of transport, and levels of production and

At the same workshop, however, the GEMS representative stated that, while the pilot project was a success in providing a baseline for the global distribution of oil and in testing the methods and the usefulness of the IGOSS network, for all the money spent it simply yielded commonsense knowledge and data, i.e., that petroleum pollution was present along shipping routes. From UNEP's point of view, according to him, the continuation on a permanent operational basis of this type of monitoring was not warranted, as it would not provide additional information of 'sufficient environmental significance and interest'. Accordingly, he said that UNEP was ending its support, owing to financial constraints and because it wished to concentrate the limited financial resources available in its Regional Seas Programme, although it would continue to follow with interest any work by IOC at the global level.[12]

UNEP's position was shared by many who felt that the pilot project produced masses of data of little interest or usefulness and that its principal contribution had been to demonstrate the viability of such monitoring schemes based on the willingness of many countries to engage in the collection of data. The differences that surfaced reflected the different organizational outlook, mandates, and priorities of UNEP and IOC, as well as WMO, the different coalitions of governments in control of each, and their varied domestic constituencies. Also, they reflected disagreements among the scientists involved.[13]

IOC had been established in 1960 as a subsidiary body within the framework of UNESCO.[14] Founded by a few interested countries to promote and coordinate

consumption of petroleum were to change significantly. The continuation of the collection and measurement of floating tar balls was proposed. Similarly, it was proposed to continue the monitoring of dissolved/dispersed hydrocarbons, which are especially important for an understanding of pollutant pathways and effects on marine biological resources. A search for new sampling and/or observation methods with respect to oil in seawater was recommended (e.g., surface microlayer sampling, hydrocarbons at greater depths and in bottom sediments, remote sensing).

12 See *ibid.*, Annex VI.
13 For an overview of the issues, see A. Preston, 'Marine Environmental Monitoring Requirements: Integrated Global Monitoring', submitted to the Riga Symposium on Integrated Global Monitoring, held in 1979. It is interesting that the types of conclusions on the subject reached at various meetings or workshops depended considerably on who attended them, and therefore the selection of the 'right' scientists to participate was of critical importance to an organization's stand. An instance of differing views concerns the monitoring of trace metals and hydrocarbons. UNEP interpreted the results of the Bermuda intercalibration exercise as meaning that 'operational monitoring of these pollutants in the open ocean was hardly warranted at this stage, whether from the point of view of existing concentration levels or from the point of view of practicality' (IOC Workshop Report No.22, *op. cit.*). (A similar conclusion was reached by ICES regarding trace metals baseline studies in open ocean waters. It was agreed that imperfect methodologies did not allow, at the present time, useful monitoring of baseline levels and trends in trace metal pollution in the open ocean. See 'Fourth Report of the Working Group on Marine Pollution Baseline and Monitoring Studies in the North Atlantic', ICES document CM.1978/E: 6.) However, the 'Scientific Report of the Intercalibration Exercise, the IOC/WMO/UNEP Pilot Project on Monitoring Background Levels of Selected Pollutants in Open Ocean Waters', IOC Technical Series No. 22, which was prepared by the IOC secretariat and consultants as a follow-up to the Bermuda exercise, conveys a more positive impression. It states, for example, that the experiment had shown 'that sampling methods currently being used by most developed marine laboratories are adequate for trace-metal sampling in the ocean at concentrations now believed to prevail in open-ocean waters' (*ibid.*, p. 5).
14 Resolution 2.31, adopted by the Eleventh Session of the UNESCO General Conference, established the International Oceanographic Commission 'to promote scientific investigation with a view to

oceanographic research and to encourage cooperative research projects, it had to live in the shadow of UNESCO, with limited resources to pursue its ambitious tasks. The Stockholm Conference was seen as an opportunity to strengthen the hand of IOC within UNESCO, to expand its mandate and channel more financial resources to its activities. It was assumed that the governments would be inclined to contribute to pollution-related work. The new interest in the environment was seen as offering an opportunity to obtain additional resources for IOC activities and to boost the organization's role. Most of the issues concerning the marine environment are closely interrelated and, in monitoring and studying marine pollution, one must necessarily study oceanographic issues and natural processes as well.

IOC, which already had an initial blueprint for research and monitoring of marine pollution in the form of GIPME, was given a prominent place in the UNCHE Action Plan. In its recommendation No. 87, the Conference called for the improvement of the 'constitutional, financial and operational basis' of IOC in order for it 'to be able to take on additional responsibilities for the promotion and coordination of scientific programmes and services'. Several recommendations were addressed to IOC for implementation and the IGOSS pilot project was one of these recommended activities. It was considered as the first step in building up a programme of open ocean pollution monitoring and research. In general, it was expected that, in the implementation of GIPME, which approached the marine environment in a comprehensive manner, IOC would be able to rely on regular support from the Environment Fund.

An effort was also under way to change the image of IOC from that of an exclusive circle of oceanographic nations to a more representative forum that would attract the developing countries' participation on an active basis. The mass entry of developing countries generated the need for technical assistance and training. This led to the launching of the Training, Education and Mutual Assistance (TEMA) component of IOC, specifically designed to help all interested countries take part in its work and to strengthen the research capabilities of developing countries. It also showed the need to initiate programmes that would be of interest to them, and that would secure their active participation. Marine pollution research

---

learning more about the nature of resources of the oceans, through the concerted action of its members'. For a study of IOC history, functions and achievements, see H. U. Roll, *op. cit.* For the structure and workings of the Commission and its secretariat see *IOC Manual* – Part I, second revised edition, July 1982, Paris. The small secretariat of IOC (10–15 people) is provided by UNESCO, but is *de facto* an autonomous unit. (This was recognized *de jure* only in 1987, when an amendment to the Statutes of IOC was adopted by the Twenty-fourth General Conference of UNESCO, giving IOC functional autonomy within UNESCO.) In serving the Intergovernmental Commission, the IOC secretariat cooperates with the specialized agencies of the UN system (which second staff members to it) and with other organizations involved in the study of the oceans. The major programme coordinated by IOC is the above-mentioned LEPOR, which focuses on physical, chemical, geological, and biological processes of the oceans. Its aim is to increase knowledge of the ocean, its content and the content of its subsoil, its interfaces with the land, atmosphere, and the ocean floor, and to improve the knowledge of processes affecting the marine environment. The ultimate goal is effective utilization of the ocean and its resources for the benefit of humanity.

and monitoring was considered as an area that would be of interest to developing countries.

IOC was thus not pleased when doubts began to be voiced as to the immediate value of allocating part of scarce GEMS resources to open ocean study and monitoring, and when UNEP announced its intent to concentrate on regional seas activities. On the assumption that the Environment Fund was created to finance the activities of the agencies on a continuing basis, IOC felt that it was being deprived of funding to which it was rightfully entitled.[15] Its reaction was all the more anxious because of its own institutional weakness, the fact that it was underfunded (e.g., $3.5 million budget for the biennium 1988–9, of which 60 per cent for staff costs), and its dependence on outside financing, including that from the Environment Fund, for executing its various programmes. Guidelines and blueprints for pursuing study and research of open ocean pollution had been prepared on the assumption of support from UNEP. Were such support to cease or fail to materialize, many of its programmes could not be implemented as planned. In addition, as will be discussed below, it was upset by what it considered to be jurisdictional trespassing into its own domain by the Regional Seas Programme of UNEP which staked out an interesting and promising area of work with the help of money from the Environment Fund.

UNEP's changing position regarding its financial support for activities related to study and monitoring of pollution in the open ocean was influenced by a series of factors. There were doubts shared by many scientists as to the feasibility and value of such monitoring.[16] The high costs of open ocean monitoring also helped dampen the original enthusiasm. Most of all, UNEP was swayed by the widespread view that the monitoring effort should concentrate on the most critical spots proximate to the major sources of pollution, which by definition lay close to shore. It was in these areas that monitoring would yield data and make possible assessments of immediate interest to governments which could be used to launch measures aimed at controlling environmental degradation. This view was already present in the pre-Stockholm phase. It was temporarily overshadowed by the global emphasis at UNCHE and the more sweeping and ambitious concept of an integrated ocean monitoring scheme.

This opinion crystallized during the deliberations of the 1976 expert group convened to review and propose the lines of action to be followed.[17] While the

---

15 For the IOC secretariat's interpretation of IOC and UNEP roles, see doc. IOC/INF–523, Add. The choice of quotations from the constituent documents of IOC and UNEP and from other sources focuses on the non-operational, coordinating, stimulating, catalyzing, and financing roles of UNEP, and the fact that the Environment Fund should be used for financing wholly or partly programmes of general interest, such as regional and global monitoring and environmental research.

16 Among other factors, the level of most pollutants in the open ocean is so low that the most sophisticated methods and instruments must be used to detect them. With a few exceptions (mostly manufactured pollutants, such as organochlorine compounds and radionuclides), most pollutants appear as natural constituents of the open ocean and it is difficult, and often impossible, to distinguish between a natural fluctuation of their concentration and what is introduced by humans. See A. Preston, *op. cit.*

17 See 'Report of the WMO/IOC/UNEP Preparatory Meeting of Experts on a Monitoring System for the Assessment of Background Levels of Selected Pollutants in Open Ocean Waters', WMO,

expert group formulated a programme for study and monitoring background levels of selected pollutants in open ocean waters, it also endorsed UNEP's focus on regional seas, which began to take shape with its Mediterranean initiative. UNEP's choice was of a pragmatic and priority character, namely, where it would be most important and useful to channel its limited resources and organizational effort at that particular juncture. To employ UNEP's own terminology, it was concluded that work related to study of open ocean baselines and open ocean monitoring should be carried out as a *Level Two* activity by the interested governments and organizations but without its financial support.

The shift was made easier by a series of factors that were specific to UNEP. The decision to initiate a Regional Seas Programme modelled on the Stockholm scheme of assessment/management, and to concentrate UNEP's marine activities in this sphere, as well as to tap the funds from the oceans budget line (a management activity) for purposes of marine pollution monitoring and thus supplement the GEMS budget line (an assessment activity) and leave more money for other monitoring clusters, determined a great deal of what happened eventually. Owing to the lack of appropriate expertise and interest among the small staff of GEMS PAC, and the limited funds available for monitoring activities in need of support, it was decided by the UNEP management to concentrate the efforts of GEMS PAC on climate-and health-related monitoring, and on natural resources monitoring, and to allocate the money in GEMS budget line primarily to these activities. The responsibility for marine pollution was devolved to the Regional Seas Programme, which was beginning on an experimental basis in the Mediterranean Sea and which was the only place in UNEP where expertise related to marine environment was to be found. The monitoring activities were to be charged to the oceans budget line, which at that time was larger than the allocation for Earthwatch as a whole (e.g., $3.0 million vs. $2.2 million in 1976, and $4.0 million vs. $2.6 million in 1977).

Both institutionally and substantively, this was a significant move. Marine pollution monitoring became the responsibility of a programme activity centre specialized in marine environment, with relatively significant resources at its disposal, an operationally ambitious approach, and a mandate concerned primarily with management, with monitoring and assessment playing a supporting role. This called for a different, action-oriented posture *vis-à-vis* the outside world and the cooperating institutions, in contrast to the more passive, primarily financing role common in other GEMS clusters. This pattern provides a useful insight and contrast in UNEP's role performance.

In the new framework, marine pollution monitoring was explicitly linked with and subordinated to specific management applications, and was thus distanced somewhat from the primarily scientific and research concerns dominant in the work of IOC. It was directed at coastal and territorial waters, where problems were more serious and where corrective and regulatory action was possible, rather than at areas beyond national jurisdiction. Its purpose was to tie in with the current concerns and

Geneva, 23–5 March 1976, and 'Joint UNEP/IOC-WMO IPLAN Meeting of Governmental Experts on a Monitoring System for the Assessment of Background Levels of Selected Pollutants in Open Ocean Waters'.

worries of the governments and to involve as many countries as possible. The shift in UNEP's attitude was also encouraged by the lack of interest in open ocean monitoring that it detected among its broad constituency.[18]

As the IGOSS petroleum monitoring pilot project demonstrated, in spite of the efforts by IOC to involve many countries in its execution, it was the oceanographic nations that did the major share of the work. The participation of developing countries was little more than symbolic.[19] Indeed, unlike health- and climate-related monitoring, where the system build-up was achieved essentially by linking national facilities and activities into a global network and by relying on specific domestic interests for motivation, study and monitoring of open ocean pollution involved territory beyond national jurisdiction. This effectively precluded the participation of all but a handful of developed nations that had the necessary capabilities and resources for, and special interest in, this type of monitoring.

UNEP's choice to concentrate on the regional seas and coastal areas in the initial stages was pragmatically inspired, as illustrated by the case of the Mediterranean. This, of course, did not mean a loss of interest in open ocean research and monitoring, as will be described later in this chapter.

POLLUTION MONITORING IN REGIONAL SEAS: THE MEDITERRANEAN MODEL

Monitoring marine pollution in regional seas and coastal waters has some practical advantages over monitoring open ocean. For example, the levels of concentration of various pollutants are generally higher than in the open ocean, and usually easier to detect and measure. In many instances, it is possible to identify point sources of pollution. The link between monitoring data and assessment, on the one hand, and action management, on the other, is realizable, which makes monitoring potentially of direct value to governments and thus commands their greater interest. The monitoring packages can be tailored to suit specific situations in a given body of water, and to correspond to the needs and interests of the countries involved, to their capabilities, and to the resources available. By focusing on the interests and concerns of groups of countries, it becomes easier to involve them and to mobilize

18 See the statement made by the UNEP representative to the third session of the Working Group of GIPME, Malta 1979 (doc. IOC/WG-GIPME-III/3). See also doc. IOC/INF-372, Annex I, Table 10, dated Oct. 1978, which contains a survey of the views of governments regarding a proposed programme for monitoring background levels of selected pollutants in open ocean waters. Of the 42 governments replying, only 7 indicated a readiness to participate (Argentina, France, the Federal Republic of Germany, Japan, Norway, the US, and the USSR).

19 One of the activities in which the developing countries took part was the counting of tar balls on the beaches: this was a coastal activity that could just as well be carried out as part of regional seas efforts. Their scientists were also invited to oceanographic vessels for training, which was not always relevant to their needs upon returning home. By mid-point in the pilot project, 12 developing countries were formally taking part. Only India had reported some data. See doc. IOC/INF-372, annex I, tables 11, 12. This was in marked contrast to the mass participation of developing countries in various Regional Seas Programmes. During preparations for UNCHE, developing countries asked for assistance with marine pollution monitoring, especially for technical and scientific training of personnel, provision of equipment and facilities, and the establishment of regional training and monitoring centres. See doc. A/CONF.48/IWGMP. I/5, Annex X/Rev.1.

financial resources, as shown by the willingness of governments to establish special trust funds for various Regional Seas Programmes. By mounting such programmes in critical areas around the globe, one can enlist the participation of most countries and achieve worldwide coverage.

The logistics of regional seas monitoring are less complex than in the case of monitoring the open ocean and it is comparatively easier to bring on-shore infrastructure into play. Moreover, the tasks of intercalibration, improving analytical and sampling procedures, and obtaining comparable data are less difficult to organize and carry out in a geographically delimited setting. Various regional efforts at developing and testing methodologies and techniques, data processing, and assessing monitoring data can complement one another. They can be transferred between regions, thus contributing to a more rational effort and use of resources on a global scale.

UNEP's venture into the regional seas was in response to the Stockholm Conference. In its recommendation No. 92, the Conference called upon governments to adopt national measures for the control of sources of marine pollution and to coordinate their actions regionally and, where appropriate, globally. This reflected the view that the main causes and manifestations of marine pollution and degradation were to be found in areas adjacent to land and in the enclosed and semi-enclosed seas, and that these represented the foci for corrective and protective action and management. UNEP also correctly surmised that most governments were interested primarily in the seas within their jurisdiction or adjacent to their territorial waters. In fact, as it turned out, governments were chiefly concerned with the waters very close to the shore and showed far less enthusiasm and willingness to pay for monitoring the more distant open waters even in the regional seas context.

UNEP drew on the research, monitoring, and assessment elements of GIPME that were relevant to regional needs.[20] It combined these with management and regulation objectives to produce regional action plans responding to the specific needs of different regions. It tried to link science and research-oriented needs, on the one hand, with practical and immediate concerns, on the other hand. It drew heavily on the experience and approaches already developed in the framework of ICES, which were reflected in the regional conventions for the North Sea and the Baltic that were concluded in response to UNCHE though outside the framework of UN.

The Mediterranean Action Plan was the first and remains the best known step in this direction. It grew out of the fact that UNEP was aware of an institutional vacuum and the absence of a mechanism for accommodating the desire of the countries bordering on the Mediterranean to work collectively in order to prevent further degradation of this regional sea. During the preparations for UNCHE, several of these countries had proposed that scientific and technical cooperation be undertaken, including the monitoring of pollution and the establishment of joint

---

20 For GIPME content, see 'Comprehensive Plan for Global Investigation of Pollution in the Marine Environment and Baseline Studies Guidelines', IOC Technical Series No. 14, UNESCO, 1976. Its overall objective is to provide a scientifically sound basis for the assessment and control of marine pollution.

monitoring networks. A number of preprogamming activities were carried out, involving the World Health Organization (WHO), WMO, the Food and Agriculture Organization (FAO), the International Atomic Energy Agency (IAEA), and IOC. These prepared the ground essential for the follow-up action.

The Mediterranean Basin posed a difficulty as concerns the institutional focus. It was parcelled up among three UN regional economic commissions, as well as between the three WHO regional offices. Their territories did not coincide. A further complication was that four countries on the northern reaches of the sea belonged to the European Economic Community. Also, an integrated approach to environmentally sound management of this sea spanned the substantive responsibilities of the specialized agencies. However, none of them was able to offer the right mix of qualities required for taking a lead role in this kind of activity. This presented UNEP with an opportunity. Relying on its broad jurisdiction and the money from the sizeable oceans budget line, spurred on by the Mediterranean governments and in cooperation with the specialized agencies, it took the lead role in launching and implementing the Mediterranean Action Plan.

The model for the UNEP regional seas and coastal areas programme was thus initially developed and tested in the Mediterranean Sea. It is structured according to the UNCHE assessment/management scheme. Its viability is based on intergovernmental negotiations, leading to policy agreement on action plans, buttressed by legal instruments (e.g. conventions and protocols) binding governments to take part and to act.

The monitoring of marine pollution has an important role to play in regional seas action plans. Tailored to specific problems and management needs of different regions, monitoring is supposed to provide inputs for decision-making and, in general, to upgrade the knowledge and understanding of human impacts on the regional seas and their health. Operational monitoring has been carried on longest and practical experience gained primarily in the Mediterranean.[21]

The monitoring-assessment component of the Mediterranean Action Plan (i.e., the Coordinated Mediterranean Pollution Monitoring and Research Programme), dubbed MED POL, was initiated in 1975. Its purpose was to support the implementation of the Barcelona Convention for the Protection of the Mediterranean Sea. The aim of Phase I of MED POL was to formulate and carry out a coordinated monitoring and research programme; to analyse sources, amounts, levels, pathways, and trends of pollutants (oil and hydrocarbons, heavy metals, polychlorinated hydrocarbons, total coliforms) in biota, seawater, and sediments; to study the effects of pollutants on ecosystems; and to help riparian countries to participate in the programme (eventually, 84 national centres from 16 states took part). A series

---

21 For an initial review of various monitoring activities, see 'Achievements and Planned Development of UNEP's Regional Seas Programme and Comparable Programmes Sponsored by Other Bodies', UNEP Regional Seas Reports and Studies, No. 1, Geneva, 1982, pp. 5–9. Outside UNEP, monitoring of marine pollution in the Baltic, northeast Atlantic, and the North Sea has been actively and routinely carried out pursuant to the respective regional conventions and under the technical auspices of ICES. For a review of ICES coordinated monitoring programmes, see 'ICES Cooperative Research Report Series'.

Marine pollution monitoring 135

of pilot projects, coordinated and paid for by UNEP, were executed by the specialized agencies.[22]

Considerable financial resources were expended on this regional programme. The total contribution by UNEP to MED POL, through June 1980, was estimated at $4,432,790. Another $534,000 came from the Mediterranean Trust Fund.[23] In contrast to the global setting discussed in climate-and health-related environmental monitoring, in a regional situation this amount of financial support from UNEP had significant impact, especially on efforts to involve the developing countries. For example, international funding made it possible to buy and service sophisticated and costly equipment for many laboratories in the developing countries bordering the Mediterranean.

MED POL Phase I laid the foundation for a longer-term monitoring and assessment programme, projected to cover the period 1981–91. The follow-up MED POL Phase II was supposed to generate indicators of the effectiveness of the measures taken and to provide periodic assessments of the pollution situation in the Mediterranean Sea. It was intended to promote the establishment of an operational pollution monitoring and assessment programme for the pollutants monitored in MED POL Phase I and to link it directly to decision-making and management.[24]

22 Three of these projects involved monitoring: MED POL I – baseline studies and monitoring of oil and petroleum hydrocarbons in marine waters (IOC/WMO/UNEP), with UNEP contributing $223,862 of the total $335,462; MED POL II – baseline studies and monitoring of metals, particularly mercury and cadmium, in marine organisms (FAO/UNEP), with UNEP contributing $660,701 of the total $744,454; and MED POL III – baseline studies and monitoring of DDT, PCBs, and other polychlorinated hydrocarbons in marine organisms (FAO/UNEP), with UNEP contributing $367,118 of the total $429,935. Additional projects in MED POL Phase I, which provided direct support for monitoring, included such topics as the coastal transport of pollutants, coastal water quality control, biogeochemical studies of selected pollutants in the open waters, the role of sedimentation in pollution, pollutants from land-based sources, intercalibration of analytical techniques and common maintenance services, and introduction of pollutants into the sea through the atmosphere. See document UNEP/WG.46/3, which offers a review of MED POL Phase I for the period Feb. 1975 – June 1980. Also see 'Coordinated Mediterranean Pollution Monitoring and Research Programme (MED POL)–Phase I: Programme Description', UNEP Regional Seas Reports and Studies, No. 23, 1983, and 'Regional Seas Programme: Compendium of Projects', UNEP Regional Seas Reports and Studies, No. 19, Rev. 1, 1984. (Note that discrepancies occur in project cost figures cited in different UNEP documents. They are accounted for by continuing adjustment and uncertainty of total expenditures.)

23 See doc. UNEP/WG. 46/3, which offers a review of MED POL Phase I for the period February 1975 – June 1980.

24 The following activities were envisaged: monitoring of sources of pollution; monitoring of coastal waters for a series of variables in water, in sediments, in biota, in suspended matter, and on the shore (the list of variables was to be expanded within three years to include additional parameters whose monitoring becomes mandatory under the Convention and the related Protocols); monitoring of reference areas that are not directly influenced by pollutants in order to provide information on general trends; and monitoring of environmental media that play a role in the transport of pollutants to the Mediterranean Sea, including the atmosphere. Twelve research projects were to be carried out, including studies of basic oceanic processes and on the toxicity, carcinogenicity, and epidemiology of selected pollutants. For the original draft proposal of MED POL Phase II, see documents UNEP/WG.46/4 and 5. See also 'Long-Term Programme for Pollution Monitoring and Research in the Mediterranean (MED POL)–Phase II', UNEP Regional Seas Reports and Studies, No. 28, 1983, pp. 5–11. The Contracting Parties to the Barcelona Convention endorsed the proposal. See document UNEP/WG.62/7.

Many of the issues involved in trying to set up and carry out marine pollution monitoring had to be faced in MED POL. The Mediterranean Basin encompasses both developed and developing countries. One of the main objectives of UNEP was to help build the capabilities and secure the full participation of developing countries in MED POL and to ensure the balanced development of the monitoring/assessment programme.

Through various projects, it has provided the monitoring equipment to many laboratories, has offered prompt and cost-free maintenance and repair services for sophisticated monitoring equipment (e.g., atomic absorption spectrophotometers, gas chromatographs, fluorimeters), has provided free samples and reference materials to laboratories that need them, etc. These services, which may appear minor at first sight, often determine the outcome of efforts to involve the developing countries in monitoring and assessment activities. To rely on commercial servicing of equipment would, in most of these countries, mean no servicing at all (either because of the costs involved or because of poor servicing facilities by companies making and selling instrumentation). UNEP therefore has paid for an expert, housed in the IAEA Monaco Laboratory of Marine Radioactivity, who goes anywhere within the region to service the equipment on twenty-four hours' notice. Often, a laboratory may not be able to perform an operation for lack of a simple and cheap chemical, which it cannot obtain because of foreign currency restrictions or administrative complications; UNEP distributes these chemicals free of charge in order to avoid such bottlenecks.

As regards training, UNEP has tried to ensure that the technicians working on MED POL are trained on matters relevant to the programme within the region, using instruments and working under conditions similar to those prevailing in their own countries. This is in contrast to the usual practice of training technicians from developing countries at institutions in developed countries, using sophisticated equipment and under conditions that cannot be replicated back home. Upon returning to their countries, they are usually frustrated by the lack of opportunities to apply their newly acquired knowledge, and become discouraged by the conditions and difficulties encountered in this type of work.

Special attention has been paid to the simplifying and adapting of methodologies and reference methods to use with instrumentation available in developing countries, to the experience and capabilities of trainees, and to the limited financial resources available. The basic objective is to obtain an acceptable quality of results using different methods and instrumentation (operational documents contain a minimal work programme that is obligatory for all, and an extended programme that can be performed by the more advanced research centres). The quality of data and the work done by laboratories is checked periodically through intercalibration exercises executed by the IAEA Monaco Laboratory of Marine Radioactivity.

Achieving participation and self-reliance is costly, involves a good deal of frustration, and is slow. However, it is also realized that unless the effort is made and money is spent to secure developing countries' participation, progress will not be made. Practical experience indicates that, until the interest in the subject matter becomes institutionalized and a part of the regular budget activity, a country's response is often determined by subjective factors, such as the motivation and calibre of those directly involved and the setting in which they work. The problem is that many countries show interest only if they have an opportunity to receive financial support, and that the money available is far from sufficient to meet the needs and demands: for example, the initial cost of equipping a simple laboratory for monitoring mercury has been estimated at $500,000 to $1 million.

Progress was manifested by the increasing participation of developing countries' laboratories and institutions in MED POL and their growing capability to carry out monitoring and assessment operations. For example, at the start of MED POL Phase I, only one or two developing countries' laboratories were capable of routine monitoring of mercury in fish and water. By the end of Phase I, at least one laboratory in each developing country taking part was capable of performing this operation. Their involvement kept the overall quality of the MED POL effort on the side of the lowest common denominator and has prompted critical remarks from oceanographers.[25] For UNEP, these were unavoidable trade-offs if developing countries were to become involved and be provided with the necessary tools and capabilities. Improved monitoring capability and quality are a function of time, the resources invested, the infrastructure set up, experience gained, and the level of interest and commitment generated and maintained.

A simpler and quicker way of obtaining adequate monitoring results, of course, was to have the advanced countries' laboratories and technicians do the monitoring on behalf of the developing countries, a solution advocated by some. This approach was ruled out, however. The build-up of national capabilities and direct involvement of developing countries in monitoring and assessment was considered a necessary condition for their participation in the management component of the Mediterranean Action Plan.

The Mediterranean experience also shows how important is the role of practice in determining the usefulness of a given type of monitoring and the approaches to be taken, as well as in applications of monitoring data and assessment to management and decision-making. For example, experience contributed to modification of initial monitoring objectives concerning mercury.[26] Similarly, in carrying out

---

25 One of the arguments used is that 'no data are better than bad data', and that poor data should not be tolerated by the scientific community. Interestingly, some of the industrialized countries taking part in the Mediterranean Action Plan (e.g., Italy, Greece, or Spain) are also not exempt from such criticisms. Their contribution in terms of quality is considered as falling in the intermediate zone, between the developing countries and the more advanced industrialized countries, in this instance France.

26 As had already been discovered through basic research studies, MED POL monitoring confirmed that natural sources are the main cause of mercury pollution in the Mediterranean. Out of approximately 500 tons of mercury reaching the Mediterranean annually from all sources (largely through the atmosphere, volcanic activity, and rivers), only 10 tons come from land-based sources of pollution via outfalls. It was thus confirmed that expensive standards and measures to control direct discharges of mercury would not contribute significantly to the overall quality of the Mediterranean Sea. Actual monitoring also showed that continuous and costly monitoring of mercury was not really necessary and that periodic baseline measurements would be adequate. The overall conclusion drawn was that direct mercury pollution caused by human activity was primarily a local problem in areas close to major settlements and production facilities, and should be monitored and dealt with locally. See 'Assessment of the Present State of Pollution by Mercury in the

various pilot projects, doubt has arisen as to the cost-effectiveness of monitoring approaches or the usefulness of continuous monitoring as opposed to periodic sampling in providing data of value for decision-making and increased knowledge. As in other clusters of GEMS, stress is being placed on the design of sampling and measurement strategies as a critical factor in obtaining the required data with least effort and cost and with a maximum degree of certainty.

In general, the MED POL experience has proved the soundness of the UNCHE assessment-cum-management scheme and, more specifically, the important role that the monitoring of marine pollution can play in assuring this linkage. The data concerning trends and baselines of marine pollution in the regional seas and their usefulness in checking the value and status of corrective and regulatory measures undertaken make monitoring an indispensable tool. For example, the Protocol for the Protection of the Mediterranean Sea Against Pollution from Land-Based Sources was endorsed by governments, after a delay, with the help of data and technical information assembled and produced by MED POL on origins and amounts of pollutants entering the sea from land-based sources.[27] Without the quantitative back-up and overview, the ratification of the Protocol would probably not have been possible.

The demonstration of the usefulness of monitoring has helped persuade governments to continue to support and participate in this type of work. Its direct link with management has played an important role in generating this interest and commitment. The real challenge is how to achieve the linkage between assessment and collective action, and how to generate the type of monitoring data and assessments that would be most conducive to attaining this objective. Indeed, an important and difficult part of the work remains to be done after the monitoring and assessment stage. This was shown by the difficulties experienced and by the failure of the governments in 1984 to agree on standards of water quality for the beaches in the Mediterranean Basin. The proposed standards were worked out on the basis of the MED POL results.[28]

The Mediterranean Action Plan is based on the full and active participation of governments. They assume the obligation to undertake monitoring when they become parties to action plans and related international agreements.

Unavoidably, the involvement of governments can lead to the politicization of technical matters, interference in scientific work and limitations on the freedom of inquiry and action of scientists; it can also influence the design of monitoring schemes and choice of variables to be monitored, the selection of experts, and the nature of the assessments made. However, since the responsibility for action rests with the governments, and since monitoring and assessment are organically linked

Mediterranean Sea and Proposed Control Measures', doc. UNEP/WG.91/5. While there was nothing scientifically novel in these findings, they were valuable because they helped educate all those involved and especially the governments highly alarmed about mercury pollution. These findings also led to revision of monitoring and management strategies.

27 For the initial set of data, see doc. 'Pollutants from Land-Based Sources in the Mediterranean', UNEP Regional Seas Reports and Studies, No. 32, 1984.

28 See 'Assessment of the Present State of Microbial Pollution in the Mediterranean Sea and Proposed Control Measures', doc. UNEP/WG.91/6.

with management, it is necessary to work and interact with their representatives and heed their views and guidance. An excellent monitoring/assessment package, devised and executed by experts and scientists only, usually does not carry sufficient weight and influence, especially in such politicized undertakings as regional seas action plans.

In MED POL, therefore, the governments have been closely involved in all stages of the process, including assessment. In this sense, it has differed from the more scientifically independent approach pursued in health- and climate-related environmental monitoring, where reliance is placed principally on scientists and the direct involvement of governments is kept at a minimum. Notwithstanding this direct participation of governments, the Regional Seas PAC (RS PAC) has also maintained the capability and freedom to arrange for more independent assessments in instances where it has deemed this necessary. Accordingly, relying on the flexibility provided by the financial resources at its disposal, it has sought to produce independent assessments of the health of the Mediterranean Sea by commissioning experts of its own choice. Such assessments are supposed to be repeated every five years. While the governments can make comments on these assessments, they do not take an active part in their preparation.

The final point that needs to be made has to do with the institutional role of RS PAC in mounting and implementing the Mediterranean Action Plan and MED POL. It has interpreted and played its coordinating/catalytic role in an activist manner. True to the letter of UNEP's programmatic approach, the projects and operational activities were carried out by the specialized agencies and other cooperating organizations. RS PAC, however, has not been simply a dispenser of funds to cooperating organizations. It has been very active in promoting and running the whole programme itself, has relied on its own forces to prepare the programmes and to carry out and guide the assessments and, in general, to direct all the policy, substantive, and technical aspects of MED POL. This was contrary to the view of the specialized agencies, which at the outset argued against UNEP taking any part in the substantive coordination and management of the programme, and opposed the proposal that it should chair the regular coordination of MED POL. However, the UNEP initiative and activism met with a positive response from governments. This was an indication of the flexibility of its terms of reference, provided the governments are convinced by the value of its work. It is also a recognition of the breadth and nature of the subject matter, which placed UNEP in a unique position to coordinate this effort, as well as of the success of RS PAC in carrying out its mission.

The implementation of the Mediterranean Action Plan, which was the only major regional seas activity in the early days, was helped greatly by the PAC status given to the office in charge within UNEP. This gave it greater administrative, operational, and policy independence and manoeuvrability. It was also helped by its Geneva location. This was of special value for running the Mediterranean programme in a nearby region. On the operational side, the word processing equipment, which at that time was still not available in Nairobi, gave it an important advantage. Also, the location freed it from some of the bureaucratic and other constraints on action associated with being housed at UNEP Headquarters.[29] It was also helped by the high motivation of its small team. A critical ingredient of success was the willingness to go beyond the standard restrictive and minimalist interpetation of UNEP's role, to rely as much as possible on its own forces and abilities, and to utilize imaginatively and sometimes aggressively the opportunities for action inherent in the subject matter and in the institutional mandate of UNEP.

MARINE POLLUTION MONITORING IN GEMS: FACTORS AT PLAY

The customary gap experienced in other GEMS clusters between objectives and expectations, on the one hand, and what is available in terms of financial resources, technical capabilities, scientific knowledge, infrastructure, and ultimately the governments' commitment to take action, on the other hand, has also been manifest in the case of marine pollution monitoring. The post-Stockholm experience has served to scale down and modify the practical goals to match the resources available for their implementation, and to take into account the practical, technical, and scientific difficulties, as well as the new knowledge that has been acquired.

The question of financing played a key role. The inadequate funding available to IOC for some of its programmes – and its expectation of obtaining the needed support from UNEP, which, in turn, was trying to ration its money and allocate it according to its own vision of priorities for action – was at the base of institutional friction between the two organizations. The shortage of funds in the GEMS budget played a role in UNEP's decision to redeploy marine pollution monitoring to the Regional Seas Programme. In this manner, it relieved some of the pressure on the GEMS budget line and tapped other resources for purposes of marine pollution monitoring. However, it also affected the direction and content of the marine pollution monitoring activities. In the Regional Seas Programme, the nature of monitoring was determined by management needs and the physical and development characteristics of the regions under study. It was to be expected that, in this situation, study, monitoring, and assessment of the open ocean would be assigned lesser weight.

To a considerable degree, the very flexibility and success of the Regional Seas PAC in mounting and implementing MED POL stemmed from the availability of a comparatively large, although not necessarily adequate, amount of financial resources to support various monitoring activities. The funds contributed to MED

---

29 The RS PAC, which was set up in 1977 when the global Regional Seas Programme was initiated, was preceded by a unit in charge of the Mediterranean Action Plan. Geneva was chosen as a temporary, compromise location for this unit until the coastal states could agree on its permanent site. The RS PAC remained in Geneva because of the logistical and operational advantages that it enjoyed there. One of the reasons given for this was the availability of word processing equipment, which greatly increased the throughput of the small office and which at that time was still not available in Nairobi. The RS PAC was moved to Nairobi in 1985, after a prolonged controversy as to whether or not it should remain in Geneva. By then, the name of the Programme Activity Centre had been changed to Oceans and Coastal Areas (OCA PAC). The responsibility for the coordination of the Mediterranean Action Plan was deployed to a special office in Athens, established by the contracting parties. UNEP's partial disengagement was in line with the basic premises of its catalytic role.

POL by UNEP in the early period 1976–80 ($4.4 million) were, for example, twice what it had contributed to climate-related monitoring during that same period.

The tension between the scientific quest for knowledge and the more practical concerns and uses of monitoring data figured significantly in the case of marine pollution monitoring. The wish of most governments to concentrate on the situation close to shore made for a link between management and monitoring and resulted in the fashioning of the latter to suit the needs of the former. This trend was strengthened institutionally as well, through the launching of the UNEP Regional Seas Programme, a management activity, with marine pollution monitoring and assessment representing an integral part of the package.

Focusing attention and the available resources on regional seas and coastal areas detracted from monitoring and assessment of pollution in the open ocean, which is beyond national jurisdiction and hence considered by most governments to be of secondary interest. Yet, from the global perspective the quest for basic knowledge about ocean processes and mechanisms, including pollution problems, was as important. However, for practical and institutional reasons the separation did occur. Unlike climate-related monitoring, where management and scientific needs could not be easily separated, in the case of marine pollution this same distinction tends to be accentuated because of the possibility to delimit the problem territorially and to separate national jurisdiction from what lies beyond. The focus on regional seas and coastal areas was given support by the findings of a 1982 study by the Group of Experts on the Scientific Aspects of Marine Pollution (GESAMP).[30] It concluded that, at least in the short run, the open ocean was not seriously threatened by pollution and that there was no justification for the alarm about the health of the oceans voiced at the time of UNCHE.[31]

North–South differences have also had an impact on the direction of marine pollution monitoring in GEMS. With very few exceptions, developing countries were not likely to take a meaningful part in study and monitoring of open ocean waters. The open ocean was accessible mostly to countries with adequate resources and scientific capabilities to carry out the task. Since one of UNEP's major policy objectives was to involve developing countries in monitoring and assessment activities, it was obviously much more inclined towards the regional seas approach. This was a promising way to engage their interest and to respond to their immediate interests and needs. Its policy choice was to contribute to global marine pollution monitoring through the participation of as many countries as possible, as an

30 GESAMP, or the IMO/FAO/UNESCO/WMO/WHO/IAEA/UN/UNEP Joint Group of Experts on the Scientific Aspects of Marine Pollution, was created in 1969 upon the initiative of the various sponsoring organizations within the UN system. It is a multidisciplinary group composed of distinguished scientists acting in their individual capacities. It has played the central role in the study and analysis of marine pollution. It meets annually, while its working groups, which deal with specific topics, meet as required. UNEP's financial support for some of its working groups has been of great importance for GESAMP. For a review of GESAMP's activities and work, see V. Pravdic, 'GESAMP: The First Dozen Years', UNEP, 1981.
31 GESAMP, 'The Health of the Oceans', UNEP Regional Seas Reports and Studies, No. 16, 1982. The same report was also printed by UNESCO as 'GESAMP Reports and Studies No.15, the Review of the Health of the Oceans', UNESCO, 1982.

alternative to a system of relying on the sophisticated facilities and ships of countries with oceanographic research capabilities.

The regional seas model met with the approval of developing countries also because it is based on the principle of territorial exclusiveness, i.e., the action plans are limited to riparian countries only. This gave them the pretext to keep out other countries, mainly the USA and the USSR, whose presence was viewed with suspicion in many regional groupings.[32] It is no surprise that the two oceanographic powers were critical of the territorial exclusiveness of the regional seas approach and have tried to have open access to various action plans.[33]

They have also advocated and promoted activities related to open ocean monitoring.[34] In UNEP's Governing Council, however, they did not meet with much success, since the majority of the governments were not prepared to approve the funding for this type of monitoring. In fact, some of the tension between IOC and UNEP was due to the dominance of the oceanographic nations in the former and of a different line-up of forces and interests in the latter. Also, owing to the divergent attitudes of the IOC and UNEP's domestic constituencies it happened occasionally that the delegations of the same country took different positions in the two organizations. Indeed, the position of one country's delegation within UNEP has differed from one session to the next, depending on which national institution happened to be represented or to have drafted the position paper.

Finally, the institutional dimension has played a role in shaping this activity in GEMS. How do the international organizations involved define their objectives and who controls them? What are their capabilities and how are they used? It is worth repeating that a good deal of commotion has been caused by the comparatively weak institutional position of IOC, which in some ways resembles that of UNEP itself. IOC efforts to achieve greater latitude and institutional independence within UNESCO and to obtain financing for its ambitious blueprints and operational programmes inspired its demands on UNEP.[35] The nature of UNEP's

32 A typical instance of the sensitivity of a regional grouping occurred in the Mediterranean. An attempt by UNEP to spend some of its non-convertible currencies by leasing an oceanographic vessel from the USSR failed because of the opposition of most signatories to the Barcelona Convention. Even the compromise proposal to lease the vessel without the crew was turned down. Note that the Black Sea is not included within the scope of the Mediterranean Action Plan, even though it belongs to the 'Mediterranean Region' on the IOC map. One of the reasons for this is that most riparian countries did not wish the USSR to be involved in the Action Plan. In contrast, IOC functions on the principle of universality, with all interested countries having free access to any regional activity.

33 As the 200-mile zone was more generally adopted, the oceanographic nations saw their access to these waters curtailed. The regional action plans of UNEP reinforced this trend. With the oceanographic vessels playing an important role in tracking satellites and missiles, free access to the 'regional seas' and complete 'freedom of scientific research' were undoubtedly important, especially if such access could be legitimated through participation in a UN-sponsored activity.

34 For example, see the report of the eighth session of UNEP's Governing Council, which notes that some countries criticized the lack of UNEP activities related to open-ocean monitoring (doc. A/35/25, para. 329–30).

35 The action paper submitted by the IOC secretariat to the twelfth session of the IOC Assembly in November 1982 noted UNEP's tendency to set up new structures rather than to help develop and work through the existing ones. It called on UNEP to consider using IOC and its subsidiary bodies and regional mechanisms for planning and implementing marine science components of the Regional Seas Programme. It also called for UNEP financial support to the Working Committee of

mandate, its internal set-up, personalities involved, and the choices it made in the development of this programme (in part made possible by the absence of a specialized agency for oceans and the resulting institutional leeway) also played a role in what transpired. Having started a successful, quasi-operational initiative of its own, which was meeting the demand of the Governing Council and of its constituency, the UNEP secretariat was unlikely to yield and miss an opportunity for further substantive, policy, and institutional self-assertion.

TOWARDS COMPREHENSIVE GLOBAL MARINE POLLUTION MONITORING

Almost two decades after UNCHE, the international community was still far from the operational global network of marine pollution monitoring envisaged originally. As described above, the efforts to move in this direction were halting and controversial. The reasons for this, as reviewed, include the geopolitical and physical peculiarities of the marine environment, the differing interests and ways in which individual countries relate to oceans and seas, and the scientific, technical, and financial challenges posed by the monitoring of the oceans. Besides, there was the perennial issue of relations between the international organizations involved, their mandates and characteristics, financial resources available for international programmes, and so on.

It is to be regretted that some of the worthwhile activities had to be neglected, ultimately because of the lack of the minimal financing that was needed. And it would have been preferable to have had an agreed blueprint and a better coordinated process. Still, it is the results that matter, and the fact is that, in spite of tension and institutional uncertainty, the progress recorded during this period is quite significant. The matter has moved to the point where the framework and components of a global observation system of the marine environment are known and broadly agreed to, and the setting up of such a system has become a feasible proposition.

An important early achievement was to link monitoring and assessment with management and action by focusing on regional seas and coastal areas, where some of the most serious cases of environmental degradation occur and whence significant pollution loads spread to the open ocean. This helped to telescope the process and to bring management and regulation into the picture early. Such a linkage would have been hard to achieve had open ocean pollution study and monitoring been concentrated on. Regional seas and coastal areas also provided the opportunity for meaningful involvement of developing countries, thus fulfilling one of the objectives of GEMS. Moreover, the focus helped to bring more governments into

GIPME and to MARPOLMON (doc. IOC-XII/8). The draft resolution submitted by the IOC secretariat included all these points and also called for increased UNEP support for the assessment of background levels of selected pollutants in open ocean waters, which is supposed to provide basic information for the regional seas programme (doc. IOC-XII/DR.4). The UNEP Governing Council, with a different coalition of governments in control and with UNEP's programme priorities in mind, was non-committal in its response to the IOC initiative. It took note of the past cooperation between UNEP and IOC and called for future cooperation within the available resources. (See decision 11/7, part IV, in UNEP/GC.11/18, Annex I.)

the act and to generate additional financial resources, an important ingredient for successful action.

The modular or 'jig-saw' approach that is inherent to the regional seas programme, combined with the open ocean baseline studies and monitoring – provided that the coordination and intercalibration of methods and efforts is attained – makes it possible to contribute to a build-up of a global monitoring/assessment network that would provide a quantitative and qualitative overview and assessment of the health of significant areas of world oceans and seas.[36]

This is why one of the more important challenges in the period to come is not just to build up various Regional Seas Programmes into functioning and efficient undertakings, but also to coordinate these and synthesize their findings, keep them in line with the overall framework and objectives, and ensure that they evolve into the building blocks for a global marine pollution monitoring and assessment system. In part, this is a question of effective logistics and cooperation, which can be attained with adequate capability at the centre of the system, and the provision of resources and mechanisms for that purpose. Among other things, this entails the dissemination and application of compatible methodologies, data formats, data processing, storage and exchange mechanisms, organizing coordination and interaction among regional seas programmes, intercalibration exercises among various monitoring programmes, working on comparability of data.[37]

With respect to open ocean pollution monitoring, the experience with the IGOSS pilot project, ICES activities, the work done by oceanographic nations, activities within the IOC including various scientific plans and methods developed within the framework of GIPME,[38] the work carried out by GESAMP to develop scientific principles for global and comprehensive marine pollution monitoring, and so on, all contributed to an improved understanding of what is involved and what is required to put such monitoring into practice. Although the feasibility of such

---

36 GESAMP's terms of reference require it to prepare periodic, updated reviews of the state of the marine environment as regards marine pollution, i.e., the above cited 'Review of the Health of Oceans'. Precisely because of the lack of systematic monitoring and monitoring data in most places, especially as regards the open ocean, the assessments in this review were mostly tentative and qualitative. (Note that the first effort of this kind, commissioned by UNESCO, was undertaken by a single scientist. See E. D. Goldberg, *The Health of the Oceans*, UNESCO, Paris, 1976.) For the same purpose GESAMP has been using the mechanism of a working group on the state of the marine environment (its Working Group 26), which consults a wide network of peers and relies on regional task teams.

37 OCA PAC has been working in this direction through organizing meetings on monitoring and assessment methodologies, developing common procedures, producing intercomparable results, planning to designate data centres where monitoring data from all Regional Seas Programmes would be deposited, appointing and circulating its own staff within various bodies running different regional seas action plans, organizing technical assistance missions by including experts from the more advanced regional sea schemes to share their experience, organizing intercalibration exercises involving different programmes (e.g. MED POL and ICES), etc. Interestingly, various Regional Seas Programmes invariably show an insular tendency, and an agent is needed to keep them in line and to maintain an awareness of the other programmes and of the global framework. For a set of initial recommendations of how to deal with these problems, see Report of the Meeting of Governmental Experts on Regional Marine Programmes, January 1982, doc. UNEP/WG.63/4.

38 For a review of these activities see 'The Marine Pollution Research and Monitoring Programme of the Intergovernmental Oceanographic Commission', doc. IOC/INF–523, *op. cit.*

monitoring with available techniques continued to be doubted and the feeling of urgency was reduced with the general acceptance of the view that the pollution threats to the open ocean were not as imminent or serious as it was assumed in the early 1970s, the matter remained very much alive, within UNEP and outside.

While UNEP may have appeared to have distanced itself from open ocean concerns, it had in fact maintained an important role. This was done mainly through cooperation between OCA PAC and GESAMP, with the former providing financial support for the work of GESAMP and taking an active part in some of the activities of this group. Also, the development of a global ocean monitoring system continued as a topical item in relations between IOC and UNEP, which began gradually to improve and assumed a more clearly defined form as a consequence of several decisions by the governing bodies of the two organizations and intersecretariat agreements on cooperation.[39]

Pollution of the open ocean was a topical scientific and management issue that could not be wished away. Plans and blueprints for open ocean research and monitoring – most notably embodied in GIPME – were being pursued actively by IOC with the objective of providing a scientifically sound basis for the assessment and control of marine pollution. And the interest of the oceanographic nations continued undiminished. It was primarily the initiative of the USSR that maintained open ocean pollution monitoring on the agenda, in spite of the continuing reserve on the part of UNEP. It is the USSR initiative that led to the convening of a Symposium on Integrated Global Ocean Monitoring (IGOM), in Tallinn in 1983. It was organized with financial backing from UNEP in the framework of GEMS, even though UNEP continued to express scepticism about the idea.[40]

---

39 See decisions 11/7 of the UNEP Governing Council and resolutions XII–20 and XII–21 of 1982, and XIV–15 of 1987 of the IOC Assembly. The two secretariats thereafter signed several *aides-mémoire* and, in 1987, a memorandum of understanding on cooperation between UNEP and IOC. The memorandum lists both areas of ongoing cooperation and those of potential future cooperation. For a review of cooperation between IOC and UNEP, submitted by the IOC secretariat to the UNEP Governing Council in March 1988, see doc. IOC-INF 743. UNEP had joined as co-sponsor of three expert groups within GIPME, though not GIPME itself, namely, Group of Experts on Methods, Standards and Intercalibration (GEMSI), Group of Experts on Standards and Reference Materials (GESREM), and Group of Experts on Effects of Pollutants (GEEP). The work of these three groups is essential for methodologies to ensure global comparability and good quality of monitoring data. For the background on IOC activities and its vision of the future, see *Ocean Science for the Year 2000*, UNESCO, Paris, 1984, a report prepared by the Scientific Committee on Oceanic Research and the Advisory Committee on Marine Resources Research.

40 For the summary report and recommendations of the Tallinn Symposium, see document GESAMP XIV/8. For the abstracts of the papers presented, see 'The First International Symposium on Integrated Global Ocean Monitoring', USSR, Tallinn, Oct. 2–10, 1983, Abstracts, Gidrometeoizdat, Moscow, 1983. See also report on IGOM presented to the International Symposium on Integrated Global Monitoring of the Biosphere, Tashkent, USSR, 14–19 Oct. 1985, in its 'Proceedings', EPMRP series, No.45. For the UNEP view, see 'Prospects for Global Ocean Pollution Monitoring', UNEP Regional Seas Reports and Studies No. 47, 1984. It contains *inter alia* a feasibility study of using BAPMoN stations for monitoring background open ocean pollution and a study on the prospects of global ocean monitoring. The latter in essence argues that projects may be undertaken for research purposes, but that considerable further work and development of well proven methods are required before operational monitoring is attempted. It concludes that although 'continuous open ocean monitoring is neither feasible nor justified in terms of pollution problems, specific wide-scale research or survey projects are a viable alternative' (*ibid.*, p.5).

146 *The quest for world environmental cooperation*

IGOM is a component of the integrated global monitoring of the environmental pollution concept promoted by the USSR, which is discussed briefly in chapter six below. The IGOM plan highlights the need for a unified approach to assessment. It notes that world ocean pollution is becoming global in nature due to large-scale ocean water circulation and atmospheric transfer and deposition of pollutants, and that persistent low-level pollution is present in those areas of the ocean where different water masses converge. It calls for the concepts of assimilative capacity and biological pollution indicators to be used as the basis of a comprehensive plan for integrated global ocean monitoring. It proposes the establishment of polygons for combined ecological investigation and integrated global ocean monitoring. To make all this possible, it proposes the adoption of organizational (institutional) principles of the comprehensive plan.[41]

The Tallinn Symposium recognized the scientific importance of working on the establishment of an integrated approach to ocean monitoring. It concluded that it would still be premature to launch such a programme in view of the existence of methodological and technical problems and gaps in the knowledge about the oceanic processes, and that considerable additional preparatory work would be required. The matter was passed to GESAMP to examine the scientific justification and to prepare an agreed strategy for the development of integrated ocean monitoring as a part of GEMS, including methodological feasibility and the structure and support that might be required.[42]

It was felt that integrated ocean monitoring should be developed slowly, in close contact with the basic research. In this context, the Symposium also recommended that basic oceanographic research should be pursued within GEMS. This should be done concurrently with and related to the existing more management-oriented strategy concentrated on the regional seas, which itself was recognized as a vital element in the achievement of integrated global ocean monitoring. The Symposium stressed the assessment of interaction between ocean and other compartments of the environment and the role of the ocean in geochemical, geophysical, and biological processes. And it proposed that UNEP should support baseline studies of selected pollutants in open ocean waters, as well as the expansion of the 'mussel watch' sentinel organism monitoring programme into a global activity and tool to be used when needed.[43]

---

41 For excerpts from a comprehensive plan for IGOM submitted by the USSR, see document GESAMP XIV/8, Annex II, *op. cit.*

42 For the terms of reference of the GESAMP Working Group on Integrated Global Ocean Monitoring see 'Prospects for Global Ocean Pollution Monitoring', *op. cit.*, p.49. Among other things, the Group was to review observations and measurements, the availability of sampling and analytical techniques that can be expected to provide data on a global basis, and the observations and sampling strategy.

43 The possibility of using bivalve molluscs as sentinel organisms was discussed first at a Global Oceans Research Workshop in Ponza as early as 1969. Mussels were chosen because they extract all the major types of pollutants (including halogenated hydrocarbons, heavy metals, petroleum hydrocarbons and transuranic elements), are a widespread species around the globe, concentrate pollutants by factors of a thousand to a hundred thousand, and apparently lack enzyme systems to metabolize most contaminants (doc. IOC/WC-GIPME-V/8, p.7). On 'mussel watch' see E. D. Goldberg, ed., *The International 'Mussel Watch'*, National Academy of Sciences, Washington, D. C., 1980, and by the same author, 'The Mussel Watch Concept', *The Siren*, March 1984. For a

After two years of study, the GESAMP Working Group on Integrated Global Ocean Monitoring concluded that IGOM is both desirable and feasible, and required pilot-scale implementation. It offered a plan and a timetable for its implementation.[44] It did not recommend 'continuous global and intensive monitoring of the open oceans'; rather, it proposed an open ocean pilot international research monitoring programme, as a means of developing gradually the necessary capabilities and of contributing to a baseline for future reference. At the same time, giving a nod to the regional seas approach of UNEP, it recognized the potential of the regional approach to evolve into a system yielding a global picture, while dealing with priorities of given regions. And, in fact, it proposed two components for IGOM, i.e., global ocean monitoring and regional/coastal programmes.

However, many of the methodological, practical, and technical problems remained unsolved.[45] Also, the high cost of ocean research and monitoring was an important obstacle to implementation. Thus, reservations were expressed about a number of assumptions made by Working Group 24. On the whole, its proposals failed to convince GESAMP, which felt that the scientific rationale was not complete and that it was still premature to initiate a large-scale global ocean monitoring programme of a large number of parameters.[46]

The GESAMP action notwithstanding, consensus had been crystallizing round the idea that such a system was needed and that its build-up had to start eventually. Once the operational decision was taken, it could rely on the definition of IGOM agreed to by Working Group 24, i.e., 'an international multidisciplinary programme for systematic surveillance and evaluation of those characteristics of the world ocean that may be altered as a consequence of human activity'. In IGOM, monitoring was defined as a 'systematic time sequence of observation of the marine environment conducted in order to detect changes, if any, from a given baseline,

---

critical view of the idea see H. H. White, 'Mussel Madness', *ibid.*, May 1984. In preparing for the global expansion of the 'mussel watch' and owing to the serious logistical difficulties likely to be experienced in organizing it worldwide, including the sampling, packaging, and shipping of frozen mussel tissues, at one point attention was also given to 'artificial mussel' technique, but without much practical success. For this purpose, sheets of urethane foam and amberlite resin were proposed. Placed in seawater, they would be collected after a given period and mailed in special envelopes to the central analytical laboratory.

44 See the Report of the GESAMP Working Group 24 on Integrated Global Ocean Monitoring (IGOM), held in Batumi, USSR, in 1985 (doc. GESAMP XVI/7). The approach that was recommended largely resembled the strategy pursued in other GEMS clusters. It recommended criteria for selection of geographic areas for global open ocean research and monitoring; it proposed observations in the near-surface atmospheric layer, in the surface water microlayer, of the water column, of particulate fluxes, and of the sea bottom; it recommended applications of remote sensing; it proposed a sampling strategy and intercalibration, quality control of data, and data processing and storage in existing mechanisms and centres.

45 For example, improved techniques were needed for supply and analysis of trace contaminants, and for identifying ecological changes due to anthropogenic impacts; methods for measurement of fluxes had yet to be developed, etc. An emerging issue had to do with the need to integrate and interpret data collected over large space and time scales. See doc. GESAMP XVI/7, *op. cit.*

46 See GESAMP Reports and Studies, No.31, 1987, Report of the Seventeenth Session, Rome, 30 Mar. – 3 Apr. 1987, pp.7–9. For the summary of the report of the second meeting of the Working Group held in Moscow in November 1986, see Annex VII, *ibid.*, pp. 27–9. The report was not endorsed by GESAMP for publication in its series.

and to identify temporal and spatial trends', its overall objective being to determine to what extent 'the marine environment is impaired by certain human activities'.[47] A step in this direction was made when UNEP approved the 'mussel watch' pilot project in 1987, a challenging undertaking conceived and devised by scientists.[48]

The USSR initiative and the IOC position were directly responsible for forcing the integrated ocean monitoring into the fold of GEMS. Ultimately, however, it was the very nature of the issues and global oceanic processes under study and observation, including worldwide transport of pollutants through air and water movement, that called for greater complexity, integration and comprehensivness of research, monitoring and assessment activities. The world oceans system, similar to the world climate system, requires a systemic and global monitoring and assessment approach. And, similar to climate-related monitoring, in the case of marine pollution monitoring much more needs to be known about global oceanic processes to be able to place anthropogenic impacts in a proper perspective and assess their role in the marine environment. Thus, oceanographic research gained greater legitimacy in GEMS and the dichotomy between management needs and science was softened. Similar to the understanding that crystallized in climate-related monitoring and gave rise to WCP and later to IGBP, marine pollution monitoring and assessment had to be placed within the broader context of global environmental change, interaction between environmental media, including fluxes of energy and materials, impacts on world climate, changes in marine ecosystems, etc. What is required, therefore, is an integrated ocean research and monitoring system.

There is another parallel with climate-related monitoring, namely, scientific and technological advances hold the key to making IGOM, or any kind of world ocean watch, operational. The technical potential, where it exists, is still to be applied systematically; where it does not exist, it remains to be designed and developed. Satellite systems, and in particular a new generation of oceanographic satellites planned for the 1990s, promise to overcome some currently intractable problems related to the cost, logistics, and methodological and technical difficulties involved in monitoring pollution in the open ocean, especially the need to cover large areas.

There has been relatively little application of satellite remote sensing to chemical oceanographic research and marine pollution monitoring on a global scale. There have been no satellites designed for oceanographic research after the ill-fated and short-lived SEASAT was launched by the US, and specialized sensing tech-

---

47 *Ibid.*, p.27.
48 UNEP contributed $115,000 of the total cost of $430,000 for a one-year pilot project, with the rest mainly in kind from IOC, from the Chesapeake Bay Laboratory (which acts as the secretariat for the pilot project), the International Mussel Watch Commission, etc. The long-term purpose is to determine levels of biocides and pesticides (PCBs and DDT) in mussels, and to establish a global data set that would make it possible to identify trends and the state of the health of the marine environment and mount corresponding measures for its protection. Three interlinked regional networks were planned (Atlantic, Pacific and Indian Oceans), each with its own analytical centre. Such worldly difficulties as obtaining permission of governments to get samples of mussels from their territorial waters had to be sorted out. Transporting samples in liquid air, organizing a bank of samples for future analysis, and assuring comparability of samples from different parts of the world were among the more technical hurdles that had to be dealt with.

nologies have not been developed.⁴⁹ While in the past the countries with the necessary capabilities have not done much to develop such tools systematically, it is also true that many of the remote sensing technologies already in existence and devised for military and intelligence purposes could result in important spin-off applications for marine pollution monitoring.⁵⁰

Backed up by oceanographic vessels and other devices providing sea truth, remote sensing is likely, in the years and decades to come, to help fulfil many of the objectives of comprehensive ocean pollution monitoring that were first envisaged at the time of the Stockholm Conference. It will be aided by other technological advances that will help overcome various practical problems. For example, drifting buoys are being developed that will relay monitoring data to satellites. Among the problems that will have to be solved is the development of appropriate, long-lasting instruments that could function for a period of five years or longer, on their own. The fact that there would be no need to recover or service the buoys, combined with the on-line transmission of data via satellites, would completely change the economics of ocean monitoring and make regular surveillance of the world's oceans possible.

The growing interest in the Earth as a system and the appreciation of the many roles of the world oceans in this system have generated adequate policy interest and allocation of resources for monitoring of oceans from satellites. Thus, for example, the Earth Observing System (EOS), cited in chapter four above, has as one of its major components routine and continuing measurement of physical, chemical and biological properties of the oceans.

While waiting for satellite monitoring to materialize and to place monitoring and assessment of the health of the oceans on a solid and durable footing, periodic assessment of the kind initiated by GESAMP will continue and will no doubt improve in comprehensiveness and coverage as years go by.

In the longer run, such global assessments, firmly grounded in the data generated by marine pollution monitoring and oceanographic research, are bound to play a critical role in the evolving interrelationship between humans and the marine environment, and in the informed and controlled uses of the oceans' capacity to degrade and disperse waste materials generated by society.⁵¹ By providing the hard

49 See N. Andersen, 'The Potential Applications of Remote Sensing in Chemical Oceanographic Research', ONR West 80–1, Office of Naval Research, Scripps Institution of Oceanography, June 1980. Also see K. H. Szekialda, 'Uses of Remote Sensing in Global Pollution Monitoring Systems', IOC/WC-GIPME-II/9. Remote sensing has been used for monitoring physical oceanographic phenomena. There has been some experimental work related to pollution (e.g., ocean red tides and harmless algal blooms, classifying eutrophied waters, spectral signatures of sewage plumes, oil spills surveillance). In general, in order to achieve adequate capability, especially to distinguish human-induced changes from the natural variations, special tools and sensors need to be developed and incorporated in multiple sensing systems.
50 An example of this is the application of laser technology to antisubmarine warfare. It could also be used for monitoring the characteristics of the vertical column of the ocean. Some laser systems have shown the ability to detect extremely thin layers of organic substances on the surface of the sea. (Andersen, *op. cit.*, p.59). However, many of these capabilities remain strictly in the military domain.
51 See A. Preston, 'Use of Oceans for Waste Disposal', *The Siren*, September 1983. He addresses the issue of 'controlled exploitation of the ocean's capacity to degrade and disperse waste materials' as an 'underutilized aspect of ocean resources', in part because of overconcern for the marine

scientific data essential for understanding the situation and for well-informed decision-making and management, monitoring of the marine environment should help to control anthropogenic inputs, to watch over the health of the oceans and seas, and to ensure that their regenerative capacity and resilience are wisely used. The work done since UNCHE, while it has gone some way towards setting up the capability that is required, represents only the beginnings of a slow process of learning, scientific progress, and international and indeed institutional cooperation.

How to rationalize the whole undertaking, including the need to overcome frequent institutional and political obstacles that add to the already formidable scientific and operational challenges, how to sort out the differing roles and capabilities of countries in the global system, and how to make policy and management use of the scientific and technological potential is a challenge faced by the international community and a matter ripe for policy review in the framework of GEMS. In its recommendations, the GESAMP Working Group 24 touched on a key aspect when it stressed the need for an 'effective mechanism' for the coordination of the global open ocean research and monitoring programme, to be responsible for all aspects of IGOM, including the control of data, and their processing and evaluation. It further proposed that a scientific coordination board be set up for IGOM. For the rest, the basic model followed by UNEP and GEMS was endorsed, i.e., national focal points, responsible centres in charge of intercalibration and data processing, etc.

Monitoring and assessment of the global marine environment are indispensable both for building up necessary knowledge and for devising strategies for rational management and use of the world's oceans. The oceans represent not only one of the last economic frontiers for humankind, an important buffer for environmental impacts, and the ultimate sink for most human wastes, but also a vital component of the geosphere–biosphere-climate system and thus of the very habitability of the Earth.

---

environment. The current definition of marine pollution, formulated by GESAMP at the time of UNCHE, states that it represents 'the introduction by man, directly or indirectly, of substances or energy into the marine environment (including estuaries) resulting in such deleterious effects as harm to living resources, hazards to human health, hindrance to marine activities, including fishing, impairment of quality for use of seawater, and reduction of amenities'. See GESAMP 'Health of the Oceans', *op. cit.*, p.9. This definition implies that introduction of substances or energy that does not result in deleterious effects should not be considered as marine pollution. Thus the issue becomes what is an 'acceptable introduction' of substances and energy into the marine environment. This 'broadmindedness' has been criticized by those who prefer to see any kind or amount of substances or energy considered as creating marine pollution. The original definition, however, has withstood the challenges, with GESAMP arguing that there was no scientific evidence calling for its change.

6 Natural resources and integrated monitoring in GEMS

Increasing pressures on the natural environment, careless uses and often predatory exploitation of natural resources, and the degradation of ecosystems were among the priority concerns of the Stockholm Conference. Emphasis was placed on the effects of these trends on the prospects for sustainable development of the regions affected, and on the biosphere, weather, and the climate. The impacts of pollutants on non-human targets, their deposition in and impacts on ecosystems, and circulation within and among the various compartments of the environment were highlighted as well. The Conference made specific proposals for monitoring the natural resource base, i.e., trends in soil degradation, capabilities, and regeneration; basic data on forests, changes in biomass, and changes having an impact on the environment; world genetic resources; world fishery resources; and animal species endangered because of their commercial value. It also recommended monitoring of hazardous compounds in biological and abiotic materials.

As discussed in chapter two, during the initial deliberations in the UN Environment Programme (UNEP), an effort was made to limit the fledgling Global Environment Monitoring System (GEMS) to monitoring of pollution and pollution effects only.[1] At the first Governing Council, however, developing countries argued forcefully that monitoring of the status of natural resources was of direct concern to many of them, and thus should not be sidelined. Their demands led to the reshuffling of the proposed priorities and the recasting of the initial orientation of GEMS. The monitoring of desertification and of the degradation and depletion of soil and living resources such as forests, grasslands, wildlife, and aquatic ecosystems, i.e., monitoring of the degradation of the natural resource base caused by agricultural and land-use practices, was thus included among the priority objectives of GEMS.

The events at the first Council and the 1974 Intergovernmental Meeting on Monitoring, discussed in chapters one and two above, gave greater policy weight

---

1 During the preparations for the Conference, monitoring of the status of natural resources received less attention and figured less prominently in the proposed system than monitoring of impacts of pollution on non-human targets and ecosystems. As an example, see the US paper on monitoring (doc. A/CONF.48/IWGM. I/INF.10). In the context of terrestrial monitoring and of a global network of ecological baseline monitoring stations – which it proposed should be established to identify long-term changes of major significance for the flora and fauna of the world – the paper refers essentially to pollution-related problems.

to this type of monitoring. The desire to attract developing countries and to show that something concrete was being done in GEMS regarding problems of interest to them resulted in prompt action. Two project proposals by the Food and Agriculture Organization of the UN (FAO), intended to develop pilot methodologies for soil degradation monitoring and tropical forest monitoring, were approved. FAO, with its long experience of studies and research on soil degradation and tropical forests, and having aired these project proposals already at the UN Conference on the Human Environment (UNCHE), was ready to start immediately. These two projects formed the basis of the natural resources monitoring cluster of GEMS. Later, its scope was expanded to include the monitoring of arid and semi-arid rangelands and monitoring of the status of species of wild flora and fauna.

As a result of modified priorities, the monitoring of pollution impacts on ecosystems was relegated to the background. The idea of supporting this type of monitoring within UNESCO's Man and Biosphere Programme (MAB), based on an initial blueprint elaborated with some financial assistance of UNEP, had to be abandoned.[2]

The decision taken by UNEP not to assist financially projects related to monitoring pollution impacts on ecosystems was due not to loss of interest but to the need to decide how to allocate the limited resources. Moreover, in striking a balance between the interests of developed and developing countries' and between pollution-oriented and non-pollution oriented activities, leaving out this type of monitoring seemed a natural thing to do. Monitoring pollution impacts on ecosystems was heavily research oriented and mostly of concern to industrialized countries. It was thus assumed that these nations would continue to do the relevant work domestically, as well as within the framework of MAB. The withholding by UNEP of financial support for MAB project 14 on research into environmental pollution and its effects on the biosphere prevented it from getting off the ground

---

[2] The role of MAB in the study of pollution effects and the monitoring of pollutant levels could not be settled at the first session of the MAB Council in 1971 and the matter was deferred until after UNCHE. Recommendation No. 80 of the Stockholm Conference called, among other things, for more knowledge about the inputs, movement, residence time, and ecological effects of critical pollutants, and suggested that the MAB programmes be used to monitor the accumulation of hazardous compounds in biological and abiotic materials, as well as the effects of such accumulation on the reproduction and population size of selected species. The report of the Interagency Working Group on Monitoring, which served as the background document for the 1974 Intergovernmental Meeting, recommended that a task force headed by UNESCO should meet to consider the feasibility of monitoring the impacts of pollution on the structure and functioning of terrestrial and freshwater ecosystems, the research needed in this respect and the role of MAB. (See doc. UNEP/IG. I/2, p. 40). A task force was convened to decide what should be done in relation to research into the effects of pollution on the structure and functions of terrestrial and freshwater ecosystems, and what methodologies should be used in research and monitoring relating to the above effects. The work of the task force, with UNEP's support, culminated in a meeting in Moscow in April 1974, which proposed the establishment of a new MAB project, No. 14, entitled 'Research on environmental pollution and its effects on the biosphere'. See 'Task force on: pollution monitoring and research in the framework of the MAB programme, final report', MAB Report Series No. 20. On development of MAB project No.14, see 'Task force on project 14: research on environmental pollution and its effects on the biosphere, final report', MAB Report Series No. 32. See also 'Task force on: criteria and guidelines for the choice and establishment of biosphere reserves', MAB Report Series No. 22, recommending that some biosphere reserves in undisturbed natural ecosystems should serve as baseline areas for research and monitoring activities in GEMS.

as planned and hoped for. However, this did not affect the process of establishing a global network of MAB biosphere reserves, which could be used for a variety of environmental monitoring needs, were a required decision to be taken.

This was the starting point, then. In the pages that follow, monitoring the status of natural resources and monitoring of pollution impacts on ecosystems are treated separately. The latter is subsumed within the broader framework of 'integrated' or 'multimedia' monitoring.

NATURAL RESOURCES MONITORING

As in the case of health-related and climate-related monitoring, one of the initial concerns of the GEMS natural resources monitoring cluster was the development of the requisite methods and tools. Yet, it was quite different in the sense that there was no real global network build-up. The reason was that the priority areas chosen, namely tropical forests and resources of arid and semi-arid lands, were mostly in developing countries. This is where the major degradation was taking place and where the effort to obtain the primary data had to be concentrated. For the developing countries, these resources are of practical economic and often strategic importance. Understandably, most of them were not willing to have these resources and their status known, especially to their neighbours and economic rivals. Thus, although the global aspects of natural resources monitoring and assessment remained an important concern in GEMS, for practical and political reasons the primary focus was on the domestic needs of developing countries.

In brief, the UNEP-supported projects had two main initial objectives:

- To devise, test, and in due course apply methods and tools for monitoring the status and degradation of natural resources in developing countries; and
- To get these countries interested and involved, and to provide them with the methodological packages and assistance they needed to undertake natural resource monitoring domestically.[3]

Monitoring soil degradation

Despite the fact that political concern had been voiced about soil degradation – its nature, total area and distribution – many critical areas, mostly in the developing countries, had not been adequately explored. Estimates were approximate and qualitative, and there was no standardized methodology or uniform criteria to assess such degradation. The purpose of the GEMS pilot project on soil monitoring, executed by FAO and UNESCO, was to develop a methodology applicable at all scales, from continental down to the local ranch or farm, to devise a set of criteria to measure and monitor the extent, intensity, and rate of soil degradation (defined

---

3 For the general background of this monitoring cluster, see 'Monitoring for the Assessment of Selected Critical Environmental Problems Related to Agricultural and Land-use Practices', Report of the UNEP/FAO Government Expert Group on Environmental Monitoring of Soil and Vegetation Cover, March 1976, Rome, 1976.

as a process that lowers the capability of soil to produce), and to explore the feasibility of a comprehensive soil monitoring system.[4]

Focusing on the Middle East and Africa north of the Equator, the project resulted in a pilot methodology (i.e., guidelines for standardized data collection, processing, classification, and retrieval) and reviewed the possible contributions of remote sensing to this type of monitoring. To illustrate the outcome of the study, three maps of this area were prepared, which showed soil degradation and degradation risks (limited to soil erosion by wind and water, and salinization and alkalization). Since the maps at this scale (1: 5,000,000) are not of much practical use, in order to ration the limited funds available it was decided not to aim for a global coverage, for which 17 maps were needed. After all, what really mattered was the pilot methodology.

The methodology was to be tested in different regions, on different scales and under widely differing conditions.[5] After the testing phase, the operational and sustained monitoring of soil change and modification was supposed to be initiated on a worldwide scale, with special emphasis on selected critical areas. However, owing to the impulse of the UN Conference on Desertification, the emphasis shifted to monitoring desertification. Efforts to develop appropriate monitoring methods were not very successful and continue to be tested. Some of the problems encountered were due to the fact that the experts called in to advise were not familiar with the process of desertification as it occurs in Third World regions.

Monitoring tropical forests

The degradation of tropical forests was one of the hotly debated issues at UNCHE. Because of the lack of quantitative data and precise knowledge on the status of processes taking place, emphasis was placed on the need to go beyond 'guesstimates' and to undertake systematic monitoring and assessment of tropical forests. This was not an original or novel idea. As early as 1951, the FAO Conference had recommended that information on world forest resources be published at five-year intervals. Relying on answers to questionnaires sent to governments, FAO published the first 'World Forest Inventory' in 1953. However, on account of differences in national objectives, needs, characteristics, levels of development, and technical knowledge, and in the absence of an agreed vegetation classification that can be applied worldwide and especially in the tropical areas, and of a common methodology, and because data did not add up and were often inaccurate, and because

4 For the background of this project, see 'A World Assessment of Soil Degradation – An International Programme of Soil Conservation', report of FAO/UNEP expert consultations, Rome, June 1974. Among the degradation processes that can be monitored are water and wind erosion, excess of salts, chemical degradation (acidification, toxicity), physical degradation (decrease of permeability), and biological degradation (decrease in humus).

5 See 'A Provisional Methodology for Soil Degradation and Assessment', FAO, Rome 1979. The FAO national constituency was invited to use the provisional methodology for soil degradation assessment and to comment on it to FAO. International consultations were held to evaluate the methodology and the results of its application in the field and to make recommendations for follow-up action. One of the problems that had to be faced was the practical difficulty of devising a single method that would be useful on different scales.

Natural resources and integrated monitoring 155

many governments did not even bother to respond, the inventory never materialized. Although it was continued for the temperate zone, it was eventually abandoned on a global scale, leaving a critical information and knowledge gap regarding tropical forests.

Recommendation No. 25 of the Stockholm Conference renewed the call for surveillance of the world's forest cover. A pilot project on tropical forest monitoring executed by FAO was launched in the framework of GEMS. The project was aimed at preparing a pilot methodology and guidelines for tropical forest monitoring, exploring possible uses of remote sensing, and proposing modes for computerizing and analysing the data.[6]

A pilot methodology was developed, based on aerial photography and remote sensing.[7] It was to be tested, further refined, and adapted for use in the three main tropical forest regions. Eventually, an institutionalized global system of tropical forest monitoring was to be set up, composed of regional and sub-regional forest monitoring units, acting as focal points for the initiation and execution of monitoring in the groups of countries concerned.[8]

However, in the overall review of programme priorities carried out by UNEP at that time, this objective was demoted from high to low priority status, which weakened the drive to make such a system operational. In view of the rapid degradation and depletion of tropical forest, and in order to provide the international community with an assessment of the existing situation, UNEP and FAO launched a project in 1978 aimed at pulling together and interpreting the data already available. The first desk-top inventory was thus produced in 1981.[9] It is to

6   For the background of the project, see 'Report on the Formulation of a Tropical Forest Cover Monitoring Project', a FAO/UNEP publication, Rome, 1975. Togo, Benin, and Cameroon were selected for the pilot project, Africa being the region with the largest proportion of cloud-free and usable satellite imagery. Moreover, the possibility existed of extending the pilot project to link up with the monitoring of adjoining arid and semi-arid rangelands. The project was to prepare for the extension of monitoring to the whole tropical and sub-tropical region.

7   Monitoring tropical forests by remote sensing from satellite has helped to generate a global view of their status. However, remote sensing is still far from being able to provide all the information needed. An agreed standard vegetation classification is required, which could easily relate to what is seen on satellite imagery. It would dispense with major problems of interpretation as to what a tropical forest is. There is also a need for sensors that can penetrate clouds to monitor more effectively those areas in the tropics that are mostly under cloud cover. Furthermore, while remote sensing can detect complete removal of forests, and with some difficulty secondary regrowth or plantation replanting after deforestation, it cannot detect selective logging in mixed-species forests. Yet, this happens to be an important aspect of tropical deforestation. In the future, it should become possible to monitor differences in the extent and nature of the canopy, which would make remote sensing more useful.

8   'Global Environmental Monitoring System, Pilot Project on Tropical Forest Cover Monitoring, Benin–Cameroon–Togo, Project Implementation: Methodology, Results and Conclusions', FAO, Rome, 1980. See appendix I of the methodology document, *op. cit.*, for the proposed follow-up activities. See also T. J. Synnott, 'Monitoring Tropical Forests: A Review with Special Reference to Africa', MARC report No. 5, Chelsea College, London, 1977, and A. W. Mitchell, 'Reaching the Rain Forest Roof', a handbook on techniques of access and study in the canopy, University of Leeds, 1982. Both of these studies were executed in the framework of GEMS. It is also worth noting that the Mitchell study was a by-product of a GEMS-sponsored meeting on tropical forests held in Leeds. This meeting and the associated publications on problems of tropical forests started the process that eventually led to the adoption of the Tropical Forests Action Plan, coordinated by FAO.

9   Among the data studied in the preparation of the first assessment were pre-investment surveys,

be updated every five years, in parallel with the efforts to establish the monitoring system. Monitoring data generated by the system will be incorporated and utilized as and when they become available. In the longer term, the two efforts are to merge, as the monitoring system becomes operational and is gradually expanded to include sub-tropical and temperate forest zones.

As the environmental concerns regained momentum and as deforestation in the tropical areas was becoming one of the key public concerns, especially in so far as it is thought to have a bearing on global atmospheric and climate change, the lack of precise quantitative and qualitative information on the state of tropical forests became apparent. The policy debates and the efforts to mount action were undermined for lack of internationally authoritative data and assessments. The enormous difference between various estimates reduced the credibility of the arguments and confused the public. Undoubtedly, the absence of coordinated and sustained effort at the global level was responsible for the lack of the necessary capability and data.

Relevant work was proceeding, however, mostly through efforts to apply remote sensing techniques from satellites to tropical forest cover monitoring.[10] Indeed, the problem spurred so great a proliferation of projects that by the end of the 1980s an important issue became how to devise a coordinated approach, link existing projects and make data and assessments available. It is in this context that in 1989 UNEP took the initiative of convening an expert meeting on forest cover monitoring.[11]

The meeting outlined the basic challenge as being to maintain the originality, independence and scientific independence of existing projects, while at the same time integrating the data available and assuring communication and cooperation among these programmes.[12]

small-scale UNESCO vegetation maps, LANDSAT data (interpreted by using what had been learned in the pilot project on tropical forest monitoring). The assessment includes both natural forests and plantations, thus making it also useful in the study of the global carbon dioxide budget. See Jean-Paul Lanly, 'Tropical Forest Resources', FAO Forestry Paper 30, Rome, 1982. This volume represents the synthesis of the survey. See also the three technical reports for the three regions, 'Tropical Forest Resources Assessment Project: Forest Resources of Tropical Asia; Forest Resources of Tropical Africa; Forest Resources of Tropical America', FAO, 1981. For an executive summary of the main report see 'The Global Assessment of Tropical Forest Resources', GEMS PAC Information Series, No. 3, Nairobi, April 1982. (The potential of these pilot activities is well illustrated by the fact that once the project was terminated, its manager was invited by the Government of Brazil to give advice on how to set up and operate Brazil's forest inventory and national monitoring programme.)

10 GEMS PAC has thus supported work on the use of coarse spatial resolution of the NOAA weather satellites for analysis of global vegetation. See C. O. Justice, J. R. G. Townshend, B. N. Holben and C. J. Tucker, 'Analysis of the phenology of global vegetation using meteorological satellite data', *International Journal of Remote Sensing*, Vol. 6, No. 8, Aug. 1985. The capability of NOAA weather satellites to distinguish forest from non-forest was discovered accidentally. These satellites pass over the entire globe every single day, which gives more chance for cloud-free imagery in the tropics. With the resolution of more than one kilometre, the weather satellite imagery needs to be supplemented by more detailed images of LANDSAT, or similar satellites. It has been argued that a systematic assessment of the rates of deforestation in the tropical areas could be carried out for about $5 million per year. 'Monitoring the fate of the forests from space', *Science*, 17 Mar. 1989.

11 See 'Report of a Group of Experts on the Scientific Aspects of Landcover Monitoring – Informal Technical Meeting on Forest Cover Monitoring', Geneva, 16–17 Jan. 1989, GEMS Report Series No. 7, Nairobi, January 1989.

12 See annex, *ibid.*, for a listing of various activities which include a programme on monitoring

It is the GEMS' Global Resource Information Database (GRID) – which will be discussed in chapter eight below – that was chosen to act as a central data archiving and disseminating facility, building on its tropical forest mapping project in West Africa.

Monitoring arid and semi-arid rangelands

The degradation of arid and semi-arid rangelands in the tropics and sub-tropics represents another important international and national concern. The pilot project on the monitoring and assessment of this type of natural resource degradation was initiated by UNEP and FAO as a logical complement to and extension of the soil and tropical forest monitoring projects.[13]

A methodology had already been developed, tested, and applied in East Africa, with monitoring undertaken from satellites, on the ground, and from low-flying light aircraft guided by radar to keep constant height above ground level.[14]

The purpose of the pilot project was to demonstrate the method's applicability to West African conditions, to refine it into a tool for the repeated inventorying and monitoring of tropical rangeland systems, and to prepare the ground for its application in other parts of the world. The primary aim was to provide national decision-makers with a tool for the assessment and management of their countries' rangelands. The longer-term objective was to lead to a global system of rangeland monitoring based on standardized methodologies and yielding comparable data. The monitoring and assessment methodology and the institutional package refined in the course of the pilot project have thus been promoted for country use by UNEP. As one of the follow-ups, and in response to the recommendation made by the UN Conference on Desertification, a set of coordinated national projects on the ecological monitoring of desertification in Argentina, Bolivia, Peru, and Chile was proposed. Of these, only the project for Peru materialized.

vegetation in tropical and sub-tropical areas by the Joint Research Centre of the European Community.

13 For some of the background, see 'The Ecological Management of Arid and Semi-Arid Rangelands in Africa and the Near East', report of expert consultations, Rome, May 1974, a FAO/UNEP publication, Rome, 1974. See also 'The Ecological Management of Arid and Semi-Arid Rangelands in Africa and the Near and Middle East (EMASAR), the formulation of an international cooperative programme', report of an international conference, February 1975, a FAO/UNEP publication, Rome, 1975. Senegal was chosen for the pilot project because it is in the Sahel and because of a number of agricultural projects in the zone, the facilities, the historical and current data available, etc.

14 Human observers and photography were used for surveys from the aircraft. With the help of the global navigation system, the transects flown by the aircraft were located, so that the same transects could be flown on future occasions. This made survey repeatability possible. For the description of the monitoring and assessment methodology developed in Kenya Rangelands Environmental Monitoring Unit (KREMU), see H. Croze, M. Norton-Griffiths and M. D. Gwynne, 'Ecological monitoring in East Africa', *New Scientist*, Vol. 77, 2 Feb. 1978, pp. 283–5. The three 'platforms' complement each other and each type of monitoring has its own uses, advantages, and disadvantages. The ground work is difficult, time-consuming, and expensive, is limited to a few sample areas, and is not very useful for the study and overview of large areas. It is crucial, however, for generating information needed by local decision-makers, for studying processes, and for providing ground truth for remote sensing.

Monitoring endangered species of wild flora and fauna

The last GEMS activity dealing with the monitoring of the status of the renewable natural resource base concerns the endangered species of wild flora and fauna. Responding to the recommendation of UNCHE for the establishment of species monitoring, UNEP started to support financially the activity of the International Union for Conservation of Nature and Natural Resources (IUCN) related to Red Data Books. Specifically, it supported efforts of IUCN's Conservation Monitoring Centre to develop a computerized 'Red Data Base'. It is also involved in the major restructuring of all the databases held by the Centre in order to make these an integral part of GEMS. One of the basic purposes of this activity was to place the effort on a more solid scientific foundation and to make the data more accessible to decision-makers and managers and easier for them to use. It helped strengthen the centre and its conservation database. The centre was eventually renamed the World Conservation Monitoring Centre, jointly sponsored by UNEP, IUCN and the World Wide Fund for Nature (WWF).

In addition, a database study of the African elephant was undertaken using GRID capabilities, a study which responded to a major public concern and yielded data valuable to conservation efforts.[15]

Factors at work

Progress has been made since UNCHE in the field of natural resource monitoring. In respect to each problem area, work has been initiated, pilot methodologies have been developed and readied for testing and application, and the foundations for a systematic global monitoring effort concerning the status and degradation of the natural resource base are slowly taking shape. Within the existing constraints, the developing countries' desire to give priority to this type of monitoring has been satisfied. In the initial ten-year period, through 1983, a comparatively large amount of resources was allocated by the Environment Fund to this activity, i.e., $3.28 million.[16] The total was boosted by resorting to other budget lines in addition to the GEMS allocation, a precedent of sorts, as this practice was commonly discouraged in the early days of UNEP.[17] Indeed, this made it one of the better endowed

---

15 See 'The African Elephant', UNEP/GEMS Environment Library, No.3, UNEP, Nairobi, 1989.
16 In the case of the soil degradation monitoring and assessment project, UNEP contributed $541,500 (out of a total of $1,005,300) for the period June 1975 – May 1978. In the case of tropical forest monitoring, it contributed $24,295 (out of a total of $33,295) for the initial formulation of the project; $637,582 (out of a total $1,037,582) during the period Sept. 1975 – Dec. 1978 for the pilot project itself; and $239,000 (out of a total $302,000) during the period Oct. 1978 – May 1980 for the assessment study. In the case of rangelands monitoring, UNEP contributed $1,562,680 (out of a total $2,083,680) for the period Oct. 1979 – Nov. 1983; for the project on the use of modern remote sensing techniques in monitoring desertification of natural resources in Egypt, it contributed $150,000 (out of a total $2,140,710) for the period Apr. 1981 – Dec. 1982. In the case of the monitoring of endangered species, it contributed $127,500 (out of a total $633,000) for the period Jan. 1981 – Dec. 1983.
17 Thus, use was made of soil, tropical woodlands, and ecosystems budget lines. In UNEP, up to then, a rigid distinction had been maintained between budget lines and programme activities, mainly for administrative reasons. Thus, in a sense, the tail came to wag the dog. This administrative rationale

GEMS clusters, which was in part a reflection of the high operational costs involved in devising pilot monitoring methodologies in developing countries, a category of expenditures for which the Environment Fund was not really equipped, or intended.[18]

As in other GEMS clusters, funding from UNEP was virtually all that was available in cash, thus limiting what could be done. The projects were carried out with resources that were modest compared to what was really needed.[19] The gap was exaggerated by the fact that this was a monitoring activity in which only developing countries were taking part. As a result, adequate counterpart support and participation was usually not forthcoming. Moreover, the whole effort was slowed down and made more difficult by the fact that there was little knowledge of the status and dynamics of the renewable natural base in the tropics and sub-tropics. The existing knowledge about soil and forests in temperate zones was not transferable. Research and monitoring of the characteristics and behaviour of ecosystems in tropical and sub-tropical zones was sporadic and not advanced, with serious gaps in the knowledge and understanding of the conditions and processes involved, in particular on the regional and continental levels.

For the most part, the effort to mount a global and coordinated monitoring programme had to start virtually from scratch. In order to develop knowledge about the renewable natural resource base and to monitor it, a corresponding scientific and institutional infrastructure is needed. In many places such infrastructure is weak or even non-existent; building it up is not a simple task, and requires major

---

went against the need to integrate activities and fostered programmatic and in-house sectoralization. It introduced rigidity and prevented interdisciplinary projects and activities from being initiated. In this instance, the use of other budget lines was made possible because the available funds could not all be expended on management activities and projects. Substantively, the rationale was that monitoring/assessment represented a precondition for developing management strategies, especially since the knowledge possessed about the ecosystems in question was generally weak. Gradually, multi-line projects supported from a number of budget lines became more common in UNEP practice. For example, the *World Resources* report, which is published biennially by the World Resources Institute (WRI) and the International Institute for Environment and Development (IIED), is given financial support through a GEMS project which draws on several budget lines in UNEP. For the report, which uses some of the GEMS data, see *World Resources 1988–89*, Basic Books, New York, 1988.

18 UNEP funds were used mainly for catalytic purposes, such as consultants, travel, publications, expert meetings. In the case of the Senegal project, however, UNEP also paid for personnel ($996,180 for a project manager and the range resource ecologist), for aerial reconaissance ($157,000), for equipment ($337,500), etc.

19 A good illustration of this is the pilot project on tropical forest monitoring. FAO had originally prepared a comprehensive five-year project at a total cost of $9 million, of which UNEP was supposed to contribute $5,743,000. Among the items that UNEP was to pay for were experts ($1,438,000), travel ($437,000), aerial surveys in 45 test areas ($1,238,000), and training ($382,000). Also, $470,000 was allotted to research on applications of the automatic interpretation of SLR and satellite imagery. In case this comprehensive version of the pilot project was not acceptable to UNEP, FAO had prepared a fall-back position, which would confine the project to the African region only. In this version, UNEP's contribution was estimated at $2,802,000, with personnel costs amounting to $779,000, travel to $225,000, and contractual services to $1,100,000, of which $468,000 was for aerial surveys. The total cost of this pilot project was to be about $5 million. In the end, as noted above, the pilot project was trimmed down to $1,058,000, with UNEP contributing $557,000. See 'Report on the Formulation of a Tropical Forest Cover Monitoring Project', *op. cit.*, pp. 36–50.

160 *The quest for world environmental cooperation*

financial inputs, training, time, institutional decisions, etc. Ultimately, of course, this depends on the willingness of individual developing countries to take the necessary action and to persist in developing and using this type of capability. The fact that it is of direct importance to their economic development process and is obviously of economic value ensures their interest and commitment, at least in principle, and makes the prospects for this dimension of GEMS appear promising.

Nevertheless, the need remains for international encouragement and support for such activities, both in the form of seed money and payment of operational costs. The cost of monitoring is determined by the level of detail and the quality of the information required, and the objective in any given situation should be to gather only the data that are necessary. This is something that will have to be decided iteratively, depending on local conditions, priorities, and capabilities. In the long run, the key to the success of these monitoring efforts will be whether they prove to be useful and economically valuable to individual countries.

The difficulties that have been encountered in financing and launching the pilot monitoring projects in Latin America illustrate some of the underlying problems.[20] This is an issue that will have to be faced as the methodologies become ready for use, as the operational phase of natural resource monitoring is entered, and as attempts are made to involve a large number of developing countries.

As noted above, GEMS had to reckon with the sensitivity of developing countries to any seeming interference from outside with their sovereign right to use and manage their natural resources. This is why the GEMS strategy has been to encourage the creation of national monitoring capabilities and to develop standardized monitoring packages that could be used by countries domestically, if they choose to do so. For obvious reasons, it has played down the causes of natural resource degradation, which often have to do with internal political, economic, and social issues that countries are not eager to have discussed or known outside.

Remote sensing of natural resources

Remote sensing from satellite plays a key role in the prospects of natural resources monitoring. Although its full potential still remains to be developed and applied widely, observation of natural resources from space, in conjunction with ground

20 Initially, three projects were proposed by GEMS PAC (Argentina, Bolivia, and Peru), using the methodology developed in Senegal as a a model. (As an example, see the Argentina project proposal in doc. DESCON–3/6, Sept. 1980).The projects were to be implemented by the countries concerned, with the greater part of the external cost being met by outside donors through the DESCON facility. This facility was created after the UN Desertification Conference to help to finance relevant national and transnational projects. In the case of Bolivia, donors were supposed to contribute $1,201,300 (or 54.5 per cent of the total budget of $2,203,550) over a period of five years; in the case of Argentina, $515,000 (or 38.8 per cent of the total of $1,328,350) over a period of four years; and in the case of Peru, $704,700 (or 57.7 per cent of the total amount of $1,221,700), also over a period of four years. (For a summary of project proposals see doc. DESCON–4/WP3, Nov. 1982). In spite of a great deal of effort by GEMS PAC to get these projects going, as already noted above, only the project in Peru materialized. In Bolivia, the Federal Republic of Germany had agreed to fund the project; however, after a new government came to power in La Paz, it withdrew its offer of help. The initiative to have a project in Chile yielded eventually the integrated monitoring project discussed later in this chapter.

truth against which to calibrate observations, has already become an indispensable tool with many current applications. These include crop conditions and crop forecasting, soil moisture, insect infestations, animal health, deforestation.

The important practical advantages of remote sensing from space, as used for purposes related to monitoring and assessment of natural resources, include the capacity to delimit rapidly and accurately surface cover and vegetation classes; improvement of soil maps; usefulness for assessing characteristics and trends of large areas and for showing extensive features that cannot be detected from the ground; ability to survey repeatedly a given situation and establish a time series needed to determine trends and change; regularity and uniformity of observation; ability to sense visible and non-visible frequencies; digitized data that lend themselves easily to storage, handling, and retrieval, to transmission, to conversion into images, to analysis and interpretation by computers.[21]

In addition, from the point of view of GEMS, remote sensing of natural resources has a dual function and offers a way out of some operational and practical predicaments. It makes it possible to separate the more strictly management-oriented and decision-making uses of natural resources monitoring at the national level from its broader applications to study and assessment of global trends and phenomena. Thus, it makes accessible to developing countries a tool for survey and assessment of their natural resources, a shortcut to monitoring that the great majority of them would not or could not undertake on their own with the traditional means and methods. Besides, from a space platform it offers a global view of the status and degradation of natural resources worldwide, a macro-view using low resolution power and a close-up view of 'hot spots' using progressively higher resolution power that is becoming available. Remote sensing from space can thus potentially substitute for and perform the role of a global monitoring network independent of individual countries' responses, to survey a series of parameters and processes simultaneously. On the ground, such a monitoring network could not be expected to come into being in the foreseeable future, for the reasons already mentioned above.

The pioneering Earth Resources Technology Satellite (ERTS) was first launched by the US in 1972. Later renamed Land Remote Sensing Satellite (LANDSAT 1), it introduced the practice of civilian applications derived from remote sensing of the Earth's surface. In part inspired by the notion of public service and interest, strongly research-oriented and proclaiming the principle of 'open skies', LANDSAT was run on an experimental basis by the National Aeronautics and Space Administration (NASA). Following the pattern set by US meteorological satellites in the Television Infra-red Operational Satellite (TIROS) series that offered free access to their data and thus led to the construction of ground stations in many countries to receive their signals, LANDSAT set the initial tone by offering liberal access to its data at a comparatively low cost. One of the explicit objectives of this policy was to placate developing countries, many of which were alarmed by the fact that foreign satellites orbited over their territories and peered into their

21 For a review of the current applications of space technology to resource management, see P. S. Thacher, 'Space technology and resource management', *Journal of International Affairs*, vol.39, No.1, Summer 1985.

resources without their consent. Thus, developing countries were encouraged to use LANDSAT data, often with direct support from the US Agency for International Development (USAID), in particular for conducting natural resources inventories. At least a dozen built ground receiving stations, while many more invested in the training of technicians and in the infrastructure needed to use and interpret LANDSAT-generated remote sensing data.

Developing countries and other users of LANDSAT services, in the US and elsewhere, were taken by surprise when in 1979 a directive by the US President announced the intent to transfer the responsibility for LANDSAT from NASA to the National Oceanic and Atmospheric Administration (NOAA) in the Department of Commerce. This marked a turn in the US policy and a signal of the intent to commercialize surveillance of natural resources from space, a trend which became especially pronounced with the arrival of the new President in the White House. It triggered an intense debate in the US over access to and the uses of remote sensing services and data, a debate which continues one decade later.[22]

The reasons for the US action were many and complex. They included the pressure for federal budget cutting, which reflected negatively, in particular, on public services that by definition did not pay for themselves. There was also the ideological preference for diminishing the role of the government and for moving to the private sector activities it had traditionally performed. The accent was placed on profitability, and the notion of public subsidy was criticized. There was also a strong responsiveness of the political party in office to the pressure from industry eager to gain a foothold in space and in an activity with many potential applications. These included the gathering of economically valuable information and intelligence worldwide, which is of special importance to transnational corporations and other actors with a global reach. The Department of Commerce, to which was assigned the task of operating and commercializing LANDSAT, is required by its terms of reference to motivate industry and to turn over to it those activities and services that industry can perform reasonably well.

The changes taking place in the US highlighted the growing political and practical importance of remote sensing of natural resources and of the Earth, and in fact signalled the beginning of a new space race. France joined when its government announced the intent to develop and launch the first fully commercially oriented satellite, known by its acronym SPOT (Le Système Probatoire d'Observation de la Terre). The first in a series was launched in 1986, attracting attention by the high resolution of its images of earth (10 square metres per pixel in black and white and 20 square metres in colour), as well by the very high prices of its products and services in the commercial market.[23]

The controversy and uncertainty over the US civilian remote sensing pro-

22 See D. Deudney, 'Space: The High Frontier in Perspective', Worldwatch Paper 50, Washington, D. C., August 1982, pp. 26–34; and 'Imaging the Earth (I): the troubled first decade of LANDSAT', *Science*, Vol. 215, 26 Mar. 1982. See also 'Remote sensing for Third World satellite data', *New Scientist*, 22 Oct. 1987, and 'US draws a veil over open skies', *op. cit.*, 5 Nov. 1987. For an early US discussion of problems concerning developing countries, see 'Resource Sensing from Space, Prospects for Developing Countries', National Academy of Sciences, Washington, D. C., 1977.
23 See 'Spot the commercial advantage', *New Scientist*, 24 Mar. 1988.

gramme continued throughout the decade. The future of LANDSAT remained in doubt, the transfer to the private sector being opposed in the US Congress. The two remaining satellites in orbit were operated and their products commercialized worldwide by a private Earth Observation Satellite Company (EOSAT), which owns the data while the US government pays for the operational costs. EOSAT was awarded a contract in 1985 by the Department of Commerce to assume the operation of existing satellites and to develop new ones under the Land Remote Sensing Commercialization Act of 1984. However, LANDSAT 4 and 5, which remained the property of the US government, ran into difficulties as NOAA could not afford the $20 million needed to monitor and keep on track the two satellites. To many this demonstrated the failure of the LANDSAT commercialization policy of the 1980s, and the need to reconsider the approach, possibly through a cooperative system between governments whereby they would reduce the administrative and operational costs by sharing them.[24]

In the meantime, work is proceeding on the LANDSAT 6 spacecraft, which will carry on board improved observation technologies, such as the Enhanced Thematic Mapper. It is scheduled to be launched in 1991. The US Congress has voted funds to subsidize part of the costs for the construction of the satellite and of the necessary ground systems, and to keep the US in an increasingly competitive race that is eroding its supremacy in this domain. Furthermore, the Earth Observing System (EOS), discussed in chapters four and five above, has as one of its major objectives the study and monitoring of land processes and terrestrial ecosystems.[25]

Other countries have also entered the competition for a view from space. Japan, Canada, and the European Space Agency have plans to launch natural resource sensing satellites. In addition, Brazil, India and Indonesia, three large developing countries that find remote sensing essential for their development, plan to launch national satellites, and thus to counter the monopoly of developed countries and also to create opportunities for South–South cooperation in this domain.[26] Indeed, one of the lessons of the 1980s that was not lost on the developing countries concerned the risks and costs of being dependent on a single developed country for these strategic services.

The mounting pressures for commercialization of the remote sensing of the Earth's surface, and the mounting costs of space technology services, have paralleled the technology's increasing power and are evidence of its political, strategic and economic value for the management of natural resources, and for other uses such as military and economic intelligence and news gathering and dissemination.[27]

24 'LANDSAT: Drifting toward Oblivion?', *Science*, 24 Feb. 1989, and 'LANDSAT Wins a Reprieve', *op. cit.*, 17 Mar. 1989.
25 *Earth System Science, a Closer View*, NASA, Washington, D. C., 1988.
26 Indonesia, in cooperation with the Dutch, plans to launch in 1991 the Tropical Earth Resources Satellite (TERS). It is intended to service the equatorial-belt countries, and to provide data on natural resources, weather, pollution, and natural disasters. Brazil, with its focus on the Amazon region, and India, which has already launched experimental satellites in its Bhaskara series, plan a regular remote sensing service by the early 1990s. 'Space scramble for the long view', *South*, June 1984.
27 See 'Space Cameras and Security Risks', *Science*, 27 Jan. 1989. The usefulness of SPOT's products for military intelligence, and its efforts to commercialize them for this purpose, gave rise to concern

More powerful and versatile observing instruments with high spatial resolution and fine spectral resolution over an increasing number of bands, new and powerful software packages and analytical systems to interpret, integrate, and use the data and to give them varied applications, and more powerful hardware to handle and process the data all promise to overcome many of the practical problems and to extend the reach and usefulness of monitoring from space. Improved communications and geostationary relay satellites promise greater speed of transmission and instantaneous availability of data worldwide, and *de facto* real-time monitoring of the globe's surface. Indeed, EOS, planned for the mid-nineties, offers a glimpse of things to come.

These are valuable developments. However, in a highly unequal world and in the absence of agreed principles and of an international regime for the peaceful uses of the outer space, with mounting pressures to commercialize space and the ballooning costs of remote sensing products and services, the prospects are not bright for the developing countries, for international programmes, nor indeed for scientific users. Will these services be available on a regular basis only to those who can afford them? Will scientific research suffer and will remote sensing data become a luxury inaccessible to most developing countries, strapped as they are for hard currency and with many needs competing for scarce resources?[28]

The underlying issues, of course, are at the heart of the North–South development gap and of the debate on how to reconcile the broader interests of humankind with the more narrow interests of given actors on the contemporary scene. They highlight the dichotomy between particular interests and the strictly economic considerations, on the one hand, and the more general objectives of seeking knowledge and promoting sound uses and management of the Earth's resources on the other.

This is a matter that should be internationalized and linked with the agenda of the more traditional negotiations between the North and the South. After all, the natural resources of developing countries are vital to the well-being of their peoples and their development and play an important role in the state of the global environment. These resources are exposed to great pressures generated by the underdevelopment of these countries and their status and situation in the interna-

---

in the US about a 'free-wheeling, uncensored surveillance of the Earth from space', and about the spread of this capability and the resulting 'strategic transparency' becoming accessible to many governments and other actors.

28 One of the major attractions of LANDSAT data in the early days was their reasonable cost. Thus, in the 1970s, in the case of rangelands monitoring in East Africa, the cost of LANDSAT data per square kilometre amounted to $0.01. This compared with $100 for the collection of data by the ground team. See H. Croze and M. D. Gwynne, 'Rangeland monitoring: function, form, results', in 'Selected Works on Ecological Monitoring of Arid Areas', GEMS/PAC Information Series, No.1, April 1980. The LANDSAT services costs increased steeply in the 1980s, however. The first act by EOSAT was to quadruple the price of each LANDSAT scene and to increase to $600,000 the annual fee for ground stations mainly installed in developing countries. 'LANDSAT: Drifting toward Oblivion', *Science, op. cit.* In 1981, a single multispectral LANDSAT picture cost $200; in 1987 the price was $730. Each frame from the Thematic Mapper from LANDSAT 4 cost $2,800 in 1983; the price rose to $4,400 in 1985, and so on. *New Scientist, op. cit.*, 22 Oct. 1987. SPOT prices were much higher, in some instances as much as nine times that of a similar product from LANDSAT.

tional economy. They are often mismanaged or destroyed for lack of information, knowledge, and needed finance – to the detriment of these countries and of the international community at large. These countries need easy access to the tools and information that would help them to manage their natural resource base effectively. This is another issue that needs to be addressed comprehensively and in a timely manner in GEMS and Earthwatch where, in spite of its obvious relevance, the question was not a subject of policy debate.

One of the obvious ways to try to overcome the many issues that have ensnared this important capability and to harness it for common purposes would be to launch specially designed research and monitoring satellites that would be internationally funded and managed. This is not a novel idea. In the early days of GEMS, Japan had informally suggested to UNEP management that it might be interested in equipping and launching a satellite for GEMS. Had there been enough persistence and imagination to probe and keep alive this idea rather than to be discouraged by the earthly obstacles and views that considered it as premature and impractical, it is certain that useful discussion could have taken place in the years that followed. Such maturation of ideas is essential in international organizations. And there is nothing inherently impossible about an international venture of this kind, although it would have to deal and live with many underlying conflicts of a political and economic nature, and military and intelligence considerations.

The experience with meteosats and communications satellites can provide valuable precedents and lessons, and points in the direction of a joint international space surveillance agency to coordinate, manage and perform these increasingly important services. Certainly, this is a critical issue not only as it concerns the future development of GEMS, but also as it relates to increasing international policy and public concern over global environmental degradation and climate warming.

It is in this broader policy framework that such pilot initiatives as the International Satellite for a Land-Surface Climatology Project (ISLSCP), which is part of the World Climate Research Programme, will find their important place. With UNEP having provided through GEMS the initial support for this cooperative project, its basic objective is to improve the understanding of the processes involved in the interactions between the biosphere and the atmosphere, and to clarify the role that changes in the land-surface characteristics play in climate and climate change. Among other things, it should help to establish the global extent and trends of desertification and deforestation.[29] A complementary approach would be to use remote sensing data in international data information centres, such as GRID, and make them available free of charge to users who otherwise would not have access to them.

Summing up

In concluding this brief review of natural resources monitoring in GEMS, it can be

---

29 See F. Becker, H-J. Bolle and P. R. Rowntree, 'The International Satellite Land-Surface Climatology Project (ISLSCP)', ICLSCP – Report No.10, Free University of Berlin, Berlin, 1988.

said that the progress has consisted mostly of preparing the ground for future action, for learning, and for acquiring practical experience. The initial UNEP estimates of how long it would take for such monitoring to become operational proved wide of the mark, underestimating the complexity of the task and the difficulties of attaining it under existing conditons, with the limited means available, and lack of interest and willingness to follow up. Tropical forest monitoring is a good example of this. It is said that it will take an additional 10 years or so before a truly comprehensive and reliable assessment of the state of tropical forests becomes available. This means year 2000, or 30 years after UNCHE and 50 years since the first step was taken by FAO in 1951, probably too late for saving a good part of tropical forest which, if some current projections prove correct, will have disappeared by then. Yet, the necessary tools for this purpose already exist.

On the whole, however, in the context of GEMS' evolution, and given the existing conditions, this has been a rather successful set of activities. The development of monitoring methods, and their field-testing and application, have produced a tool of potential value for those developing countries interested in sound management and use of their natural resource base. Global and regional views are emerging thanks to remote sensing, which, as it evolves, holds the potential to leapfrog many of the current obstacles and yield an operational global system of monitoring the status of natural resources.

Institutionally, this happens to be a monitoring activity in which GEMS PAC has played a direct and active role in conceiving, managing, and implementing various projects and activities. One of the factors that contributed to this role was the availability of the in-house expertise in the subject, gained and applied in Africa where this type of monitoring is finding its principal applications. No doubt, the location of UNEP headquarters in a geographical region where natural resources are exposed to strong pressures also played a role, illustrating local influences on programme development. The pattern set was different from the one in climate-and health-related environmental monitoring, where the agencies dominated the action, and more like the one set by Oceans and Coastal Areas Programme Activity Centre (OCA PAC) in the Regional Seas Programme.

Most importantly, it is the experience with natural resources monitoring that has helped GEMS PAC to evolve the idea of what eventually became the Global Resource Information Database (GRID), a major innovation in the framework of GEMS and Earthwatch (see chapter eight below). Suffice it to say here that its establishment as a data management mechanism was in part motivated by the need to link the expanding databases in GEMS, to make the monitoring data more accessible to potential users, and to promote assessment. Remote sensing applications to the monitoring of natural resources, and other technological advances, mostly in computer hardware and software, were among the key factors that made it possible to entertain the idea of GRID to begin with.

INTEGRATED MONITORING

The notion of an integrated or 'multimedia' approach to environmental monitoring is inherent in the systemic properties of the environment. The idea of an integrated approach to environmental monitoring was advanced as early as the proposal to hold the Stockholm Conference.[30] It was discussed during the Conference and was implicit in its recommendations to monitor pathways of pollutants and ecosystems' response to change. It was proposed in the early background documentation for GEMS. Thus, the SCOPE 3 report argued that GEMS should be designed in such a way as to make the study and analysis of interaction between media possible. It was included in the framework of MAB Project 14. In 1974, the World Meteorological Organization (WMO) Executive Committee proposed to the governments that they should, for scientific reasons and in the interests of cost-effectiveness, establish stations that could monitor several media simultaneously, and include the monitoring of pollution in the soil and biota. More specifically, it was proposed that the Background Air Pollution Monitoring Network (BAPMoN) stations should monitor pollution not only in the atmosphere but also in other environmental compartments. In 1978, the same Committee stated that such monitoring in BAPMoN would be useful not only for the study and understanding of climate but also for increasing knowledge about the exchange of pollutants between different compartments of the environment, and about geocycles in general. The concept of integrated monitoring was also studied by many scientists in their individual work or in the context of their countries' national programmes.[31]

Within GEMS and UNEP, the idea of an integrated approach to monitoring was kept alive and carried forward mainly thanks to the initiative of the USSR, which assigned to it high priority and pressed for the inclusion of corresponding activities within the system.[32] This is the reason why 'integrated monitoring' of environmental pollution was included among GEMS' priorities. Starting in 1976, many discussions and consultations were held in an effort to give greater precision and operational meaning to this concept.[33]

30 See B. Lundholm, 'Global Baseline Stations', Swedish Ecological Research Council, Stockholm, 1968. He argued for a global network of baseline stations (sampling sites) to monitor biotic as well as abiotic factors in the environment that were related to pollution.

31 See, for example, G. B. Morgan, G. B. Wiersma and D. S. Barth, 'Monitoring in Biosphere Reserves for Regional Background Levels of Pollution', in 'Proceedings of the US–USSR Symposium on Biosphere Reserves', Moscow 1976. Also see R. E. Munn, 'Basic Principles and Siting Criteria for Multi-Media Monitoring', and Yu. A. Izrael, 'Main Principles of Monitoring the Natural Environment and Climate', both papers presented to the UNEP/WMO International Symposium on Global Integrated Monitoring of Environmental Pollution, Riga, USSR, December 1978. (The Munn paper is reproduced in 'Selected Works on Integrated Monitoring', GEMS/PAC Information Series, No.2, April 1980).

32 For a description of the USSR approach see 'National Programme for Comprehensive Monitoring of Environmental Pollution in the USSR', presented to the WMO Expert Meeting on Siting Criteria, Mainz, 1976, reproduced in 'Selected Works on Integrated Monitoring', *op. cit.* Also see, *ibid.*, the USSR proposal entitled 'Programme for the First Stage of the Conduct of Comprehensive Global Monitoring of Background Pollution of the Environment', 1979. Further see Yu. A. Israel, *Ecology and Control of the State of the Environment* (in Russian), Gidrometeoizdat, Leningrad, 1973, on the concept of monitoring humans' impact on the biosphere and on ecological monitoring.

33 See 'Long-term Ecological Monitoring in Biosphere Reserves', a report on an international workshop, held in Oak Ridge, Tenn., under the auspices of UNESCO and UNEP, published by the US National Committee for MAB, July 1978; 'Interagency Consultations on Monitoring in Biosphere Reserves and Integrated Monitoring', Geneva, 1979 and Nairobi, 1980, reproduced in

The official definition of integrated monitoring has been 'the repeated measurement of a range of related environmental variables or indicators in the living and non-living compartments of the environment, and the investigation of the transfer of substances or energy from one environmental compartment to another.'[34] A good deal of ambiguity and terminological looseness occurred, however.[35] In part this was a consequence of the lack of experience and the inappropriateness of standard frames of reference and institutions for handling the processes and phenomena that were to be monitored and assessed. The definitional quandaries became magnified when attempts were made to devise concrete approaches for integrated monitoring. Seven years passed between the first interagency consultation on the subject and the definite formulation of the first pilot project on integrated monitoring. Here, a distinction is made. 'Classical ecological monitoring' focuses on ecosystems and ecosystem dynamics, and involves monitoring of such factors as biological rates (biomass, numbers, species, etc.); spatial, temporal, and demographic distributions; and biological processes (birth/death rates, rates and pathways of nutrients, and energy flows). 'Integrated monitoring' refers to systemic monitoring concerned with levels, pathways and flow rates of pollutants within, through, and between ecosystem compartments (i.e., air, water, soil, plants, and animals), and is also concerned with background pollution in the environment as a whole.

Among the many reasons for the delay were the absence of a single agency capable of assuming the systems initiative that was required and the usual problems encountered in mounting joint interagency activities. The responsibility of UNESCO for the biotic aspects and of WMO for the abiotic aspects resulted in such institutional dilemmas as which body would convene a meeting on integrated monitoring. Initially it also led to the submission of two separate project proposals to UNEP, one by UNESCO on biotic aspects, the other by WMO on abiotic aspects. The two projects were to be carried out concurrently, in the same biosphere reserves, and by the same groups of national scientists. The institutional divisions, which tend to become pronounced when it comes to operational activities and

---

'Selected Works', *op. cit.* ; 'Final Report of the Expert Meeting on the Operation of Integrated Monitoring Programmes', WMO, September 1980; 'International Symposium on Integrated Global Monitoring of Environmental Pollution', Tbilisi, Oct. 1981, Abstracts, Gidrometeoizdat, Leningrad, 1981. A selection of the papers presented at Tbilisi was published as a special issue of *Environmental Monitoring and Assessment*, Vol. 11, 1982, No. 4.

34 See 'Report of the Interagency Consultations on Integrated Monitoring', Nairobi, 14–16 January, 1980, in 'Selected Works on Integrated Monitoring', *op. cit.*

35 In some instances 'ecological' and 'integrated' monitoring have been used interchangeably. In one of its submissions, the USSR referred to 'integrated ecological monitoring of natural environment pollution', while the 1979 interagency consultations on monitoring in biosphere reserves and integrated monitoring speak of 'ecological monitoring' as a means of providing information on the state, trend, variability, and functioning of entire or selected components of production systems. The UNEP/UNESCO workshop held in 1978 used the expression 'long-term ecological monitoring' in biosphere reserves, which referred to 'selected chemical, physical, biological and anthropological variables being systematically observed, measured and interpreted for defined purposes', such purposes being directed towards a description of the state of the environment, the identification of trends in it, and the assessment of pollutant effects.

sharing the available funds, were superimposed on the broader conceptual, scientific, and practical difficulties inherent in this type of monitoring.

Also, there was no great enthusiasm or support for integrated monitoring among the broad GEMS constituency, and the whole idea had to fight an uphill battle. The approach was considered as too theoretical and research-oriented. The subject did not figure clearly as one of GEMS' priorities, even though it was related to most of them. Most of all, this type of monitoring was complicated and difficult to carry out. Another background factor had to do with the traditional political mistrust of many towards initiatives by the USSR, often suspecting ulterior motives. This was the case in spite of the fact that the US National Academy of Sciences was also championing integrated monitoring and the two countries had cooperated on the idea within their bilateral programme on biosphere reserves and indeed within GEMS. As with integrated ocean monitoring, the USSR managed to maintain the initiative in the framework of GEMS mostly thanks to the roubles it had contributed to the Environment Fund that were used to organize the Riga, Tbilisi, and Tashkent meetings on integrated monitoring.

At any rate, the fact that the phenomena and processes under study did not recognize institutional and disciplinary barriers, continuing scientific work, the growing conceptual and practical appeal of the idea, and the pressures for the formulation and execution of an interagency programme in this field all led to the eventual formulation of the first pilot project on integrated monitoring, to be carried out in the temperate mixed (coniferous/deciduous) forest ecosystems. These forest ecosystems were chosen because of their proximity to areas of intense industrial activity and their exposure to pollution, and because they were the most extensively studied and monitored and best understood type of forest ecosystem.

The pilot project, to be executed jointly by UNESCO and WMO with the financial support from UNEP, was supposed to start in two biosphere reserves, one in the USSR (Berezinskyi) and one in the US (Olympic National Park). Eventually, it was to link up with a similar biosphere reserve in Chile (Torres del Paine), as a test site for the monitoring methodologies. Background monitoring was to be carried out in remote areas to determine baseline levels of pollutants in the various environmental compartments; at the same time, monitoring was to be carried out in areas affected by pollutants.

The purpose of the project was to determine the feasibility of integrated monitoring on a global basis, and to develop methods for such monitoring, including siting criteria, sampling and analysis procedures, and methods for data analysis, synthesis, and evaluation. The long-term operational objective was to establish routine integrated monitoring. Significant research efforts were called for, involving such questions as the transfer rates and pathways of various pollutants and the performance of biota and ecosystems exposed to them. The pilot project consisted of two sub-projects. The biotic sub-project was oriented more towards classical biological and ecological monitoring and was aimed at helping to form an understanding of the functioning and dynamics of the ecosystems observed and serving as a backdrop for the abiotic project. The latter, in turn, was to focus on the levels, pathways, and flow rates of selected pollutants within, between, and through

170 *The quest for world environmental cooperation*

various compartments of the ecosystems. The eventual aim was to understand and monitor the effects of pollution on ecosystems and the response of the latter, and how pollutants behave in ecosystems.

The USSR, however, withdrew from the pilot project because of the participation of Chile, though it continued its work on integrated monitoring in close collaboration with the organizations involved. In the framework of the pilot project, at three monitoring sites (two in the US and one in Chile), measurements are taken of trace elements, sulphates, nitrates, trace gases such as ozone and sulphur dioxide in air, precipitation and deposition, surface waters, soil, forest litter, and various types of vegetation. In addition, ecosystem functions are monitored, including productivity, litter decay rates, and nutrient flux in soil.

It is on the basis of the experience acquired during these pilot activities that a comprehensive manual on integrated monitoring was prepared, reflecting the state of the art in integrated monitoring, providing the conceptual underpinnings, and giving operational guidance, techniques, and step-by-step procedures.[36] Moreover, in the CMEA region the USSR initiative led to the establishment of an incipient regional integrated monitoring network.[37]

Integrated monitoring had been assigned a peripheral place among GEMS activities, mostly on account of the choices that had to be made as to where to apportion the limited resources available, which were continuously shrinking owing to the falling dollar and the declining relative value of the resources available in the Environment Fund. However, the idea was nurtured to the point where its importance for GEMS and its feasibility became generally accepted. The work done and the experience gained constitute the kernel of a systems approach to global background environmental monitoring carried out at the micro-level.[38]

It has kept alive and embodies some of the original notions that had given rise to the very idea of GEMS, namely the need to observe and understand physical, chemical, and biological processes that constitute the human environment and life on Earth, their interactions, changes taking place in the system, and the role of anthropogenic impacts and how the natural environment responds to these. It has also advanced the concept of a global integrated background monitoring network, tentatively dubbed GEMS/IBM, using a unified programme and unified methods. Such a global network would encompass initially the three pilot sites and the regional monitoring system of the CMEA countries, and it could link up with MAB biosphere reserves chosen as sites for integrated monitoring.[39] In addition, it could

36 F. Ya. Rovinsky, USSR, and G. B. Wiersma, USA, 'Procedures and Methods for Integrated Global Background Monitoring of Environmental Pollution', EPMRP Series No.47, August 1987.

37 See 'Report on the Status of the System for Integrated Background Monitoring in East European CMEA Countries', and also the report of the Second UNEP/CMEA Meeting on Co-operation in the Field of Integrated Background Monitoring, held in Geneva, 16–20 Nov. 1987, where the earlier document was distributed.

38 For the monitoring programme for the stations, see 'Procedures and Methods for Integrated Global Background Monitoring of Environmental Pollution' *op. cit.* It gives the basic and optional programme and the observation frequency for ambient air, atmospheric precipitation and deposition, surface waters, soils and bottom sediments, and biota.

39 As noted above, MAB Project 14 did not get off the ground as originally hoped. It was limited to a few meetings organized under MAB auspices and inventories of relevant research projects carried

include those baseline stations of BAPMoN that can perform expanded monitoring programmes in several environmental media, including the ocean where possible, thus providing a link with the marine environment.[40] Within the global system, regional sub-systems would operate, backed by regional centres that offer guidance to the stations and carry out sample analysis and data quality assessment on a regional basis.

In sum, then, while peripheral to the mainstream of GEMS activities and concerns, research and practical work useful for global integrated background environmental monitoring have been accomplished. This makes it possible for an idea first mooted more than twenty years earlier to move several steps in the direction of global implementation.

The integrated monitoring is important for the general concept of GEMS because of its emphasis on the interrelationships among its various clusters and on joint interagency programmes, and because it links logically with the aims of the International Geosphere–Biosphere Programme (IGBP), outlined in chapter four above. Cross-fertilization of methodologies and approaches should take place. Suitably adjusted monitoring of background environmental pollution and its ecological consequences should be undertaken in various ecosystems, including both the industrialized and the developing countries.

Integrated monitoring can provide a useful tool for assessment of acid rain and its effects on land and associated ecosystems, regionally and globally. Indeed, as will be seen in the next chapter, based on the Swedish national monitoring programme (PMK), the notion of integrated monitoring has come to play an important role in the regional effort led by the Economic Commission for Europe (ECE) to understand and cope with the effects of transboundary air pollution. Similarly, integrated monitoring could help forge a link between ecosystems monitoring and climate-related and ocean monitoring by focusing on the cycling of gases and particulate matter through various compartments of the environment. It could link up with the GEMS HEALTH monitoring network and use its data for study of pollution impacts on non-human targets. Lastly, it could draw attention and yield some evidence concerning certain phenomena that were not given much attention in the early phases of GEMS, notably soil pollution and its effects on the recycling of nutrients in the food chain.

Although monitoring-related research is likely to continue as the principal activity, and although participation is likely to be limited to a few countries with

out by different countries. See 'Progress Report on MAB Project 14' (doc. MAB/ICC–6/19), July 1979. For some earlier proposals on the use of biosphere reserves for monitoring, see G. T. Goodman, 'Outline proposal for a collaborative (MAB 14/WMO) project on pollutant dynamics and biotic response in ecosystems', reproduced as an annex to J. F. Farrar and J. R. Thompson, 'Monitoring Requirements for Biological Systems', Mainz Report, in 'Selected Works on Integrated Monitoring', *op. cit.* Also see G. B. Wiersma and K. W. Brown, 'Recommended Pollutant Monitoring System for Biosphere Reserves', submitted to the Second Conference on Scientific Research in the National Parks, San Francisco, Calif., Nov. 1979, reproduced in 'Selected Works on Integrated Monitoring', *op. cit.*; and G. B. Wiersma, 'An analysis of the global biosphere reserve system for use in integrated monitoring', Sept. 1981, a report prepared for GEMS.

40 For the siting criteria for monitoring stations, see 'Procedures and Methods for Integrated Global Background Monitoring of Environmental Pollution', *op. cit.*, pp. 24–6.

172 *The quest for world environmental cooperation*

the required experience and capabilities, an essential tool is in the making, with a marked potential for management applications to current and emerging problems. By reason of the subject matter, the systemic properties of ecosystems and the nature of their response to and interaction with human activities, it has been possible to overcome the many practical and institutional obstacles in the way and to get GEMS involved in this important aspect of global environmental monitoring.

THE PROSPECTS FOR THE MONITORING OF NATURAL RESOURCES AND INTEGRATED MONITORING

The progress in this GEMS cluster has been relatively slow and uneven, limited as it is to pilot projects and to devising and testing monitoring methodologies. The many contributing reasons for this have been identified above. They are the usual ones: the financial and infrastructural handicaps of developing countries and the need to work out simple, inexpensive, and easily adaptable methodologies that most of them can use; the inadequacy of international support to secure their greater involvement; the sectoralization of the environmental issues according to the mandates of the specialized agencies in the UN system and the difficulties of achieving interagency cooperation; the complexity of monitoring, magnified by the need to launch and sustain global programmes; the rudimentary state-of-the-art of monitoring techniques and technologies, including remote sensing from satellite; the problems of availability, access to, and costs of remote sensing data; the limited resources and capabilities available within UNEP to follow up and promote the various objectives of GEMS; the need to bridge the gaps between scientific disciplines; North–South and East–West tensions and controversies; and so on.

The spotty record notwithstanding, the conceptual, methodological, operational, and especially technological advances that have been made over the last two decades, parallel to GEMS and within it, offer a foundation for a systematic effort towards fulfilling the stated objectives of GEMS in the period to come. The problems, needs, and gaps are better understood. The monitoring and assessment tools to support national and international decision-making and management are being developed and, in some instances, applied, with significant implications for the development process and for relations among countries.

Natural resources monitoring has largely remained at the level of pilot methodologies that could not be applied widely at the country level owing to lack of resources, interest, institutions, and so on. Once more, however, technological advances and the systems character of the global and natural environment came to the rescue. Through GRID, a capability is emerging that both pulls together diverse types of data on the global scale, and also places them at the disposal of countries in packages tailored to their specific needs. Remote sensing and advanced data management tools offer the capability to take a global view of the Earth, of the albedo of its surface, of the status of its forests, rangelands, agricultural lands, and deserts, and in general of the functioning of terrestrial ecosystems. This macro-view of natural resources and their status is of critical importance in understanding and assessing changes in the global environment, including the climate and the atmosphere, and the role and impacts of society in such changes. In sum, monitoring natural resources and ecosystems emerges as one of the principal tools for management and decision-making, as well as for such global, long-term studies as IGBP and the efforts to observe, study, and understand the Earth as a system. Indeed, its interdisciplinary integrated observatories network – Geosphere–Biosphere Observatories or GBO network – already referred to in chapter four, incorporates many of the ideas inherent to the concept of integrated monitoring that has evolved in the fold of GEMS.[41]

Integrated monitoring, in turn, has managed to secure a niche for itself in GEMS and has contributed to launching an integrated project, with several UN agencies taking part. An operational method for its application in the field has been devised. It is intended to observe processes and terrestrial ecosystem dynamics at the micro-level, with technological advances promising to help resolve many of the existing operational difficulties, and to supplement *in situ* observations with the work in the controlled environment of the laboratory.[42] It represents a twin track to the macro-view of Earth from space and the global monitoring of natural resources. As such, it has an important role in IGBP, in particular as it concerns the understanding of the interactive nature of biota and the atmosphere. Among other things, it is essential for the study and prediction of the changes in the atmospheric chemistry, Earth's radiation budget, and climate on account of the biological processes that regulate the concentration and flux of biogenic trace gases in the atmosphere – a process increasingly influenced by human activities.

Substantively and institutionally, advances have been made through attempts at integrating monitoring objectives and activities, including both the pollution and non-pollution impacts of humans on the environment, relating global trends and processes to the micro-level, and broadening the relatively static concern about the levels of pollutants in the environment to include their pathways, vehicles and dynamics as well. All this has come about thanks to the very nature of ecosystems and the biosphere, which have been imposing an integrated framework on their human observers.

Given the circumstances, probably the only way to arrive at this stage was the roundabout and slow route that has been taken, moving from the simple to the more complex. An example of this is provided by tropical forest monitoring, where the discrete project approach used initially has evolved into a more programmatic approach, which subsumes the central information/data system, the alarm system

---

41 It proposes global observations of land-surface properties and their changes, being important for the study of terrestrial biosphere–atmospheric chemistry interactions, biospheric aspects of the hydrological cycle, and the effects of climate change on terrestrial ecosystems, all of which are affected by changing human uses. 'The International Geosphere–Biosphere Programme: a Study of Global Change IGBP, a Plan for Action', *Global Change,* report No. 4 1988, section 6.4 on geosphere–biosphere observatories, pp.103–8.

42 The so-called 'field exposure chambers' or 'ecostats' have been designed to mimic natural air pollution conditions and to carry out experiments and studies of the effects of pollution on plants and their responses. Based on advanced computer technology and sensors, these chambers are helping to overcome the logistical and cost difficulties associated with direct observations in the field, can recreate realistic exposure, and can produce ordered data. See K. Randolph, 'Using computers to isolate pollution causes', *EPA Journal,* Oct. 1985.

to draw attention to major events in tropical forests, and an integrating mechanism to deal with 'scientific, methodological and international cooperation aspects of landcover monitoring'.[43] Making further progress will also be difficult and slow, because the political, conceptual, institutional and scientific constraints have not been overcome, and because the necessary resources and effort remain to be mobilized. Yet the first steps have been taken by linking its various clusters and objectives and, one could even say, bringing it back to some of the ideas that originally gave rise to GEMS, but that could not be pursued and implemented effectively in the initial stages. Accordingly, on the eve of the twentieth anniversary of UNCHE, natural resources cum integrated monitoring is also at a point where a review and evaluation of its functions, its applications, its links with other clusters of GEMS and with other global national and international programmes such as EOS and IGBP, and its place in the future development of GEMS is called for.

43 GEMS Report Series No.7, *op. cit.*, pp. 6–7.

7 Monitoring long-range transboundary air pollution in GEMS

The precipitation of acid rain and the acidification of lakes in Scandinavia were among the reasons that inspired Sweden's proposal to convene a UN Conference on the Human Environment (UNCHE). Soon after, acid rain became one of the prominent concerns on the environmental agenda, giving rise to polemics and conflicts within and among developed countries, and to intense scientific controversy.

The effects of air-borne pollutants, and more specifically of acid rain, on soil, forests, plants and agricultural crops, lakes, fish, and materials in built structures, gave policy relevance and public visibility to the issue. For example, the massive damage to and death of forests (*Waldsterben* in German) in some parts of Europe, which was attributed to the effects of acid rain and air pollution, has become an emotional, top-priority domestic and regional policy issue. In fact, forest damage caused so great a public outcry that the governments of some important developed countries had to revise their earlier indifferent attitude.

The possible effects of acid rain and of acids in the air on human health have not attracted the public's attention so far.[1] The reason is primarily the lack of research and clinical and epidemiological studies of such health effects, especially long-term chronic changes in the lungs of children and the impairment of the body's defence mechanisms, and the resulting absence of conclusive evidence on the subject. It is certain, however, that in view of the nature of suspected health effects of acid rain and of long-range air pollution, the controversy will intensify as more evidence is accumulated and as public pressure and concern mount. Acid rain has also been referred to as 'toxic rain', to denote the fact that a variety of polluting

---

1 Some non-governmental organizations, such as Greenpeace, have been agitating for the study of and research into health effects of acid deposition and long-range air pollution. Among the concerns are the health effects of inhaling acidic substances in the air, and of the acidification of untreated drinking water supplies. A decrease in the pH of water leads to an increase in the solubility of toxic metals, primarily lead, cadmium, mercury, aluminium, arsenic, and zinc. These are leached from soil; sediments; plumbing materials, including pipes made of plastic and asbestos cement; water tanks, etc. One of the resulting problems is the increasing mobilization of aluminium in its bioavailable forms and its concentration in drinking water. Aluminium has been linked to pre-senile dementias, in particular Alzheimer's disease. See 'Acid Rain's Effects on People Assessed', *Science*, Vol. 226, 21 Dec. 1984, pp. 1408–10. On health effects of acid groundwater, see 'Transboundary Air Pollution: Effects and Control', ECE Air Pollution Studies, No.3 (United Nations, New York, 1986), pp.71–2.

substances, in addition to acids, are transported over long distances by air masses and washed out from the atmosphere.

Acid rain refers to the deposition via precipitation mainly of sulphur oxides and nitrogen oxides, both of which are released into the atmosphere on a massive scale by fossil fuel combustion. It is one aspect of the phenomenon of the release, transport, and deposition of air pollutants.[2] Weather patterns, the circulation of air masses, wind trajectories, and zones of high precipitation introduce a pattern and a degree of regularity into both the transport and the wet and dry deposition of pollutants. This means that, depending on their geographic location, configuration, and relation to major sources of air pollution, countries are likely to absorb a greater or lesser amount of air-borne pollutants from both foreign and indigenous sources. Not surprisingly, the threshold of tolerance for damage inflicted by foreign pollution tends to be low; hence the political sensitivity of the issue at the international level. It is even less surprising, then, that in many instances concern about foreign sources of acid rain has influenced policy in the matter of pollution in general.

The basis for international action against air-borne pollution is provided by national boundaries. Air pollution travelling across borders (and, by definition, national policies related to pollutant emissions and even industry siting) becomes an issue in inter-state relations, subject to negotiation. 'Long-range transboundary air pollution' has thus been used to refer to pollution the physical origin of which is situated within one state and which has adverse effects on the territory of another state, at such a distance that it is generally impossible to distinguish the share accounted for in the pollution by individual emission sources or groups of sources.

What is involved is the deliberate or unintentional externalization of environmental costs; tall stacks on coal-fired power plants alleviate local pollution by transporting it to places hundreds or even thousands of miles away. Acid rain has been most intense in northern and central Europe and in eastern Canada and the United States, where over 90 per cent of the sulphur deposited originates in human activities. In Europe the controversy has been accentuated by the fact that many countries are crowded together in a region of great industrial activity and high population density. In North America, where the controversy has been just as intense, the issue of acid rain has pitted Canada against the United States and, within the US, primarily the Northeast against the Midwest.[3]

---

2   In the atmosphere, sulphur oxides and nitrogen oxides are in part transformed into sulphuric and nitric acids, among other things. They all return to ground level, either by being washed away by rain (wet deposition) or by directly adhering to various surfaces on the ground (dry deposition). Nitrogen oxides, which are mainly the product of fossil fuel combustion in electricity generating plants and motor vehicle engines, are also involved in complex photochemical processes that yield ozone and other photochemical oxidants that are especially harmful to vegetation.

3   In the case of the US and Canada, part of the common border lies mainly downwind from one of the most concentrated and diversified regions of industrial activity in the US, and for that matter in the world. Canada has felt victimized by the meteorology and geopolitics of its situation and frustrated by the ability of its big neighbour largely to ignore its arguments and demands for corrective measures. The situation does not appear to be greatly different within the US itself; the midwestern states, whose coal burning gives rise to air pollution that is considered as a major source of acidification in the eastern states, balk at the high costs that the required anti-pollution measures would entail and refuse to accept responsibility for defraying the cost alone.

MONITORING LONG-RANGE TRANSBOUNDARY AIR POLLUTION IN EUROPE

International management and regulatory actions regarding transfrontier air pollution have to be based on environmental assessment, which requires continuous monitoring of the sources and amounts of pollutants emitted, their trajectories, and deposition across borders.

When the issue was presented at the Stockholm Conference by Sweden in its national case study entitled 'Air Pollution across National Boundaries: the Impact of Sulphur in Air and Precipitation', scientific study of the phenomenon was well advanced and the concept of acid rain was one century old.[4] While the work had been done primarily at the national and sub-regional levels in Scandinavia, the matter had already spilled over into the broader international arena prior to the Conference. In the Organization for Economic Cooperation and Development (OECD), some initial pilot work had been carried out in response to the initiative of the Scandinavian countries. Their action was spearheaded by Norway, one of the principal victims of acid rain, with only 10 per cent of sulphur deposition in its territory originating from local sources.

Although UNCHE did not speak of acid rain *per se*, the matter was amply covered both in its Declaration and in the Action Plan that it adopted. This provided a policy base and gave programmatic impetus to further work on the subject. For example, principle 21 of the Declaration says that states have 'the responsibility to ensure that activities within their jurisdiction or control do not cause damage to the environment of other States or areas beyond the limits of national jurisdiction'. Recommendation No. 72 calls upon governments to coordinate with each other in 'planning and carrying out control programmes for pollutants distributed beyond

---

4 See B. Bolin, L. Granat, L. Ingelstam, M. Johannesson, E. Mattson, S. Oden, N. Rodhe and C. O. Tamin, *Air Pollution across National Boundaries, the Impact on the Environment of Sulphur in Air and Precipitation*, Sweden's Case Study for the United Nations Conference on the Human Environment sponsored by the Royal Ministry for Foreign Affairs, Norstedt and Sonner, Stockholm, 1971. The present text draws on an overview of the history of acid rain monitoring and research presented in R. F. Pueschel, 'Man and the Composition of the Atmosphere, the Background Air Pollution Monitoring Network (BAPMoN)', WMO, 1986, pp. 24–30. The term 'acid rain' was used first in 1872 by R. A. Smith in his book on chemical climatology. Smith was an English chemist who detected and analysed the phenomenon in Manchester and its surroundings as early as 1852, eventually expanding his studies to other parts of England, to Scotland, and to Germany. However, the continental scale of acid rain was not recognized for another century. In the 1940s, Swedish agricultural scientists began to study fertilization of crops by nutrients carried via the atmosphere, basing their work on the hypothesis that atmospheric transport is a major mechanism for the dispersion and transformation of chemical substances. This led to the first monitoring network for the collection and chemical analysis of precipitation. It gradually expanded to include the rest of Scandinavia, spread to West and Central Europe, and further east during the 1957–8 International Geophysical Year, making up what was called the European Chemistry Network. A further step was taken in the early 1960s, when another Scandinavian monitoring network was established to measure surface water chemistry. With the data of the two monitoring networks combined, knowledge was unified, trends and relationships became clearer, and the notion of acid precipitation and its effects finally entered the public and policy domain in the late 1960s. The bridging between scientific awareness of the problem and public and policy recognition took place when a Swedish soil scientist, S. Oden, wrote in a Stockholm newspaper about a 'chemical war' between the European states. He highlighted the regional scale of acid precipitation, and noted the existence of well-defined source and sink regions.

the national jurisdiction from which they are released'. Recommendation No. 79 speaks of the need to monitor 'properties and constituents of the atmosphere on a regional basis and especially changes in the distribution and concentration of contaminants', while recommendation No. 80 calls for research on terrestrial ecology 'to provide adequate knowledge of the inputs, movements, residence times and ecological effects of pollutants identified as critical'. Also, recommendation No. 85 speaks of the need to establish 'consultation mechanisms for speedy implementation of concerted abatement programmes with particular emphasis on regional activities'.

Following the Stockholm Conference, work in OECD was expanded and completed by 1977. Emissions of sulphur dioxide were surveyed, pH and sulphates were monitored in precipitation, and air trajectories of pollutants were studied, which demonstrated their long-range transport and transformation. The basic conclusion was that air and rain quality in every European country were measurably affected by pollutant emissions from other European countries.[5] Concurrently with the OECD work, individual countries continued to study the phenomenon. For example, Norway undertook large-scale national research to investigate the effects of acid rain on forests and fish.

Further study and action, however, called for broadening the geographical scope and inclusion of countries in Eastern Europe. Obviously, OECD was not an appropriate forum for this. The focus of action thus moved to the UN Economic Commission for Europe (ECE), where the problem of transfrontier air pollution had already been placed on the agenda in 1974 on the initiative of Norway.

The real boost for region-wide efforts, however, came from the 1975 Helsinki Conference on Security and Cooperation in Europe, which marked the high point of *détente* and the political thaw in East–West relations. In its chapter on the environment, the Conference called for cooperation between European countries in order to control air pollution and its adverse effects. Starting from principle 21 of the UNCHE Declaration, it recommended the development of an extensive programme for the monitoring and evaluation of the long-range transport of air pollutants, starting with sulphur dioxide and with the possibility of extension to other pollutants.

Agreement having been reached on policy and specific recommendations having been adopted at the Helsinki Conference (the so-called environmental basket), intensive work was mounted by ECE. It established a task force to develop a cooperative programme for the monitoring and evaluation of the long-range transport of pollutants, with Norway as a lead country. The Governing Council of the UN Environment Programme (UNEP) also asked the Executive Director to cooperate in this regional endeavour, which, it said, should be regarded as a part of the Global Environment Monitoring System (GEMS) and an important contribution to the attainment of its goals. This constituted a nod to UNEP to become involved and to provide catalytic financial support for the initial monitoring and

---

5 The findings were published in 'The OECD Programme on Long-Range Transport of Air Pollution', OECD, Paris, 1977.

evaluation of the long-range transmission of air pollutants in Europe.[6] The invitation was needed because UNEP did not show great enthusiasm for funding this project of a group of economically advanced countries that could have easily afforded the cost.

The objective of the preparatory phase of the cooperative programme was to initiate work on determining the sources and amounts of sulphur dioxide emitted, as well as fluxes across borders, and to provide a database that would eventually lead to collective action to reduce emissions causing transfrontier air pollution and acid rain. It was undertaken as a GEMS project, financed by UNEP and executed by ECE, with the World Meteorological Organization (WMO) responsible for the meteorological aspects.[7]

The project helped to establish the technical basis and infrastructure for transfrontier air pollution monitoring and evaluation in Europe. It led to the formation of a monitoring network (70 stations in 20 countries, to start with). Some of the stations were the same as those used in the OECD pilot exercise; others belonged to the existing BAPMoN network, which was already geared to monitoring variables in the atmospheric chemistry.

Standardized reporting procedures were adopted and comparable statistical analyses, interlaboratory tests, and the quality control of data were carried out. The Chemical Coordinating Centre (CCC) was established at the Norwegian Institute for Air Research to coordinate chemical measurements and analysis within the programme. The Meteorological Synthesizing Centre–West (MSC–W) at the Norwegian Meteorological Institute and the Meteorological Synthesizing Centre–East (MSC–E) at the Moscow Institute of Applied Geophysics were selected by WMO and designated to analyse air transport trajectories and meteorological data. Their task was to describe transformation and removal processes, provide country-by-country sulphur budgets, estimate the quantity of sulphur pollution that crosses national boundaries and is deposited (including concentration and deposition fields), and develop the necessary mathematical models.

The monitoring data confirmed the long-range transport and helped verify the models, while the results of scientific studies in Norway confirmed the effects of acid rain. This, combined with the change in the position of some countries, most notably the Federal Republic of Germany, which until then had refused to recognize the problem and the need for collective action, provided the necessary technical and political back-up and thrust for a legal instrument among the European states to deal with long-range transboundary air pollution. The instrument, agreed upon only after complex and lengthy negotiations, took the form of the Convention on Long-Range Transboundary Air Pollution and was adopted in 1979 at a high-level meeting within the framework of ECE. The Convention makes provision for the

---

6 For ECE task force recommendations, see doc. ECE/ENV/15, annex II. For UNEP action, see Governing Council recommendation 64(IV) in UN doc. A/31/25.

7 Entitled 'ECE/UNEP/WMO Cooperative Programme for Monitoring and Evaluation of the Long-range Transmission of Air Pollutants in Europe', the project started in October 1977. It consisted of a chemical and a meteorological part. For details see 'UNEP Report to Governments,' Nos 11, 17 and 23.

establishment of an Executive Body, responsible for supervising compliance with the provisions of the Convention and for coordination of its implementation. The monitoring and evaluation programme was incorporated into the Convention as the Cooperative Programme for Monitoring and Evaluation of the Long-range Transmission of Air Pollutants in Europe, or EMEP (i.e., Evaluation and Monitoring of Environmental Pollution).[8]

After the adoption of the Convention, and after the governments had agreed to carry out the obligations arising from it to the maximum extent possible pending its ratification and entry into force, the second measurement phase was initiated. The objective was to consolidate what had already been done, to continue the work that was programmed, and to prepare the monitoring and assessment network for full-scale operation. More specifically, the aim was to lead to the improvement and expansion of the standardized procedures for sampling and analysis, better and expanded geographical coverage, and the addition of other pollutants to those already being monitored. Provision was also made for a link-up with the research and monitoring effort on acid rain being carried out bilaterally by the US and Canada.[9]

During this period further advances were made with a view to making the infrastructure fully operational; obtaining monitoring data and calculating country-by-country budgets for emissions, depositions, and transboundary fluxes; building and testing models concerning the long-range transport of pollutants; reporting and exchanging data and information; increasing the number of monitoring stations and participating countries; etc. By 1983, when the required number of signatory countries had ratified the Convention so that it could enter into force, EMEP was an established and tested regional monitoring and assessment programme. It was entering its third, more ambitious phase, ready to provide vital support to governments in the task of implementing the Convention.

The third phase of EMEP (1984–6) led to further expansion of the network (90 stations in 22 countries by 1986) and progress along the lines initiated earlier, i.e., harmonization of measurements, quality assurance, growing numbers of substances measured, etc.[10]

The fourth phase (1987–9) calls for the continuing extension of the network, for

---

8 See doc. ECE/HLM.1/2 for the text of the Convention and the report of the High-Level Meeting within the Framework of ECE on the Protection of the Environment, which adopted the Convention.

9 See the Draft Plan of Action for the Follow-Up of the High-Level Meeting, ECE doc. ENV/R.117, and the progress report, doc. ENV/IEB/R.12. Plans were made to start monitoring nitrogen compounds, since it has been estimated that 20–30 per cent of rain acidity in Europe is derived from nitrates. Preparations were also made for subsequent phases to include the monitoring of heavy metals, ozone, oxidizing agents, and chlorinated hydrocarbons. For the specific proposals on cooperation submitted in 1983 by Canada and the US, see docs EB. AIR/GE.1/R.5 and EB. AIR/GE.1/2, section IX. Their purpose was to promote the exchange of information and cooperation on joint projects in order to achieve a rapid scientific advance in this field (e.g., data quality assurance, intercalibration exercises, intercomparison of models, intercomparison of instruments).

10 For a review of the third measurement phase, see doc. EB. AIR/GE.1/R.30. Also see 'Meeting on the Assessment of the Meteorological Aspects of the Third Phase of EMEP', IIASA, Laxenburg, Austria, 30 Mar. – 2 Apr. 1987, WMO, Environmental Pollution Monitoring and Research Programme (EPMRP) Report, No. 48.

efforts to install as widely as possible the best monitoring equipment available, for improved quality assurance, for strengthening the three international centres, and in particular for expanding the list of pollutants monitored (i.e., nitrogen compounds and ground-level ozone). All this was a reflection of the greater ambitions and capabilities of an expanding programme and, even more, the recognition of the importance of chemical species other than sulphur compounds in long-range air transport of pollution.[11]

As it approached its tenth anniversary in the fold of ECE, EMEP was generating the database essential for scientific study and assessment of long-range transport of air pollution (the European frontiers happen to be incidental to the natural phenomenon) and its deposition in different forms. This knowledge, in turn, is essential for devising, agreeing on and mounting regional and national strategies and action to reduce and control transfrontier air pollution within the ECE region, and transport and deposition of air pollution in general.

The 1982 Conference on Acidification of the Environment, held in Stockholm, and the entry of the Convention into force in 1983 marked the beginning of a new phase in the evolution of this international programme. Fortified by the adoption of the legal instrument, Finland, Norway, and Sweden were quick to propose the first quantitative target: taking 1980 as the base year for calculations, the target provides that, by 1993-5, each signatory country should have reduced by 30 per cent the total annual discharge of sulphur pollutants into the atmosphere. The aim was to fill an important gap in the Convention, which did not contain concrete goals for reduction of air pollutant emissions. A good deal of opposition was expressed to the idea of targets, and agreement could not be reached. However, some Contracting Parties indicated that anyway they would initiate measures to achieve the 30 per cent target on a voluntary basis.

The process was thus set in motion. The so-called '30 per cent Club' grew as other countries joined. After two important preparatory conferences held in Ottawa and Munich in 1984, the Executive Body of the Convention at its third session, in Helsinki in 1985, adopted the 'Protocol to the 1979 Convention on the Reduction of Sulphur Emissions or their Transboundary Fluxes by at least 30 per cent'. After the ratification by the required minimum of sixteen parties, the Protocol entered into force in September 1987, providing a useful yardstick and reference for regulatory action.[12] The operational paragraph of the Protocol obliges the Parties 'to reduce their national sulphur emissions or their transboundary fluxes by at least 30 per cent as soon as possible and at the latest by 1993, using 1980 levels as the basis for calculations of reductions'. By the time the Protocol entered into force, ten parties had already reached the target; another eleven parties expressed the intention to reduce sulphur emissions by 50 per cent by the year 1995. Another four said they would reach a 65 per cent target (Austria, Federal Republic of Germany,

---

11 For the programme of the fourth phase of EMEP, see annex III of doc. EB. AIR/GE.1/8.

12 For the Protocol, see doc. ECE/EB. AIR/12. Each party is to provide information on its levels of national annual sulphur emissions and how these were calculated. EMEP should provide calculations of transboundary fluxes and deposition of sulphur compounds utilizing appropriate models.

Sweden, and Switzerland). It was also recognized that while some Contracting Parties would not sign the Protocol they would nevertheless contribute to the reduction of transboundary air pollution and continue to make efforts to control sulphur emissions.

A similar pattern was also followed in the case of nitrogen compounds. As the sulphur protocol was entering into force, the Executive Body was already engaged in the process of negotiating a protocol on the abatement of emissions of nitrogen oxides.[13] A provision is made for its continuing and regular review, including negotiation of further measures to fulfil obligations of the signatories. Such reviews – the first having been proposed for 1991 – should take into account the relevant scientific information, including the concept of 'critical loads', which is to be given high priority in research and monitoring.[14]

In this evolving saga of control and abatement of transfrontier air pollution in Europe, EMEP has been of vital importance. Without the quantitative picture, scientific information, and models that it has provided, it would have been difficult to contemplate the elaboration of the Convention and especially of the two protocols, not to mention their implementation. Monitoring of countries' compliance with targets and obligations that they have assumed, or will assume under the protocols, is coming to be seen as a key ingredient of the package and an important function of EMEP. One monitors not just for the sake of measurement and scientific study, but also as part and parcel of an action and policy-oriented long-term programme.

Another move of intrinsic significance for the management of transfrontier air pollution involved the launching of four sectoral programmes for the monitoring and assessment of air pollution effects on forests, on freshwater ecosystems, on materials (including cultural and historical monuments), and on agricultural crops.

In order to assure linkages between these programmes and EMEP, and to help assess impacts on ecosystems, including positive ones resulting from actions taken to control emissions, the parties to the Convention also initiated work on integrated monitoring.[15] By itself the monitoring of transport and deposition of air pollutants

13 See 'Draft Protocol to the 1979 Convention on Long-Range Transboundary Air Pollution Concerning the Abatement of Emissions of Nitrogen Oxides', in docs EB. AIR/WG.3/10, 12 and 14. The draft contains alternative texts of articles proposed by different countries, in particular those dealing with the basic obligations of the signatories to control and reduce emissions of nitrogen oxides and of the need to promote exchange of technologies needed for reducing these emissions from major stationary and mobile sources.

14 Nitrogen and its compounds present a much more complex scientific challenge than sulphur and its compounds. Nitrogen is more widespread, it is involved in a broader range of chemical processes, and its sources are more varied and include motor vehicles and farming. Also, the nitrogen cycle is far more complex than that of sulphur. In view of scientific uncertainties, a great deal of effort is to be devoted to strengthening the scientific basis for reducing emissions of nitrogen oxides. It is planned to identify and quantify the effects of nitrogen oxides on people and their health, on plant and animal life, and on waters, soils, and materials, as well as their distribution in sensitive areas.

15 The Swedish/UN-ECE Workshop on Integrated Monitoring (WIM) was held in Stockholm in June 1987. For the original proposal to hold the workshop, see doc. EB. AIR/GE.1/R.25. For the conclusion and guidelines on integrated monitoring adopted by the workshop, see doc. EB. AIR/GE.1/R.31. The programme was tentatively divided into components: air and precipitation, terrestrial biota, soil and groundwater, and surface water. Since it is important to have full control

was incomplete. It represented only one segment of the cause–effect chain. An integrated monitoring and assessment approach was called for, involving different media and multiple targets, to cover the emission, transport, deposition, and effects of air pollutants. As usual in the ECE region, the work did not start from scratch. The initiative for integrated monitoring came from Sweden and other Nordic countries. Based on the Swedish national monitoring programme (PMK), which had been operated and tested over a period of time, they had a ready-made model to propose to other governments.

Traditionally, the focus has been on monitoring concentrations of chemical substances in air, water, and living organisms. The Swedish integrated approach has supplemented this with the monitoring of the terrestrial environment, namely soil, plants, and animals, with the aim of charting the biological effects on fauna and flora of environmental disturbances and pollution. This is done at reference sites situated in virgin tracts, usually in a catchment area. The objective of integrated monitoring is not just to find evidence of environmental disturbances and pollution, but also to obtain the data needed to know and understand their causes.[16]

The Swedish approach was used to prepare common guidelines for the Nordic countries.[17] This in turn provided the initiative for extending the approach throughout the ECE region, with Finland acting as the lead country of the task force for the Pilot Programme on Integrated Monitoring.[18] The notion of integrated monitoring thus made yet another important entry into GEMS, this time related to a specific region and a specific need, with an operational model to follow, and with significant practical experience to draw upon. Another important activity for global objectives of GEMS was being undertaken within the ECE region thanks to the initiative of the Nordic countries.

---

of the area chosen for integrated monitoring, nature and biosphere reserves were indicated as preferred monitoring sites.

16 For a description of the Swedish system see *Monitor 1985*, 'The National Swedish Environmental Monitoring Programme (PMK)', National Swedish Environmental Protection Board, Stockholm, 1985. Starting from studies and monitoring efforts dating back to 1973, and based on a decision of the Swedish Parliament, the Board was commissioned in 1978 by the Cabinet to initiate the build-up of PMK. In Sweden, municipalities, industries, and other actors disturbing and polluting the environment can be required by law to carry out local monitoring at their own expense, in order to determine pollutant effects on the immediate environment. However, pollutants travel to remote regions where it is difficult to single out the responsible agent and make him pay for the cost of local monitoring. Therefore, the responsibility to monitor these areas has to be assumed by the community. This was the origin of PMK. Its basic objectives are: (a) to monitor long-term and large-scale changes in the environment; (b) to collect data about environmental conditions in relatively unaffected areas as a way to assess more accurately the situation in more heavily affected areas; and (c) to determine how pollutants are transported in air, in water, and in the terrestrial environment, as well as between environmental media (*ibid.*, pp.7–10). PMK also took over the responsibility for a number of investigations, some of which had been going on for decades (e.g., chemical composition of precipitation since the mid-1950s and chemical analysis of groundwater since the early 1960s).

17 See *Guidelines for Integrated Monitoring in the Nordic Countries*, 1987, a monograph published by the Steering Body for Monitoring the Environmental Quality of the Nordic Countries (MKN).

18 See 'Second Workshop on Integrated Monitoring (WIM 2)', 5–8 Oct. 1988, Finland, Workshop Report and Annexes.

THE SIGNIFICANCE OF EMEP FOR GEMS

In more ways than one, EMEP typifies what GEMS was imagined to be like by those who conceived the idea of international environmental monitoring. Or, put differently, the story of monitoring transboundary air pollution in Europe illustrates many of the factors that go into the making of a relatively successful international environmental monitoring programme.

To begin with, there is a fairly clear-cut problem at stake, in a well-defined geographical setting, and with a limited number of countries, mostly with strong economies, taking part. The motivation to act is high on account of the nature of environmental impacts. Political and economic stakes are also high, and the countries that are most affected and public opinion there provide a continuing driving force for the undertaking. In fact, their interest and concern are such that they have successfully promoted the Convention and EMEP in spite of the reservations and even strong opposition of some important countries.

The case of transfrontier air pollution in Europe illustrates a long-term process of involving governments in a joint undertaking – some uninterested, others against their will – which leads gradually but ineluctably to a new quality and higher level of international cooperation. In this process, different interests of countries and the related conflicts are played out in the scientific arena, and can lead to a slowdown and loss of opportunities to act. For example, a decade went by while bona fide scientists argued as to whether or not acid rain had negative effects on the environment. These delays were unavoidable as long as scientific knowledge was not firmly established and could be challenged, in particular, by special interests. To reach the stage of greater scientific certainty, long-term and extensive monitoring and assessment were required.

The policy objective is clear, and the basic path to its attainment was charted from the very beginning. The implementation is a matter of time and of putting together the required pieces over a period of years, indeed decades. The whole project is in many ways a textbook illustration of a process of task expansion, emergence of collective action, progressive institutionalization, and functionalism, all of which are dear to the heart of students and advocates of international cooperation.

The very nature of the problem, the expanding scientific knowledge and substantiation made possible through research, monitoring, and assessment, and the increasing and diversifying impacts of air pollution all impel the programme, generate political pressure at the grass-roots, and raise consensus and the response to a higher level. Starting modestly and via pilot activities, often preceded by work in one or a group of Nordic countries, the ground is prepared for policy agreement. This, in turn, paves the way for institutionalization, task expansion, and a dynamic system of monitoring and assessment. This pattern was followed in the case of the Convention itself, and in the work aiming at its implementation, including the growing attention paid to nitrogen oxides and the initiation of integrated monitoring activities.

Monitoring is not a 'stand-alone' activity; rather, it is a vital part of a comprehensive scheme. While linked with research and assessment, it is also closely related to management and regulatory action. This is of special importance in the case of nitrogen compounds, for in their case it is harder than in that of sulphur compounds to trace the sources of emission, to determine the chemical changes they cause in the atmosphere, to check the transport mechanisms, and especially to assess and deal with the environmental impacts of the substances deposited.

The package is buttressed by a well-defined policy and institutional agreement between the governments involved. Indeed, it is the policy agreement, hammered out in the course of lengthy and complex negotiations as part of a much broader understanding between East and West, that represents the mainstay of the whole undertaking. On the strength of this agreement, and with scientific and technical foundations built over the years, it became possible to draft, adopt and ratify the Convention, which has served as the basis for progressive institutionalization and action and which places a legal obligation and responsibility upon countries to act in a prescribed manner. Periodic renewal and strengthening of the underlying political commitment plays an important role in providing continuing impetus for the process.[19]

On the whole, then, marked institutionalization characterizes this monitoring cluster. EMEP is thus endowed with its own intergovernmental mechanism, which functions within the broader institutional arrangements established within the ECE to guide and oversee the implementation of the Convention on Long-Range Transboundary Air Pollution.[20]

The whole enterprise is rooted in a well-defined institutional pattern; it proceeds incrementally from state-of-the-art studies, and via workshops the result arrives on the table of the Executive Body. National authorities are fully involved at all stages, and a country is entrusted to lead a task force. An example of the pattern occurred when, in 1985, the Executive Body for the Convention launched the Programme on Assessment and Monitoring of Effects of Air Pollution on Forests. It designated national focal centres in each country and two international programme coordinating centres (Bratislava in the East and Hamburg in the West). It established an intergovernmental task force to oversee the implementation of the programme, with the Federal Republic of Germany as the lead country. The manual on methodologies and criteria for sampling, monitoring, analysis, and assessment of effects of air pollution on forests was prepared by the two centres and approved by an

---

19 For example, the Vienna Follow-Up Meeting of the Conference on Security and Co-operation in Europe in its agreed conclusions in January 1989 called, among other things, for further reductions of sulphur emissions beyond the level established by the Protocol and for actions to reduce emissions of other pollutants. The Meeting also agreed to strengthen and further develop EMEP through extending and improving the system of monitoring stations, further developing comparable methods of measurement, and expanding coverage to include nitrogen oxides, hydrocarbons and photochemical oxidants.

20 The Steering Body to the Cooperative Programme for Monitoring and Evaluation of the Long-Range Transmission of Air Pollutants in Europe acts as the technical and policy body within the framework of ECE. It provides member governments with continuing and regular access to the programme, while also making it possible for EMEP to reach governments (including the more reluctant ones) on a regular basis. *Ad hoc* and specialized meetings are held as necessary.

intergovernmental working group. It was sponsored by UNEP, as a contribution to GEMS.[21]

The institutionalization includes the binding commitment to contribute financially to the international aspects of the monitoring programme. Thus, continuity is assured and a longer-term strategy and perspective of institutional and technical build-up can be adopted. True, the ECE governments were compelled to face and accept, some grudgingly, the requirement to set up regular means for financing the international aspects of EMEP only when UNEP announced its intention of ceasing to provide financial support.[22]

In view of the uncertainty and insufficiency of voluntary contributions, it was decided to elaborate a permanent financial arrangement in the form of a protocol to the Convention, which itself did not provide explicitly for financing international activities. An *ad hoc* meeting on financing of EMEP was thus held in 1983 for the purpose of drafting a formal instrument to include a cost-sharing system. Further negotiations within the Executive Body led to the adoption of a Protocol on Long-Term Financing of EMEP in 1984, which entered into force in January 1988.[23] The Protocol provides for mandatory contributions based on the UN assessment rates, and additional voluntary contributions by interested parties. UNEP funding was phased out in 1984, since when the programme has been fully financed by 27 parties, with a regular budget of approximately $1 million per year (of which 30 per cent is in non-convertible currencies). The annual budget is drawn up by the Steering Body for EMEP, and is adopted by the Executive Body. Most of the budget is spent in support of the work and activities of the three international centres. It should be kept in mind, however, that, as in the case of the GEMS model, the major and real costs are incurred by the participating governments in performing the required measurements and providing the monitoring data. Governments' payments of local costs, including those of operating local monitoring stations, are not taken into account and are considered as their counterpart contributions. It was

---

21 See 'Effects of Acidifying Depositions and Related Pollution on Forest Ecosystems', in *Air Pollution Across Boundaries*, Air Pollution Series, Vol.2, 1985, UN publication, Sales No. E.85. II. E.17. (Note that the Executive Body approves all the items to be published in the Air Pollution Series). A large-scale forest damage survey covering 16 countries in Europe was carried out on the basis of this manual. See 'Forest Damage and Air Pollution', Report of the 1986 forest damage survey in Europe, GEMS 1987. Also see P. Sand, 'Air Pollution in Europe – International Policy Responses', *Environment*, Vol. 29, No. 10, Dec. 1987. Similar to the precedent set in natural resources monitoring, split-budget-line financing for this pilot project was applied, thus tapping both the Earthwatch and ecosystems budget lines. UNEP contributed $207,000 to a three-year project, starting in 1984, mostly for the operation of the two coordinating centres.

22 During the start-up phase, EMEP was financed by UNEP and the voluntary contributions of governments, with the money essentially being used for the operation of the three centres (CCC, MSC–W and MSC–E). WMO contributed in kind an estimated $25,000 per year. As an example, in 1982, countries contributed $83,336 in convertible currencies (Denmark, $5,000; Finland, $17,491; the Netherlands, $20,000; Norway, $20,562; and Sweden, $20,283) and $227,000 in non-convertible currencies (USSR), while UNEP contributed $213,000 (of which $60,000 was in non-convertible currencies), for a grand total of $523,336 (doc. EB. AIR/GE.1/R.1).

23 For the initial proposal on the subject prepared by the ECE secretariat, see doc. EB. AIR/AC.1/R.1. For the Protocol itself, see doc. ECE/EB. AIR/11. The protocol does not provide for the financing of activities outside EMEP even though they may be within the framework of the Convention.

estimated, for example, that $9.8 million was expended by the governments to operate the monitoring network during the period 1977–83.

The approach is also sufficiently flexible in the sense that possibilities exist for continuing revision and expansion of activities and institutions to meet emerging needs and to respond to new knowledge as well as to strengthen the Convention and build into it control and regulatory powers. The successive protocols that were negotiated have proved to be a convenient device for achieving this objective.

There are many actors involved, each one playing a useful role in the evolution of the programme. While the governments and political leaders are the main actors in the enterprise, they do not work alone. Scientists, who help to gather the data and interpret their meaning, play a vital role. If motivated and prepared to link science with political action, they act as the spearhead for the programme. This has been the case with the scientists from the Nordic countries, for example. They work closely with their governments, using the regional bodies, primarily the Nordic Council of Ministers, to promote the shared objectives of controlling and reducing air pollution and its movement across frontiers.[24] The scientists have also provided governments with the data and arguments necessary for establishing and agreeing on quantitative targets. For example, on the basis of their work, it was possible to attain a consensus in 1982 on the threshold limit on sulphur deposition in sensitive surface waters, i.e., 0.5 grams of sulphur per square metre per year. A similar limit for nitrogen is under consideration. Such thresholds can be used to establish targets for reduction of emissions. Indeed, research, monitoring, and assessment in the framework of the Convention is necessary for arriving at policy agreement on measures to be undertaken, and linking quantitatively the agreed acceptable deposition and the emission levels.

The acid rain issue has also given rise to a good deal of popular reaction and participation, with the non-governmental groups and organizations playing an important role. The role of public opinion and participation is bound to intensify, especially once governmental action and achievement of agreed targets are supervised on a routine basis and the results of such monitoring are inserted into national and regional policy and the democratic process. This applies, in particular, to the countries of Eastern Europe where this supervisory activity is bound to get linked with domestic pollution issues and where in the past the absence of environmental information was one of the major grievances of the population.

The problem of transfrontier air pollution is by definition multifaceted, and requires appropriate study, assessment, and responses. Moreover, if one is concerned with the cause–effect chain and the need to identify and tackle the problem at its very roots and to understand the effects fully, a multidisciplinary, integrated approach is called for. It is broader than EMEP *per se*, includes media other than air and precipitation, and calls for coordinated monitoring and modelling of

---

24 The scientists play an activist role and lobby with the Nordic Council for financial support, and for diffusing region-wide given concepts (e.g., 'critical load'), models, and techniques. For a Nordic mix of science and advocacy, see the report of the Nordic Council of Ministers to the Nordic Council's International Conference on Transboundary Air Pollution, entitled *Europe's Air – Europe's Environment*, Stockholm, 1986.

emission, transport, deposition, and effects. This requirement gave rise to work on integrated monitoring, which was included in the workplan for the implementation of the Convention. The workplan provides, among other things, for research and exchange of information on the effects of sulphur compounds and other major pollutants on human health, agriculture, forestry, building materials, aquatic and other ecosystems, and visibility; for international collaboration and exchange of information on technologies for controlling emissions of sulphur compounds and nitrogen oxides; and for analysis of costs and benefits of sulphur emission controls.[25]

The inherent usefulness of the EMEP monitoring network and of the experience acquired became manifest at the time of the Chernobyl accident. The EMEP data and models of transboundary air pollution represented an international source of knowledge that the governments could look to for guidance in the confusion following the accident and the uncertainty as to the transport and deposition of radioactive materials. But some countries objected to the use of this store of knowledge, for fear of politicizing EMEP. While this attitude prevented EMEP's active involvement, its multipurpose potential became obvious, to be tapped at a politically more opportune moment.

In sum, then, the story of EMEP and the Convention has evolved largely according to a scenario that corresponded to the main assumptions of UNEP's programmatic approach. The high levels of development and the character of the ECE region made it possible for the model to function satisfactorily. For example, there was no need to provide support for national activities. As a consequence the pressure on the available international funding was eased, total requirements were reduced and greater catalytic effect was achieved with less money. Also, with the required infrastructure, technical and scientific capability and resources available in the participating countries, it was possible to carry out intercalibration and data quality assurance with comparative ease and efficiency. UNEP's direct role was rather marginal, although its basic inspiration was there all along and its financial support proved to be crucial in the take-off stages.[26]

The *Level Three* script, in particular, evolved largely as UNEP would like to see it. For a relatively modest US $1.4 million allocated to EMEP and without a major effort on its part, significant progress and returns were achieved.[27]

The countries taking part in the programme are all developed, a circumstance which in itself makes it easier for the basic assumptions of the UNEP model to

---

25 The workplan (doc. ECE/EB. AIR/I, annex III) was adopted at the first session of the Executive Body held in 1983. It is updated annually.

26 Substantive and technical contribution from UNEP to EMEP was primarily the consequence of the fact that the Deputy Director of GEMS PAC had the required expertise and interest, and was posted at WMO in Geneva, where he could interact with the ECE secretariat.

27 According to the *UNEP Report to Governments* (Nos 11, 17, 23), over the period Oct. 1977 Dec. 1983 UNEP allocated approximately $1.39 million of the total $3.3 million expended internationally for activities related to the monitoring of transboundary air pollution. UNEP provided $750,467 of the $1.1 million for the chemical component of the system. It provided $330,000 of the total $815,000 for MSC–W, with Norway contributing the remaining $485,000. And it provided $309,500 of the total $1.36 million for MSC–E, with the USSR contributing $1.05 million.

function successfully. It relieves EMEP of many difficulties common to other GEMS clusters. This is not to say that EMEP has been completely free from the North–South syndrome, which plays such a central role in other GEMS activities. Indeed, some of the countries, mostly those on the southern flank of the ECE region, did not seem to have easily available resources for this activity, and they were not in a position or not interested enough to participate at the desired technical and scientific levels. This slowed down the work and had the least-common-denominator effect.[28] However, the lag was narrowed as time went by and their fuller participation was secured. The initial lack of interest on the part of these countries may have been caused by the excessive attention focused on acid rain effects seen essentially through the prism of the European North and its ecosystems. As the total effect of transfrontier pollution, including that on buildings and historical monuments, and the regional situation came to be better understood, these countries also became more interested and increased the level of their participation.

This recital of the positive aspects of EMEP may appear too complimentary to those involved in its everyday, often frustrating reality. A student of GEMS may get carried away in the assessment of EMEP in the light of what has transpired in other monitoring clusters. The listing, however, was not meant to comment on the effectiveness or indeed on the quality of EMEP's outputs and performance; rather, it was meant to highlight a pattern and the underlying factors and approaches that gave it an in-built advantage compared with other monitoring activities reviewed in this study. In fact, EMEP and the whole effort to reduce air pollution under the auspices of the Convention are burdened by many shortcomings and commonplace difficulties of implementation. The East–West rigidities of intergovernmental relations in ECE, common during the period under review, were manifest. It has not been possible to standardize the sampling equipment and analytical techniques used.[29] And, from the point of view of those directly affected by air pollution, the pace of progress has been painfully slow, notwithstanding the 10 per cent reduction in sulphur dioxide since the measures were initiated. Some of the countries that are the chief 'culprits' in the regional transfer of air pollution have resisted collective

---

28 The Chemical Component of EMEP consisted initially of the Minimum Measurement Activity (MMA) and an Extended Measurement Activity (EMA). There was a considerable difference in their scope. For example, with respect to gases MMA provided for the monitoring of sulphur dioxide only, while EMA also included nitrogen dioxide, nitric acid, ammonia, and ozone. In the case of precipitation, the difference in scope was even greater. After EMEP had been operating at MMA for six years, some countries were still not able to comply with its requirements. While some countries opposed suggestions that MMA be expanded, others felt that it was too restricted and could be substantially expanded with the help of new techniques (doc. EB. AIR/GE.1/2, p. 6). The secretariat's proposal for MMA in EMEP's third phase contained significant additions (doc. EB. AIR/GE.1/R.3, p.4). It was turned down by the Steering Committee, however, and MMA was left unchanged. The possibility was left open for including in MMA, on a voluntary basis, other variables such as the nitrate ion, conductivity in precipitation, and heavy metals (doc. EB. AIR/GE.1/2, annex II, p.2). However, the initial slowdown did not prevent steady and rapid expansion of the variables monitored in EMEP, to meet the needs and to reflect the advances being made. See the programme for the fourth phase of EMEP (1987–9), which includes monitoring of photochemical pollutants, for example (doc. EB. AIR/GE.1/8, annex III).
29 H. Dovland, 'Monitoring European Transboundary Air Pollution,' *Environment*, Vol.29, No.10, Dec. 1987.

action, for example, Poland, UK, and US in the case of ratification of the sulphur dioxide protocol. Furthermore, bureaucratic tensions and irrationalities are as common as anywhere else. For example, potential conflict is brewing over finance between those concerned with air pollution monitoring and the newcomers just starting to monitor the effects of air pollution. The former fear that new types of monitoring might impinge on the limited financing available for EMEP.

All these shortcomings, however, are of marginal importance. A mix of political factors and the symbolism of *détente* of the 1970s between the East and the West embodied in the Convention, the air pollution effects in the ECE region and the popular reaction, the scientific explanation and demonstration of the nature and extent of these effects, and the institutional back-up have all contributed to the attainment of the regulatory stage of the control of transboundary air pollution. This was no mean achievement on the global scale as well, considering that this region accounts for two-thirds of the total measurable air pollution load emitted by human activities into the atmosphere.

The prospects for EMEP, and generally for monitoring and assessment within the framework of the Convention, are favourable. The foundations established and the model approach tested over a period of fifteen years provide the countries of the ECE region with a solid instrument and capability essential to confront the shared problem of air pollution. This problem is becoming more complex and serious as the years go by. To deal with it, ever greater and diversified uses of EMEP and other monitoring capabilities within the Convention are called for. This will require further and continuing work to improve the data, methods, and institutions used. It will also mean broadening its scope, away from exclusive concern with acidifying compounds to include other pollutants affecting the region, most prominently ozone and other photochemical pollutants.

EMEP has proved its usefulness to countries and governments. It has generated data, models, and knowledge showing how sulphur compounds are transported and deposited across Europe. In due course, it will generate similar data and models for nitrogen compounds and hydrocarbons, making it possible to fill gaps in knowledge and to devise, adopt, and implement region-wide control strategies. It has shown the adaptibility and flexibility needed to meet new requirements, as exemplified by the manner in which the critical issues of nitrogen oxides are being broached, opening up a whole new area of activity and collective action. In this context, it could assist other initiatives within the region, including those within the European Economic Community (EEC), to reduce air pollution from mobile sources, mainly motor vehicles.

The quality of EMEP outputs has steadily improved, as have the quality and scope of data that it is obtaining from participating countries. Notably, there has been an improvement both in the emission data supplied by countries of Eastern Europe and in the number and distribution of monitoring stations on their territory.[30] The participation of these countries is bound to become more active and

30 The countries of Eastern Europe have supplied only the figures for total annual emissions, on the basis of which spatial distribution of pollutants emitted was estimated. The monitoring stations in the USSR have been located almost exclusively along its western borders. In the case of nitrogen

open under the impacts of political changes on their environmental policies and traditional secretiveness surrounding the state of their environment.

As noted above, any closer association of EMEP with monitoring of radionuclides has been ruled out in order not to politicize excessively the implementation of the Convention. The 'piggy-back' capability is inherent to EMEP, however, and it would be technically possible and rational to utilize it also for purposes of monitoring accidental releases of radioactivity into the environment, especially during the acute stage of possible future accidents. Norway, in fact, has used EMEP sites and experience for national monitoring of radioactivity and analysis of precipitation.

EMEP has obvious hemispheric and global implications. Today, acid rain is no longer a local or regional phenomenon only. The model and the experience are obviously usable in other parts of the world where long-range transport of air pollution is becoming serious. The close operational linkages and technical cooperation between EMEP and the bilateral US–Canada activities, including interlaboratory comparison and quality assurance, imply the emergence of an intercontinental network. Indeed, models developed in EMEP have been used to estimate trans-Atlantic sulphur transport. The BAPMoN data and the assessment of mechanisms of transport of pollution have demonstrated the global scope of the problem.[31] There have been requests from the Asian region for methods of acidification measurement. Also, the EMEP grid extends over the Mediterranean, the North Sea, and the Baltic, linking it with monitoring of marine and air-borne pollution deposited in these regional seas. On the whole, then, one could envisage a modular approach, similar to the Regional Seas Programme, applied to the problem of air pollution and its long-range transport and distribution on the hemispheric scale.

Finally, the question arises as to the place and role of the ECE transboundary air pollution monitoring cluster in GEMS. Quite obviously, lessons of general value can be learnt, as indicated in the preceding paragraphs. The model approach has also had some impact on the issue of transfrontier water pollution in Europe, one of the more stubborn items on the environmental agenda of ECE, concerning which little progress has been made over the decades. Thus, consideration has been given to initiating the monitoring of transfrontier water pollution.[32]

---

compounds, these countries have shown greater willingness to transmit more detailed data, possibly because in the case of these pollutants their contribution to transboundary pollution was comparatively smaller.

31 See C. C. Wallen, 'Sulphur and Nitrogen in Precipitation: an Attempt to Use BAPMoN and other Data to Show Regional and Global Distribution', EPMRP series, No.26, 1986. Note the existence of undecided background deposition of sulphur given in EMEP figures, ranging from 3 per cent to 50 per cent, and as high as 89 per cent in the case of Iceland. This sulphur is found in air masses that have not travelled over sources of emissions recently. This 'old' sulphur can originate from natural sources, from other continents, and also from within the ECE region. Dovland, *op. cit.*

32 The political controversy surrounding the issue of transfrontier river pollution in Europe was so great – one has only to think of the Danube and Rhine – that UNEP, in spite of interest, sidestepped the matter in the post-Stockholm period. Within GEMS the issue of river quality and river discharge into the ocean was left to a more favourable moment. The first important step in initiating this long process in ECE was taken when transfrontier water pollution was placed on its agenda under the

In general, the experience with EMEP and with the Convention can be harnessed to promote some of the objectives of the global system. Yet, for all practical purposes, the GEMS Programme Activity Centre (PAC) seems to have reduced contact with and interest in the evolving EMEP. Once the financing by UNEP stopped and the project ceased, the pretext and motivation for involvement seem largely to have disappeared. However, a link with the regional programme has been kept by giving some financial support to the pilot project on monitoring the effects of acidifying deposition on forest ecosystems. GEMS PAC played its well-rehearsed catalytic role by supporting financially a fledgling activity, which itself proceeds according to the established pattern. But, is this sufficient in view of the global importance of the processes in the EEC region?

In part, this detachment stems from the usual overload on GEMS PAC, its changing focus, and the lack of required in-house expertise and interest. To some extent, it is the result of the very uncertainty as to the role of UNEP, and what it should do with self-sustaining and self-sufficient activities such as EMEP. GRID, to be discussed in the next chapter, provides the natural link. Yet, the broader question remains of the direct involvement of UNEP and GEMS PAC in this important domain of international environmental cooperation. In this respect, a policy decision is called for. After all, in addition to its direct contribution to dealing with air pollution, EMEP represents an important factor in global environmental cooperation and in the development of GEMS. It offers a ready-made model, a tested methodology and accumulated knowledge that can teach a good deal and can serve as a reference for other regions. It places monitoring within the monitoring–assessment–management continuum, and it follows an integrated approach that, starting from the very nature of the phenomenon, goes beyond the single-medium orientation. Its special importance lies in the pronounced institutionalization that has turned out to be the key component in the whole undertaking and a prerequisite for attaining various scientific, practical and policy objectives.

umbrella of the Helsinki Conference Final Act. The process was given a further boost by the 1983 Madrid Conference on European Security and Cooperation. Concerning the initiation of a pilot project or projects on a monitoring system or systems in the ECE region, see 'Terms of Reference for a Programme on Monitoring and Evaluation of Transboundary Water Pollution' (doc. ECE/WATER/28, annex II). Also see the report of an *ad hoc* Meeting on Monitoring and Evaluation of Transfrontier Water Pollution (doc. ECE/WATER/AC.4/2).

8 GRID: Global Resource Information Database

The Global Resources Information Database, or GRID, is a computer-based, interactive georeferenced global database and data management mechanism. It represents an important addition and innovation in the framework of the Global Environment Monitoring System (GEMS), and is an attempt to respond to changing requirements and expectations.

A number of factors figured in GRID's creation. The people engaged in the daily operation of GEMS Programme Activity Centre (PAC) felt a need to activate, integrate, manage, use, and improve access to the growing environmental database in GEMS. There was also the challenge of how to standardize the data and how to establish relationships among the data accumulating in GEMS' five monitoring clusters. This inherently difficult undertaking was further complicated by such practical obstacles as the physical separation of databanks, by the differences in formats of data sets, and by the differences in computer languages used. Moreover, many of the data that were accumulating were for all practical purposes inaccessible to most potential users, aged rapidly, were static, and could not be manipulated and used easily without adequate staff and financial resources.

Increasingly, GEMS and the UN Environment Programme (UNEP) had come in for open criticism from governments for allowing monitoring data to accumulate without being used to produce environmental assessments. GRID was thus conceived as a mechanism to organize and tap the growing global monitoring data more effectively, to help bridge the gap between monitoring and assessment, and to transform the user-hostile raw data into more user-friendly data sets – in short, to initiate the process for making them useful for environmental and resource planning and management.

THE ORIGINS OF GRID

The origins of the GRID idea are mixed. In GEMS PAC, it had its roots in the field experience of some of its staff, first with the Kenya Rangelands Ecological Monitoring Unit (KREMU), which they had helped conceive, set up, and operate, and subsequently with the Senegal rangelands monitoring project of GEMS. They realized that environmental data sets, as well as most actions and decisions that affect such data sets, relate to particular areas of the Earth's surface, and that therefore geographical location is an effective and practical common denominator for organizing and applying environmental data.

But the key factors that made it possible for GEMS PAC staff to contemplate GRID as a way to deal with some of the predicaments and challenges facing GEMS and Earthwatch were the technical advances and the development of new technologies and automated procedures for data processing and interpretation. This made it practical and cost effective to attempt to integrate selected global data and to try to bridge the existing gap between raw monitoring data, assessment, and management. In addition, these advances made it possible to contemplate a central repository for environmental data and for information exchange in UNEP.

Of special importance was the arrival on the scene of increasingly powerful computer software packages (and the hardware to run them). The software was mainly geared to the management and use of the increasing flow of raw remote sensing data that were rapidly improving in quality and scope. It was also designed to combine and interrelate different environmental data sets and to include the more traditional types of data such as those derived from measurements and observations on the ground, geological and soil maps, aerial surveys, and climate records.[1]

When the possibility of putting into practice the idea of such a global database became feasible, GEMS PAC seized the opportunity for task expansion, to make GEMS more versatile and valuable to its users, as well as to demonstrate the usefulness of new tools and to apply them to global problems. It started by commissioning a consultants' report on the potential role of computer software and hardware for a global land/soil monitoring system. The report reviewed the state of the art and proposed a strategy to be followed by GEMS.[2]

The proposal for a database going beyond land and soil characteristics and covering a broader scope of environmental data was floated at an international expert group meeting, held at the Monitoring and Assessment Research Centre (MARC) in early 1983. The experimental georeferenced Global Information Resources System being developed by the US National Aeronautics and Space Administration (NASA) was used as a model to illustrate the efforts to bridge the gap between monitoring, assessment, and management, and to make the data available for applications at all levels, ranging from local and national to continental and global level. Because both the software and hardware already existed to

---

[1] One of these software packages, called ELAS (Earth Laboratory Applications System), was designed primarily to register remote sensing data from different sources and sensors in a common geographic base, and to integrate them with conventional topographic, soil, rainfall, and other data. It produces outputs according to user needs, including maps, graphs, tables, TV pictures, snapshots, and time series of a given situation. The data are geocoded, and the software recognizes each bit of information according to its geographical coordinates on the Earth's surface. P. S. Thacher, 'Space technology and resource management', *Journal of International Affairs*, vol.39, No.1, Summer 1985.

[2] See S. Bie and J. Lamp, 'Project criteria soft-hardware for global land/soil monitoring system', *GRID Information Series* No.1, November 1981. The report outlined the inherent possibilities of the new types of software for analysing information in soil and land monitoring databases, and that could be put to use to overcome current problems hampering GEMS' efforts in this sphere. The report was reviewed by an expert group in Kiev in 1982 and a revised version of the paper was issued in early 1983.

support a global system and to work on integrating selected types of environmental data, the group endorsed the idea and gave the go-ahead to the proposal.[3]

It is on the basis of the recommendation by the expert group that active preparations were initiated to make GRID operational within UNEP. For this, it was necessary to secure the software and hardware required for data management, as well as to obtain the financial resources and staff for GRID. Only a portion of what was needed could be secured from the overcommitted GEMS budget line. Also, when seeking additional support from potential donor countries, it was essential to ensure that their existing contributions to the Environment Fund would not be reduced by their support for GRID. The approach chosen was to seek support and contributions in kind, namely, equipment such as computer hardware, computer software, services, staff on secondment, premises, etc. The response was very favourable, making it possible to launch GRID in 1985, as a two-year pilot project.

UNEP's cash contribution to the pilot project, initially projected at $545,000, rose to $1,834,000, or 34 per cent of the total. This was supplemented by contributions mostly in kind estimated at $3.521 million, or 66 per cent of the total.[4]

The total cost of GRID of $5.35 million during the first two years made it by far the best endowed GEMS project. Nevertheless, it was felt that the funding constraints and the shoestring approach caused some 'suboptimality' in systems design and execution.[5]

NASA accounted for more than half of the external support that was secured for GRID. In fact, it is largely NASA's work and the support that it gave to UNEP that made the idea of GRID possible to begin with, and enabled UNEP to 'jump start' the whole operation. NASA's efforts have been instrumental in developing the remote sensing capabilities and applications, including the software, that were at the basis of GRID.

As mentioned in chapter four, NASA had conceived and tried to internationalize the 'Global Habitability' programme, making it one of the important proposals at the 1982 UN Conference on Peaceful Uses of Outer Space (UNISPACE), held in Vienna. It was on that same occasion that the UNEP representative also spoke of the need for geographically referenced information on the state and trends of critical biosphere variables within political units that are subject to planning and decision-making. Global Habitability did not receive approval and had eventually to be

---

3 See 'GEMS: Report on Ad Hoc Expert Group Meeting to Review Hardware and Software Criteria for a Global Resource Information Database', *GRID Information Series*, No.2, June 1983. The Expert Group used as the background document the report by Bie and Lamp, *op. cit.*, which had been expanded by UNEP to include additional types of data and objectives.
4 Several of the promised contributions enumerated in the initial project document did not materialize. Among those that did were those of the Environment System Research Institute, USA – $100,000 (database management system software); NASA, USA – $1,700,000 (mini-computers, satellite image analysing software, data transmission terminals, running and maintenance, three operator staff); Canton of Geneva, Swiss Development Corporation–$400,000 (computer and office accommodation for GRID Processor in Geneva, training fellowships); and Prime Computer Corporation, USA – $450,000 (two Prime mini-computers). Some of the additional staffing needs were met by addressing directly governments of some Nordic countries for assistance.
5 See 'Report, Meeting of the GRID Scientific and Technical Management Advisory Committee', UNEP, Nairobi, 18–21 June 1988, *GRID Information Series*, No.15, Nairobi, January 1988.

shelved. But many of its ideas were recycled in other programmes, including GRID, which offered NASA an opportunity to test and implement in the international setting some of the ideas and methods that it had proposed.

THE GRID PILOT PHASE

GRID started in 1985 as a two-year internal UNEP project, implemented and managed by GEMS PAC. Physically, GRID consists of two nodes, GRID Control in Nairobi and GRID Processor in Geneva. The former is housed within GEMS PAC. It is intended to control, manage, and coordinate operations and to ensure that these follow UNEP policies. In addition, it promotes GRID technology, offers demonstrations and acts as a regional centre for Africa and as a prototype for regional centres that are to be established in other parts of the world. The GRID Processor is housed in proximity to the University of Geneva. It is engaged principally in data capture (i.e., putting data into the computer and organizing them), collection, analysis, and distribution, and in modelling. It also organizes training courses for students from developing countries. For all practical purposes, the third node of GRID is in NASA's Earth Resources Laboratory in Mississippi, which has acted as a back-up facility, providing vital services and advice upon request.

The technological and operational core of GRID is thus located in the North. This choice was dictated by the functions of GRID, by practical considerations, and by the fact that donors made their contributions conditional on its being established in an industrialized country. In order to function properly and be fully operational, the hardware used in GRID needed to be in the proximity of a comprehensive computer centre and appropriate servicing facilities. On either count, Nairobi could not fulfil the requirements.[6]

Geneva was chosen among other candidate locations in the North (London and several other Swiss cities were also considered) because the Geneva authorities offered free premises for GRID. Geneva was also convenient because of the UN specialized agencies and the UNEP Liaison Office which the city hosted.

The pilot phase was used to set up GRID, to make it operational, and to prove its value and thus help to ensure that it would be continued on a permanent basis. The key objective of the pilot phase was to develop geographic information systems (GIS) methodologies and procedures for constructing, manipulating, and making available global environmental data sets. The effort was concentrated on gradually building a geocoded database and digitizing data, focusing at first on the African region. Efforts were also made to evolve and test global data handling methods, to train users from developing countries, and to promote GRID among the potential clientele in governments and elsewhere.

---

6 During a one-year period under review, downtime of the computer in Nairobi amounted to 44 per cent of the working days. This was due to various reasons, such as power failures, hardware problems, lack of servicing and maintenance and air conditioning breakdowns. During that same period the computer node in Geneva never experienced downtime. 'Status Report', April–September, 1986, *GRID Information Series*, No.6, October 1986, Nairobi.

Use was made of data generated by remote sensing from satellites, local monitoring in rangeland conditions, and standard types of data such as those on soil, climate, geology. Pilot methods for global data handling were developed and tested. Among GRID products were overlay maps and summary statistics. Its data sets were prepared according to various requirements and at various scales, related primarily to soils, forests, hydrology, vegetation, land use, climate, etc. The basic objective was to make it possible for decision-makers, scientists, and other potential users to have 'finger-tip access' to 'integrated environmental data sets and data management technology, using common formats and systems of storage and retrieval'.[7]

The fascination with the data management and manipulation made possible by the computers and software, especially the graphics, was great among those who saw the demonstration of GRID capability. Many people, however, were not quite sure how to apply it to their own decision-making and management needs. Often, what they found out with the help of GRID was self-evident, or could be derived without such a major technological input. This is why a special effort was made to demonstrate the potential and usefulness of GRID in practical situations and to promote the learning. A series of case studies in the field was initiated to demonstrate differing applications of the GIS technology at the national level.[8]

The pilot phase was largely successful in achieving its objectives, most of all in assuring support for the continuation of GRID. In fact, the UNEP Governing Council gave the green light for the operational phase of GRID virtually without any comment.

THE IMPLEMENTATION PHASE OF GRID

The implementation phase of GRID (1988–90), to be followed by the operational phase (1990 onwards), places continued emphasis on system build-up, data set expansion and refinement, and training. In addition, however, it is planned to have GRID contribute to global research and assessment, and to build up a global network through the establishment of regional and national nodes and through the setting up of global telecommunications links.

UNEP continued to spend more on GRID relative to other GEMS activities, yet this covered only one-third of the actual cost, and outside support remained vital.[9]

7 For a description of the early concept of GRID, see a brochure entitled *GRID*, UNEP, Nairobi, 1985, a GEMS publication. For a summary of the pilot phase, see 'GRID Pilot Phase 1985–87: Final Report', *GRID Information Series*, No. 14, January 1986, Nairobi.

8 Among the case studies that were carried out were: analysis of environmental changes due to forest cover degradation in a watershed in northern Thailand; a water quality study in Lake Geneva in Switzerland; a survey of natural resources in Uganda; and the African elephant database project. A regional application was also attempted in a study of the impacts of climate change in the Mediterranean.

9 The UNEP contribution for 1988–9 was estimated at $1.66 million, or 34 per cent of the total of $4.91 million. The supporting organizations and governments are supposed to make counterpart contributions of $1.20 million in cash (24 per cent) and $2.05 million in kind (42 per cent) ('Report, Meeting of the GRID Scientific and Technical Management Advisory Committee', *op. cit.*). It should be noted, however, that the informal nature of some of the arrangements introduced a degree

In a sense, the situation reversed the standard UNEP experience; this time, it was the outside donors that were providing the so-called catalytic support for a UNEP-implemented activity, with UNEP committed, at least in theory, to assume full responsibility for financing it. This assumption was not realistic, at least in the short run. To be viable, the GRID undertaking had to continue as a collectively supported activity. Its real costs are high, making it impossible for UNEP to foot the bill alone. The key prospect and chance for GRID in the immediate future is to benefit from the fallout from the technological progress made in the more advanced countries and to enlist these in the service of GEMS and Earthwatch objectives.

The bulk of the substantive effort in the initial stage of the operational phase of GRID was to be concentrated on improving the methods, and on the continuing build-up and diversification of the global database and of its uses. In addition, GRID was to contribute to the management and interpretation of environmental and monitoring data (e.g., environmental impact assessments, inputs into scientific research, support for development planning). It is worth noting that, in designing the implementation phase and in reviewing the results of the pilot phase, GEMS PAC sought the advice of the GRID Scientific and Technical Management Advisory Committee, whose report was cited above.

A parallel stream of activity concerned the build-up of usership through training, pilot projects, provision of services, distribution of data by geographic area or topic of concern, and so on. Also, plans were laid to build up gradually a global network of regional centres, of which the first one was set up in Bangkok in 1989. Relying on modern means of telecommunication, it is planned that these future GRID nodes should be linked interactively and on-line.[10]

The network would not only help to establish linkages and interaction, but would also serve to decentralize data into regional databanks easily accessible and geared to the needs of users within the region. This was a less costly and more efficient alternative than trying to centralize all data and functions in a single GRID repository.

In view of the costs and delays involved in setting up a fully operational, on-line global network, and the inability of many countries to take part in and benefit from it in the foreseeable future, a back-up system is envisaged whereby individual countries will be supplied with appropriate microcomputers with geographical

of uncertainty into the project. In the case of the pilot project, for example, some countries did not honour their commitments. Because of the devaluation of the Australian dollar, two professionals who were promised by that country failed to materialize. Similarly, the change of government in Canada led to the withdrawal of an earlier promise to second staff.

10 The issue of commercialization and of communication costs remains to be settled and is likely to hinder the needed development. Direct and on-line interaction in the GRID network would have to use the existing commercial channels, but this does not seem to be a viable proposition for a public service of this kind and for the majority of countries that might not be prepared or able to pay the communications costs involved. Interestingly, in the early days of UNEP the issue of free and unimpeded global communication for environmental purposes was considered. A free channel was offered for the purpose on the experimental Franco-German satellite 'Symphonie'. A concrete platform was built at UNEP headquarters to install the required satellite dish, but it has remained empty. The initiative was ultimately torpedoed in Intelsat, mostly by those interests that stood to lose some income were communications to and from UNEP to have gone via a non-commercial satellite channel.

information system capability. In this manner, they could access GRID and its regional nodes by ordering floppy disks with the environmental and natural resources data they need for a given purpose. Such microcomputers could also serve as an incentive and focus for developing countries to start to build up their own national environmental and natural resources databases. The required hardware, which in its more sophisticated version runs up to $35,000 per unit, could be supplied to many developing countries within the framework of traditional technical assistance in view of the fact that it represents an important development planning and management tool.

An unexpected boost to GRID's operational capability, its plans for the implementation phase and its future prospects came in the form of a $6.5 million equipment donation from IBM. This included, *inter alia*, a mainframe computer, 12 office terminals and 5 graphic workstations for GRID–Geneva, and a mid-range computer for GRID–Nairobi, as well as personal computers for 15 African countries and 3 European institutions that cooperate with UNEP in building up the technical potential of developing countries. In this manner, IBM wished to make a contribution to UNEP's 'global efforts to achieve sustainable development'.[11]

GRID AND GEMS

GRID is of special interest to the analysis of GEMS and Earthwatch. It is executed as an internal project of UNEP and represents an operational activity, the first one within GEMS PAC. It is an example of an initiative taken from inside UNEP, using its financial resources to sustain and gradually build it into a major component of GEMS. It went largely against the established rules and practice, and successfully avoided the usual institutional constraints characteristic of UNEP. It is an instance of self-reliance and of getting away from the usual project-funding mode, from the grip of the specialized agencies and from total dependence on their services. It is also an example of successful catalysis, where an activity which is needed and responds to the perceptions of at least some governments is used to attract significant additional resources and support from outside.

GRID embodies increasing ambitions and represents an effort to keep up with developments and to ride on the wave of scientific and technological change by piggy-backing the needs of GEMS on the activities already taking place in the advanced scientific and industrial centres. Moreover, it is an example of the potential of linking up directly and interacting with the scientific and technological communities.

11 'IBM Makes $6.5 Million Equipment Donation to United Nations Environment Programme,' IBM Press Release, London, 23 Nov. 1988. By its articles of incorporation, IBM is supposed to set apart a given percentage of its net income each year and to give it to some worthy cause. For example, in the year preceding the donation to GRID, IBM donated equipment to the AIDS programme of WHO. In search of a worthy cause, those responsible in IBM supposedly were inspired by a favourable reference to GEMS and GRID in the report of the World Commission on Environment and Development, following which GEMS PAC was approached with the offer. The stated objective of the donation was twofold, to boost information management capability of GRID, and to assist developing countries in evolving their information management capability necessary for sustainable development.

GRID relies on advanced technological tools, direct support from outside and networking, the presence on its staff of people with the required technical expertise and experience to run the system, and in sufficient numbers to make the project credible and viable.[12] Also, it illustrates that a minimum critical mass is needed to make it into a meaningful undertaking, with modern technology stepping in to fulfil a number of functions and to supplement the reach of a small staff.

Similar to the International Geosphere–Biosphere Programme (IGBP) in the broader domain of monitoring and assessment, GRID is a manifestation of the underlying need for a global view and corresponds to demands to integrate and link data, to overcome traditional sectoralization of data, and to begin making sense of the data that have been accumulating, often in a disorderly and non-standardized manner. The inconsistencies become easier to detect and, consequently, to remedy when different data sets are brought together in a databank for use. GRID further reflects the need to put into effect the linkage between monitoring data, assessment, and management, which is one of the basic premises of Earthwatch and a key objective set by the Stockholm Conference. The fact that it embodies a capability that can make the hidden messages of the data more accessible to decision-makers and the public gives it practical value in the policy and decision-making process.

GRID provides evidence of the modern world's dependence on technological and scientific progress and related applications to make such complex and ambitious global undertakings possible. It is the technological tools that make possible observation, processing, integration, and interpretation of data, interaction with the database, a global network and communications, and so on. Indeed, many new functions of GRID, some of which at present are not even contemplated, will become possible in the years to come. For example, a global view of pollution in the atmosphere, which is missing in the database, will become possible through the application and further development of such techniques as the LIDAR, mentioned in chapter four. The difficulties experienced in the digitizing of data to make them georeferenced within the computer, which has been one of the major problems and limitations of GRID, is likely to be overcome gradually as technology advances. Also, a decentralized, interconnected global database system and a global GRID networked system will become possible thanks to evolving networked modular extendable architecture systems of computer hardware, as distinct from the traditional mainframe or mini-computer system used by GRID.[13]

Some of the institutional objectives implicit in GEMS are also dependent on technological progress. GRID itself is the beginning of a global nucleus for environmental data, where technological capabilities make it possible to rise above sectors, disciplines and institutions, and to work on an integrated approach to global data management and data assessment. As such, it justifies the operational claim

12 The GRID staff is larger than the staff of GEMS PAC. The latter consists of three professionals, while GRID Processor in Geneva has eight technical staff with another five deployed to GRID Control in Nairobi. 'GRID News', Vol.1, no. 3, 1988. Moreover, the regular GEMS PAC staff provides the coordinator for GRID and devotes a good deal of time and effort to the requirements of interacting with and helping direct and manage GRID.
13 See 'Report of an Ad Hoc Expert Workshop on GRID Systems and Software', J. O. Weber, ed., Nairobi, 14–18 Sept. 1987, *GRID Information Series*, No. 12, Nov. 1987, p.4.

by UNEP and its unique role, which cannot be contested by the specialized agencies or the governments.

Technological tools are usually developed for the needs of the North, with little concern for the special characteristics and needs of the South. How to make these tools more appropriate and accessible to the South thus figures as an important challenge for GRID. Technology once more helps to bring together North and South in a GEMS activity, by making it possible to cater simultaneously to their differing needs and interests, provided a deliberate effort is made to respond to the special situation and needs of the developing countries.

Indeed, GRID will have to face the continuing challenge of how to keep up with rapid technological advances in both computer hardware and software, which generates technical, financial, and even political issues. For example, the rapid obsolescence of equipment impairs functionality, access to service and parts, and use. Hence GRID will have to remain at the state-of-the-art level if it wishes to be viable, as well as to satisfy the users among its varied global clientele. This, in turn, shows the need to recognize explicitly its links and dependence on some countries and institutions, and to manage these relationships as an integral part of GEMS' growth.

The future of GRID is not self-evident. However, the basic need for such a global mechanism, or one like it, exists and will increase in time. Logically, it belongs within, or should be placed at the disposal of, the international community through GEMS and Earthwatch.

The IGBP Plan of Action, for example, in its proposal for data management and information systems, restates the very objectives that gave rise to GRID, i.e., being accessible to scientists and policy-makers in a timely manner and free of charge (for non-commercial uses), setting up databanks that will be left as a heritage to future generations, and having efficient data handling systems. These data management and information systems should be accessible through reliable data networks, based on state-of-the-art, interactive communication systems, to help inform about ongoing events and research findings, to help request and transmit data, and so on.[14]

In the case of the US Earth Observing System (EOS), an advanced information system is at the very core of the whole undertaking. Indeed, it is considered that the design, development and management of such an information system approaches in scope and complexity the design, development and operation of space-based monitoring systems.[15] Different satellite remote sensing programmes will generate large quantities of data, which will need to be managed and converted into meaningful information.[16]

14 See 'The International Geosphere–Biosphere Programme: a Study of Global Change, a Plan of Action', *Global Change, op. cit.*, Report No.4, 1988, pp. 79–84, 105.
15 See *Earth System Science, a Closer View*, Report of the Earth System Science Committee, NASA Advisory Council, NASA, Washington, D. C., Jan. 1988, p.37 and chapter 8 entitled 'Observing and Information Systems for Earth System Science: Characteristics and Evolution'.
16 It has been estimated that satellite monitoring of the Earth in the 1990s will generate in two weeks as great a volume of data as LANDSAT had created in its 17-year lifetime, which have swamped the existing facilities to the point that some people have dubbed the data centres as 'data cemeteries'.

It may appear quixotic to juxtapose the multibillion dollar EOS scheme with GRID. The former, however, is the reflection of what is needed if the job is to be well done. The latter is the reflection of the state of international cooperation and a kernel of a long-term undertaking. The comparative advantage on its side is its international status, which in the longer run holds the promise of a truly global capability linking up various programmes, including EOS, into an interactive global system.

For the time being, GRID has to live and evolve within the existing framework and limitations that have characterized GEMS over the years. One of these limitations concerns the problems encountered in utilizing the data from various GEMS clusters, and the many gaps that exist in the coverage, compatibility and quality of the data. Indeed, because of this and the practical considerations, the start-up GRID database seems to have come almost wholly from the US sources.[17] As the database expands and diversifies, the issue of harmonization of environmental measurement will become of central importance for environmental knowledge and assessments, for enforcement of regulations and conventions, and for effective communication with and persuasion of public opinion and decision-makers.

It is the need for harmonization of environmental measurement that gave rise to an initiative of the major industrialized countries, based on a recommendation of their June 1987 Economic Summit in Venice, asking UNEP to work on the improvement and harmonization of environmental measurement.[18] This eventually led to the establishment of a small office in the Federal Republic of Germany, called 'Centre for Harmonization of Environmental Measurement', as a project in support of and under the auspices of GEMS.

The recognition of this need at the high political level was, of course, a welcome sign of interest and an incentive to focus attention on a critical issue. Once more the response was a minimalist one. Yet, like early steps in GRID itself, it represents a beginning heralding a growing support and a focus for an integrated consideration of the issue of harmonization of environmental measurement and generation of internationally comparable environmental data. As such, it has a potentially important contribution to make to the further evolution of GRID, and of GEMS where so far this concern was dealt with only in individual clusters.[19]

The Geneva location has undoubtedly been of advantage for the growth and daily operation of GRID Processor. It has freed it from some of the usual bureaucratic and institutional constraints associated with UNEP headquarters. Also, and

See 'Bringing NASA Down to Earth' and 'Early Data: Losing Our Memory?' *Science*, 16 June 1989.

17 See global and regional data sets held by GRID in 'GRID News', Vol. 1, no. 3, 1988.
18 See UNEP Governing Council decision 14/24 entitled 'Improvement and harmonization of environmental measurement', adopted in 1987. The recommendation of the Venice International Economic Summit Conference of Industrialized Nations, held in June 1987, was based on the report of the Summit's Environmental Experts. This report defined the problem and the types of action needed, and was the result of the initiative of the Federal Republic of Germany and more specifically of one of its influential scientists.
19 See 'Report of the UNEP Expert Meeting on Improvement and Harmonization of Environmental Measurement', Munich, 9–11 Dec. 1987. The US opposed the establishment of the new mechanism on measurement as diverting resources and attention from existing bodies.

this is of particular importance for the fulfilment of its functions, its location has made it easier and possible to interface with various supporting institutions and to operate at the appropriate level of technological back-up. On the other hand, the installation of a part of GRID in Nairobi – and a considerable duplication of the systems between the two locations – was also functional. It demonstrated the possibility of a gradual build-up of a sophisticated technological system in Nairobi, and thus provided an example for the establishment of regional nodes of GRID in other parts of the developing world. As well, it helped GRID to carve its place within GEMS and UNEP. Most of all, it highlighted the role of global networking and the link up of institutions and activities in a common undertaking, as opposed to centralization.

The basic capability of GRID points inevitably in the direction of assessment. Yet, it is not at all clear who will do the assessment. Should GRID, and indeed GEMS and UNEP, remain passive organizers of the database, content to make its services available upon request to those interested? Or should they take a further step and use the experience, the unique institutional position, the analytical capability, and the database available to organize and undertake independent assessments in the context of Earthwatch?

The answer to this question thus far has been largely a restrictive one: the database is there for use by others and not by UNEP independently. In the words of the pertinent policy directive, GRID does not do assessments, it contributes to them; GEMS does not do assessments, it coordinates and orchestrates them; UNEP does not do assessments alone, but always in concert with or vicariously through other agencies. Assessments are coordinated by UNEP. Assessments have to reflect the objectives of international consensus, and hence should be concerted views on a given issue arrived at through a process of review and analysis by a group of international subject area experts.

As usual, many factors seem to have played a role in this cautious policy ruling. There is the usual reserve on taking initiatives when the mandate, capacity and resources to perform the task are limited. Then, there is the standard defensive posture and deference *vis-à-vis* the specialized agencies that continue to object whenever UNEP tries to assume an independent and more direct initiative in any domain. There is also the usual, neutral, and hands-off attitude to data and fear of controversy that their interpretation could give rise to. And there may be some reluctance to expose the weaknesses of existing and still incomplete monitoring databases, which would inevitably come to light as a result of any systematic attempts at assessment. Most of all, however, there is the defensive and cautious attitude of the international civil servants facing the governments.

Governments have not been keen to see independent environmental assessments carried out by the secretariats of international organizations, and have not had confidence in the internal capability of UNEP, or the ability of the consultants that it might enlist, to undertake such assessments. Moreover, they tend to differ as to what they wish to have assessed, how and by whom. Given the differences and potential conflicts, the governments prefer to keep assessments under their control, in full awareness of the different views and interests involved, preferably in small groups of experts appointed by themselves.

Accordingly, it should not be surprising that GRID finds itself affected by the same syndrome that was evident in the discussions of the five monitoring clusters of GEMS, where hesitation and uncertainty are exhibited precisely at the critical point when the available data needs to be assessed with a view to action.

This opportunity for a more active and independent role by international organizations should be grasped because of the nature of the environmental issues involved and because of the need for an actor other than governments to take the initiative and offer the leadership and global vision rising above particular and short-term interests. It is also a tool that will make possible assessments on different scales, from local to planetary, and thus for promoting the shared objectives of humankind first embodied in the outcomes of the Stockholm Conference.

To be successful in this task implies following a path of greater activism, which has been traditionally frowned upon and discouraged in international organizations. It certainly calls for more than the normal dose of initiative in trying to deal with the existing institutional, political, economic and mindset constraints that have been characteristic of UNEP as well. At the same time, what is required is that the governments should overcome some of their traditional sensitivities and accept that environmental issues of international concern merit new and more open treatment.

Whatever happens, with GRID in place, GEMS and Earthwatch have acquired a less ethereal character than had been theirs in the past. The role of the coordinator, the GEMS PAC officials who come and go, and the clusters of monitoring activities or projects, spinning in different parts of the UN orbit and tied tenuously to UNEP by the financial cord, are now supplemented by a data and information management mechanism, a physical nucleus on the way to becoming permanent and a global capability essential for the objectives of GEMS. GRID certainly has the potential to become an essential link and a means to make an international global interactive data and information network and an environmental monitoring and assessment system into a reality. Also, its existence and the experience that it has generated will certainly play an important role in the future evolution of GEMS, an issue that is addressed in the next chapter.

9 GEMS: inching towards a global system?

As the last decade of the twentieth century begins its course, the international community is preparing for the 1992 UN Conference on Environment and Development. Twenty years after the first conference specifically devoted to the environment, the Stockholm Conference, the subject seems, at last, to be firmly established on the international political and economic agendas.

The bursting on to the global political scene of the environmental issue and the upsurge in interest and concern are based on improved knowledge and growing scientific consensus about the role of human activities in the processes and changes on the planetary scale. The main reason, however, why the environment has become a hot topic is the successful popularization and politicization of this knowledge and of the relevant issues; both the public at large and the decision-makers are now familiar with, sensitive to and alarmed about trends and prospects for the environment, in the near future and of course the longer term.

The link has been established in a convincing manner between the global change and environmental impacts of modern civilization and the social, economic and physical prospects and well-being of the international community of nations and peoples.

The phenomena responsible for the change in traditional outlook, foremost the probable global climate warming, are also among those that inspired the very idea of GEMS. Moreover, environmental monitoring data and assessment are at the very base of the rising apprehension.

It is logical, therefore, to expect GEMS to have played a prominent role in the rise of public and policy concern. Yet, one would have a hard time to demonstrate a direct role and influence of GEMS in the formation of policy and opinion, although some of the data produced by its various clusters have certainly been used by the scientists in their analyses, and by the NGOs and the media in their efforts to enlighten and frighten into action the political establishment and the public.

ASSESSING GEMS

A hardened veteran of international undertakings has remarked that it takes 25 years for a global system or programme to mature in the UN, that being the period needed as a pilot phase for learning and institutionalization. While this estimate may appear unduly pessimistic and too long, the complexities, weaknesses and inertia of international structures and processes are such that this rule of thumb may yet turn out to be too optimistic in the case of GEMS. Unavoidably, the international endeavours seem to traverse a longer and more sinuous route to their final objective than may appear to be necessary. GEMS is no exception.

In more ways than one, as a global system GEMS was ahead of its time, preceding the real needs, the necessary political and social consciousness, the demand for its services, and the scientific knowledge and technical back-up which was essential to make it truly operational. Suddenly, however, things have begun to change. The need for GEMS and its services has become self-evident and pressing. Therefore, the question arises as to the future direction of GEMS: is it ready to carve out for itself an important niche in the changing policy and action context and will it be capable of adjusting and responding to growing demands as the international community moves into a phase of greater readiness and urgency to act on environmental challenges?

The GEMS story is largely one of a contradiction between its North-based role model and logic, on the one hand, and the broader world reality combined with the still rudimentary nature of international cooperation as embodied in the UN, on the other hand. In looking at the road travelled so far, one should not judge GEMS and its performance against an ideal model. Nor should GEMS be evaluated by reference to an ambitious yardstick derived from the examples of activities carried out and the scientific and financial resources available for this purpose in the developed North and on the assumption that the world is a homogeneous one and that the countries are all equally able or keen to take part in GEMS.

Any evaluation of GEMS should make reference to conditions and structural obstacles with which it had to contend as it evolved. These were amply illustrated in the preceding chapters, for example, the North–South development gap and the East–West split, and the political, organizational and resource constraints to which the international organizations are subject. The story of GEMS is also very much the story of UNEP, of its subjective and objective characteristics and faults.

Likewise, the ambiguity of national and international responses to the environmental problématique, unavoidably reduced the interest and drive invested in GEMS. The rival claims of pure science and the more mundane management created additional dilemmas in the evolution of GEMS.

Limitations of the technologies available, incomplete scientific understanding of the processes involved, high capital and operating costs of monitoring in a situation of national financial stringency, the lack of methods and structures to secure a binding commitment by governments to take part and contribute – all these combined to retard the growth of GEMS over the years.

Probably the single most obvious flaw of the system has been the uneven participation of member states, and the resulting major gaps in global coverage and outputs of GEMS. This was due to a variety of factors, including unfulfilled expectations and failure of direct benefits to materialize from GEMS (i.e. the usual dichotomy between scientific yields and the more direct management-related outputs), with the consequence that the initial interest and enthusiasm shown by some countries faded. It was due primarily to the differences in levels of development among countries and to the lack of financial and other resources needed to carry on the necessary activities in a sustained manner. As regards the countries of Eastern Europe and the USSR, a key factor that affected their participation in some clusters of GEMS was the excessive secretiveness of the pre-glasnost era compounded by the neglect of environmental issues which resulted in the serious degradation of both their local and regional environments; the reality in these countries, as it turned out, was very different from the positive public image they had been trying to project internationally.

Various shortfalls notwithstanding, the work done and knowledge acquired since the 1970s and the greatly enhanced observational, modelling and computational capabilities, provide a solid foundation for fashioning the existing pieces into a greater semblance of a global environment monitoring system.

The challenge at this juncture, therefore, is to invigorate and freshly orient GEMS, building on the established foundations and putting to good use what has been learned so far, taking fully into account the difficulties inherent in implementing a largely voluntary global undertaking in a heterogeneous international setting comprising many actors of different power, priority interests and capabilities.

As constituted at present, GEMS lends itself to transformation into a more structured global scheme, in line with the basic premises and objectives outlined as long ago as the preparatory process for UNCHE and sketched in chapter two above. In order to move in this direction, however, it is necessary to go beyond the individual clusters and to view GEMS as a whole, a true system which is more than just the sum of its parts.

GEMS has carried on for almost two decades without a truly in-depth effort to review its rationale, how it has fared in practice, and especially what needs to be done to mend the shortcomings and improve its performance.[1]

A great deal can be said in favour of protecting fledgling activities from excessive or destructive criticism. However, there is nothing inherently contradictory in being simultaneously positive in assessing particular actions and performance, and critical about shortcomings, as a means of improving the programme and prodding it in a desired direction. The ostrich-like approach which prevailed is in part a reflection of the rigidity and inertia of the intergovernmental deliberative process and of the usual defensiveness of the secretariats of the UN organizations *vis-à-vis* criticism and probes of their work.

This was reinforced both by the lack of major interest in GEMS' fortunes and uncertainty about how to handle environment-related matters, and by the reluctance to stir up the institutional hornets' nest of UNEP by a frank analysis of the

---

1 An in-house effort at self-appraisal of GEMS carried out by UNEP in 1986, while quite detailed, failed to acknowledge most of the key underlying issues. Among the failures and weaknesses one was singled out, i.e. overstretching of GEMS PAC's activities, resulting in its weak participation in project execution and follow-up. It was concluded that GEMS PAC staff should play a more direct and assertive role during both the initiation and development of projects, and in overseeing their management. Whether the existing staff of three professionals could do this effectively was not considered and corrective action was not taken. UNEP document 'Executive Summary of the Evaluation of the Achievements of UNEP in the GEMS subprogramme', 27 Jan. 1986.

difficulties experienced in piecing GEMS together. Indeed, many of the flaws and improvisations evident in GEMS, in addition to the systemic and other constraints characteristic of the UN, were also a direct consequence of the UNEP's institutional model, of the limited choices that were available, of the way in which the governments, the UNEP secretariat and those in charge of GEMS translated it into practice, and of how the other actors responded to its needs and how they played the roles for which they were cast in the UNEP-drawn scenario.

The unwillingness to question the build-up and performance of GEMS and the underlying causes of its difficulties was in line with the standard practice of a majority of governments in the UN to vote for ambitious and far-reaching recommendations and then to apply to their execution a 'band-aid' or 'toy gun' approach, i.e. to stipulate least cost, least manpower, least commitment and minimalist international organization for putting them into practice.

The inertia was also in part the consequence of an uneasy truce between the North and the South, noted in chapter one above, which was tacitly struck in the early days of UNEP. It was maintained, in part, by securing a friendly attitude of developing countries with the help of some financial assistance and attempts to involve them in various GEMS activities. Primarily, however, it was due to lack of organized political pressure by the Group of 77, of the kind demonstrated exceptionally at the first Governing Council session and the 1974 Intergovernmental Meeting on Monitoring, the collective failure of developing countries to consider the implications of GEMS for their development, their apparent indifference to this undertaking and a rather perfunctory role in its implementation.

The countries of the North were probably aware of the deficiencies of the system, and were certainly frustrated by the attitude and/or lack of capability of many developing countries and the resulting unevenness and lag in the performance of GEMS and the non-fulfilment of the underlying expectations of the institutional model. However, they appear to have largely contented themselves to leave GEMS to inertia, voicing disappointment and criticism but without making any sustained effort to press for greater organizational rationality and efficiency.

This relative passivity may have been inspired by unwillingness to face the true requirements of GEMS' implementation, by fear of reawakening the inconclusive debates of the kind that had taken place in 1974, and probably by the loss of patience with the inherent difficulties of the intergovernmental process in the UN and the perennial development refrain. Similarly, they wished to avoid giving an opportunity to developing countries to raise the standard question of additionality and to seek technical and financial assistance for their monitoring activities. Accordingly, the developed countries seemed to leave GEMS to its own course, and to concentrate their attention on those activities within its framework which were of interest to them, and in parallel followed their own separate paths, individually, on a regional basis and even globally in a number of environmental monitoring activities.

Then, of course, there was the usual 'conspiracy' of the specialized agencies and of the sectors and scientific disciplines that they represent. These organizations largely persisted in their attitude dating from pre-Stockholm days. They did not see any compelling reason why the existing set-up should be reconsidered. And they were quite content to run their own monitoring programmes and to receive whatever funding they could obtain from UNEP for this purpose. Their attitude was on the whole supported by the governments, which often acted in an uncoordinated manner, or even at cross purposes, in governing bodies of different organizations.

Symptomatically, only a disaster such as that at Chernobyl, which inspired a unified directive from the governments, was able to bring all specialized agencies around the table to consider acting together on an issue and yielded enough political pressure and public demand to secure a degree of meaningful coordination and cooperation. This was in contrast to the UNEP Governing Council-based, still largely pro-forma coordination practices of Designated Officials on Environmental Matters (DOEM) and the System-Wide Medium-Term Environment Plan (SWMTEP), two major vehicles devised by UNEP in its effort to give some sense of direction in environmental policy and action to the diverse and fragmented UN family of organizations.

UNEP for its part was trying to build up its global role and worldwide reputation starting from the catalytic funding and coordinating model approach, with a tiny staff of two or three GEMS PAC professionals and hamstrung by the 'non-operationality' criterion. This led it into a trap of ambitious commitments and high expectations which were not backed by adequate means to deliver. What made the situation even more frustrating was that UNEP, though aware of the gap, tried to give the appearance of normalcy.

GEMS PAC was thus involved in a global 'LEGO' exercise of sorts, but with many pieces missing and those available not fitting all that well the framework that it had in mind. By listing the many monitoring activities as belonging to GEMS, and implying that these were UNEP's own, it inevitably gave the impression of seeking credit or projecting institutional claims.[2]

This puzzles those who cannot fathom the subtle distinctions between UNEP's *Level Two* and *Level Three*, or are not quite sure as to what GEMS is all about. At the same time, many of those operationally responsible for various monitoring activities viewed UNEP's role and claims with a jaundiced eye. Some have referred to GEMS as little more than a 'label', because of what they felt was UNEP's attempt to buy a piece of action and credit for itself by insisting that publications emanating from the monitoring projects to which it had contributed financially bear prominently a reference to GEMS on the title page.

As an integrated global system, therefore, GEMS has been somewhat of a fiction. While the monitoring activities and clusters made sense by themselves, the global systemic perspective, structure and linkages were in fact largely missing. They were kept alive primarily in UNEP's documentation and public statements, while the real work and efforts in this direction were taking place elsewhere, mainly among scientists in those few industrialized countries where the necessary resour-

2 For example, a UNEP write-up on GEMS, which lists a series of monitoring activities in different parts of the world, says that 'they are examples of UNEP's Global Environment Monitoring System (GEMS) in action'. See 'UNEP Regional Bulletin for Europe', Number 7, Summer 1989.

ces, capabilities and interest existed, and in their cooperative efforts such as the International Geosphere–Biosphere Programme (IGBP).

UNEP adhered to the project-funding approach in the build-up and carrying out of activities of GEMS. However, it could not do much with the minimal financial and manpower resources at its disposal, which moreover had to be parcelled out among many different claims. Once locked in the sectoral tracks, UNEP found it difficult to get disengaged and to focus on its key comparative advantage, i.e. the mandate to take a global view and pursue an integrating approach. It was also discouraged from trying to play multiple, challenging roles inherent in its institutional mandate, including that of forging links between monitoring, research, assessment, management and regulation.

Admittedly, this chosen course was also a consequence of a deliberate choice to concentrate on the five monitoring clusters, as defined by the 1974 Intergovernmental Meeting on Monitoring, with a view to having some monitoring data ten years later and a semblance of a system. It was assumed that as networks grew, data accumulated and experience was gained, integration would be easier to achieve and the clusters would begin to join up.[3]

Besides, GEMS had no specialized intergovernmental organ of its own to give it the necessary policy focus, continuity, and a firmer sense of direction, or to provide system-wide guidance, authority and necessary support for action and coordination. The Governing Council could not fulfil this role adequately, dealing as it did with a large and heterogeneous basket of activities.

Furthermore, the minuscule secretariat in GEMS PAC, in spite of its qualities, dedication and efforts, could not project a credible posture across a wide spectrum of activities and disciplines *vis-à-vis* the scientific community, the UN system, and its national and international constituencies. Its predicament was accentuated by the impact of location, which in the case of a technical activity such as GEMS was felt more acutely as a handicap than in some other programme areas of UNEP.[4]

The very uncertainty of the whole notion and structure of GEMS and the failure of UNEP to secure a firm foothold and role were illustrated by the June 1989 action

---

3 M. D. Gwynne, 'The Global Environment Monitoring System (GEMS); Some Recent Developments', in 'Proceedings of the International Symposium on Integrated Global Monitoring of the Biosphere', Tashkent, USSR, 14–19 Oct., 1985, EPMRP Series, No.45.

4 In the case of a programme such as GEMS, the Nairobi location has been a disadvantage in carrying out the intended functions. While Nairobi had both positive and negative impacts on the evolution and effectiveness of UNEP as a global UN organization and the fulfilment of its model role, in the case of small, specialized programmes such as GEMS, the unfavourable impacts are easier to demonstrate. Were one to imagine, for example, a well-funded and strong institution in charge of GEMS exclusively, it would have been easier to deal with some of the logistics of the Nairobi location. In this important yet controversial matter of the location of a body or agency, the issue is usually presented in the light of North–South divisions, and the 'peripheral' position of the latter. In the case of GEMS, however, what matters more is proximity to cooperating organizations. In other words, Tokyo, Rome or Montreal, all of which are in the North, would not have been demonstrably more favourable locations for GEMS PAC and its staff of three professionals than Nairobi, whereas Geneva would have been. An interesting study could be made of the comparative experience of OCA PAC, and the associated politics, its management, staffing and performance, first in Geneva, then at UNEP's HQs in Nairobi. To what extent the differences were due to the actual location, as opposed to UNEP's own peculiarities, structures, personalities and style of management is a matter worth exploring.

of the Executive Council of WMO (referred to in chapter four above) to create a Global Atmosphere Watch (GAW) as an umbrella encompassing the many monitoring activities involving atmospheric composition change.

This step was quite logical and long overdue in that it grouped all atmospheric pollution monitoring under a single framework based on the BAPMoN, ozone and other climate-related monitoring experiences. It meant a belated move towards an integrated global network for monitoring atmospheric pollution. According to the institutional script, such an initiative should have come from UNEP and within GEMS. What actually happened was that UNEP was largely confined to a role of observer, whose financial assistance was sought according to the well-established pattern within the framework of GEMS.

Both GAW and IGBP initiatives are a reflection of the need and demand for a functioning and rational global environmental monitoring system. Also, they raise questions about the very rationale of GEMS, how it was translated into practice and the role of UNEP in the next stage of the global environmental monitoring system build-up.

WHITHER GEMS?

What is said above is not meant in any way to belittle the importance of the distance travelled since 1972 which was amply detailed in the preceding chapters. Thanks to the work done and lessons learned, the international community has at its disposal a good deal of knowledge and data, valuable experience and a set of monitoring capabilities and infrastructure. It is thus well poised, provided that there is the will and interest to put in place a more advanced version of GEMS, heralding a genuine system of global environmental monitoring. Indeed, many of the lessons learned over the years, in combination with the very requirements of environmental monitoring and the dynamism of the existing activities, seem to be moving GEMS in the direction of a more integrated undertaking, even though the institutional and policy realities remain essentially unchanged. One of the indicators of this change is the newly placed accent on setting up a global system for integrated monitoring, as noted in chapter six above.[5]

The future of GEMS will be shaped largely in the policy context of domestic, regional and global environment–development politics, and of North–South relations. In addition, it will be affected by the evolution of UNEP, the agencies and interagency relations. On the assumption that the need and demand for an international programme such as GEMS will give it a major impulse in the years to come, a few comments are offered below as to the possible steps which would have to be taken in order that it may fulfil the expectations.

Various shortcomings notwithstanding, it goes to the credit of the UNEP catalytic approach that the notion of a global system has been preserved and that none of the options has been foreclosed. An agreed framework continues to exist. A series of monitoring activities are operational and can be used to varying degrees

5   For a description of GEMS and its rationale at the beginning of the 1990s, see 'GEMS: Global Environment Monitoring System', Nairobi, 1990.

for assessment and management purposes. And the establishment of GRID has created an international data nucleus which could develop into the heart of GEMS as a global system.

The next phase in evolving global monitoring capabilities and assessment, however, requires a much greater sense of order, direction, and a more serious institution-building effort to match the state-of-the-art technological capabilities and the growing demand for monitoring services. It means pressing for a system that would be truly global in terms of participation, commitment and involvement of countries, and of coverage, and setting in place a network of internationally funded and operated observatories, research and data centres.

Such an undertaking presupposes a clear definition of the roles of the various participants and their obligations. It involves identifying the objectives and problems to be sorted out, and estimating the necessary financial and other resources and specifying how to mobilize these, nationally and internationally. Further, it means apportioning tasks and coordinating effectively the various components of the system. And it requires the institutionalizing of a critical and independent review of the performance of GEMS and of the uses of its products.

In other words, the piecing together of GEMS, which so far has been largely a patchwork of loosely related activities, should be superseded by a much more structured approach. It should become the subject of a global master plan and of a corresponding implementation programme, reflecting the more advanced stage in the build-up of GEMS and the growing demand for its services in science, policy, decision-making and public life. And it should be brought back into the centre of the environment–development debate where it was in the very beginning, at UNCHE and immediately thereafter.

If UNEP's role is seen as changing and evolving with the different stages of growth of GEMS and of the overall policy context, then the time has come for a reassessment and revision of its relationship with this global programme.

The undertaking is too complex and too important to be abandoned to the approach of muddling through and institutional inertia. It calls for a long-term commitment by governments and intergovernmental organizations. A global system must not have to depend on short-term and uncertain project funding. Unavoidably, GEMS should be linked explicitly with the global development and political agenda. Provided such a new start is made, there is no reason why by the year 2000 the world community should not have at its disposal an operational and relatively advanced international global environment monitoring system, which it had set as an objective some thirty years earlier.

In planning for the future of GEMS, two sets of requirements need to be combined and accommodated, i.e. those having to do with the natural environment and those having to do with the organizational, institutional and policy sphere.

It so happens that the two domains are based on different rationales and are driven by different forces. Indeed, the discrepancy between the two has been one of the chief dilemmas of GEMS, i.e. how to adapt the environmental logic to the institutional and political logic, while rationally it should have been the other way around.

Hence, one of the challenges in the years to come is to match more closely the institutional design with the social and natural systems and processes that are being studied and monitored. Favourable prospects for such an approach exist thanks, first and foremost, to technological progress.

There are three sets of variables that should be addressed in charting the future of GEMS:

- the nature of GEMS;
- the role of UNEP in the further build-up and operation of GEMS;
- the roles of other actors (i.e. governments, specialized agencies, the scientific community, the media, the public) in operating GEMS and in using its outputs.

Nature of GEMS

There appear to be four basic areas with which GEMS has been concerned so far. These need to be distinguished for operational and practical purposes: (a) the planetary or systems view of the environment; (b) human health and the environment; (c) data management as a means of integrating the system and giving it a permanent core; and (d) regulation and management, a functional task which intersects with and relies on the outputs from the first two. It would seem logical in the further evolution of GEMS that these principal areas should be recognized and used to structure the activities.

Obviously, the systems nature of the global environment exerts an integrating pressure and imposes its logic on GEMS. The integrated monitoring approach promoted actively for years mainly by the USSR in GEMS, and also embodied in IGBP and in the Earth Observing System (EOS) of the US Mission to the Planet Earth programme, are all part of the same underlying scientific quest for a global view and for an understanding of the planetary phenomena and of the role of humans in these. This approach is becoming operationally and economically possible thanks to the development of sophisticated and powerful tools for observation, monitoring and the processing and interpretation of the data.

After two decades of a parcelled-up, media-based approach forced by political, institutional, disciplinary, resource and technical considerations and limitations, the new tools and capabilities make it possible to move gradually in the direction of the original, integrated concept of global environmental monitoring.

The increasing capability to view the Earth system from space, and to monitor both the status of natural resources and the pollution, means that the international community is now in a position to evolve and deploy a potent tool for the monitoring, research, assessment and management of the global environment and of climate change. The new monitoring and data management capabilities will make it possible to bridge the knowledge and action gap between the macro-focus on planet Earth, and the micro-focus and *in situ* monitoring in geographically demarcated areas.

How to utilize these capabilities in evolving an integrated system of global environmental monitoring should be a major concern of GEMS in the period to come. Placing these capabilities, which at present are concentrated in a few advanced countries, at the service of the entire international community, and of individual countries in their national decision-making and management of the natural resource base, should also become a major policy objective in the further evolution of GEMS.

Health-related environmental monitoring offers an opportunity for GEMS to be linked with the immediate concerns and well-being of populations and individuals in all parts of the world and in specific geographic locations, and thus to be also of relevance to policies and decisions at the national level. Of necessity, it should feed on and interact with the first cluster of activities, for example, studies of ozone layer depletion, nuclear radiation, long-range transport of pollutants and like phenomena.

Inspired by the systems approach required by the nature of the environment and the social aspects of the environmental issues, the two broad clusters suggested above, i.e. global environment and human health, would make it easier to bridge the disciplinary, sectoral and institutional divides and thus move in the direction of a truly integrated global monitoring system. Adequate data management capabilities are essential for the success of this scenario. With GRID in place, an important step in this direction has already been taken.

During the formative stages of GEMS, priority was given to the setting up and efficient functioning of the various monitoring activities. The underlying objective was both to learn and to give early warning, and to manage and regulate on the basis of the knowledge thus acquired. Monitoring and Evaluation of the Long-range Transmission of Air Pollutants in Europe (EMEP), under the ECE Convention on Long-Range Transboundary Air Pollution, and monitoring in the context of the Regional Seas Programme, were the first significant ventures combining both of these objectives in internationally agreed monitoring, assessment and management packages.

EMEP, for example, involved assessment, both in monitoring the status of the environment and in monitoring the compliance of countries with given targets and obligations mutually agreed to. As time goes by, the need for similar packages will grow, linking monitoring, assessment, management and regulation into a positive feedback loop. This is another direction in which GEMS will have to move decisively.

The role of UNEP

UNEP and GEMS PAC can play a key role in the suggested scenario. To begin with, UNEP has at least two comparative advantages in its favour. First, it is the global international environmental organization and as such it provides the policy, substantive and coordinating locus for international action on the environment. Second, it has an inherently flexible mandate and institutional rationale embedded in its notion of a catalytic approach and backed by the Environment Fund, which makes it possible to re-examine its own roles and to decide on a new and more appropriate course of action in pursuit of its objectives much more readily than is the case with the more traditional UN organizations.

GEMS: inching towards a global system? 215

The question is therefore what can UNEP do most beneficially in the period to come? During the initial phase its function was primarily to provide the seed money for fledgling monitoring activities, with some $15 million disbursed from the GEMS budget line during the first decade, parcelled up among many and diverse activities, of which close to one-fourth went for the operational costs of GEMS PAC.[6]

Once these activities are initiated, however, international financing will be required principally for the funding of environmental monitoring in those countries that cannot afford the expense and the operational cost of some international activities. The GEMS budget line of UNEP's Environment Fund is not intended for these purposes. In such cases it would be more appropriate that the necessary financial resources should be provided by financing institutions such as UNDP or the World Bank, which have shown a growing interest in the environment and should assume corresponding financing responsibilities.

UNEP, in turn, should use its funds to strengthen its own efforts to coordinate and energize the system, in particular by generating financial, political, public and scientific support for GEMS. Also, it should undertake activities to promote the uses of environmental monitoring data and assessments for early warning, policy-making, environmental management and regulation.

In other words, UNEP should concentrate more on system coordination and policy, and on what the data mean, how to use them, and what to do on the basis of the findings and trends.

All this means diversifying beyond the standard, passive project-funding mode. UNEP should play a more dynamic, creative, activist and policy-oriented role, using the limited financial resources at its disposal to influence policy and catalyse

---

6 From May 1973 to December 1982, $15.7 million were committed in GEMS budget line expenditures, of which $640,000 were in non-convertible currencies (NCCs). (This is somewhat of an understatement because funds for GEMS-related expenditures coming from other budget lines are not included in the above figure. However, since the intention here is to illustrate rough magnitudes and proportions of expenditures, the difference is of minor importance.) The breakdown of the expenditure was as follows: (a) the cost of running GEMS PAC – $3.69 million; (b) climate-related monitoring – $3.32 million (plus $114,000 in NCCs); (c) health-related monitoring – $3.91 million (plus $160,000 in NCCs); (d) natural resources monitoring – $1.95 million; (e) EMEP – $941,000 (plus $255,000 in NCCs); (f) marine pollution monitoring – $356,000; (g) miscellaneous activities – $830,000, which included $541,000 for MARC in its role of support for the research function within GEMS. (These figures are based on an internal GEMS PAC balance sheet and do not necessarily tally with the figures presented in earlier chapters, which are taken mostly from appropriations entries in individual project documents.) Note that the climate-related monitoring figure also subsumes a $1.1 million contribution to WMO's coordinated Global Atmospheric Research Programme (GARP), for the execution of its first global experiment (FIGGE). This accounts for the apparent discrepancy with the figure given in chapter four, citing $2.4 million as a contribution of UNEP to climate-related monitoring activities. The project funds were used to contribute to the expense of purchasing some equipment for the participating ships. Though related to the objective of understanding better the natural climate system and general circulation of the atmosphere, and in line with the UNCHE recommendation No. 79, which called on WMO to continue its activities under GARP, this project was somewhat removed from the mainstream of GEMS climate-related monitoring activities, had little catalytic value and had no impact on either BAPMoN or WCP. It was operational and hence relatively costly. The project was approved in response to the explicit request by the US, which made its contribution to the Environment Fund conditional on such action.

action rather than spending these on what often amounts to be marginally significant financing of projects and activities.

The suggested role requires greater policy and substantive strength on the part of UNEP than in the past. This can be achieved basically in two complementary ways: first, by creating a specialized intergovernmental body for GEMS which would provide authoritative and binding guidance for the entire system and its constituent parts, and, second, by diversifying and increasing significantly the staff and resources available to GEMS PAC. These steps would go some way towards remedying two of the principal weaknesses in GEMS' build-up and performance, and also improve conditions for greater activism and leadership on UNEP's part.

There is a need to secure the involvement – indeed, to create opportunities for involvement – of governments, both of the South and of the North, to formalize their commitment, to define their roles and responsibilities. There is also a need to use their collective weight in policy making as a tool to shape and energize the system, in a manner for which the Environment Fund's resources and the rather ineffective coordination mechanisms used so far are inadequate.

An intergovernmental body, with a mandate to orchestrate and provide a permanent focus for strategy, policy, coordination, debate and evaluation of the performance of the system and of the various bodies and actors involved, could fill the existing institutional void. It could also provide greater access to GEMS for the non-governmental community, which has both the interest and a role to play in its evolution.

The risk of politicization of GEMS and conflict will consequently be greater, and both the international secretariat and the scientists will find their task harder and more complicated, and certainly more frustrating on account of the usual snags and inefficiencies of the intergovernmental discourse. However, the underlying issues, rather than being ignored and swept under the rug, should be faced and dealt with. Moreover, this is the manner in which to build a broader base to the system, to revive the sagging interest in the fortunes of GEMS, and to link the many relevant activities taking place.

One of the key objectives of such a body would be to prop up global and integrated activities of GEMS and provide the weight of policy and authority at the centre of the system. Also, it should support GEMS PAC in its role as the operational and management core of the system. More generally, it should imbue the whole undertaking with a greater sense of continuity, structure and permanence.

There is nothing inherently contradictory about a strong centre for a decentralized global system, the strength being exercised not in a hierarchical but in a substantive, policy and organizational sense. Hence, the essential counterpart of such an intergovernmental body would have to be a strong and well-staffed GEMS PAC. It should have a greater creative and throughput capacity and the ability to service the intergovernmental body in its coordinating and substantive functions. In addition, it should be an authoritative interlocutor with the governments, the scientific community and the various specialized agencies and cooperating organizations.

The strength of GEMS PAC could be bolstered by mobilizing stronger and direct

GEMS: inching towards a global system? 217

support from the scientific community. One of the approaches could be patterned along the lines of the MARC model, but with financing and inputs that would make similar relationships less of an improvisation and much more serious and able to respond to true needs.

Another possible approach would be to institutionalize an independent peer review of the system, for which purpose a scientific board and a research council for GEMS might be established.

In sum, the aim should be to create at the very centre of GEMS an authoritative, critical mass, in the political, scientific and organizational sense.

The existence of GRID is likely to influence the perception and trajectory of GEMS PAC and the role of UNEP in GEMS. GRID could gradually evolve into the central international repository of environmental monitoring data, a switchboard for a global network of specialized and regional environmental data centres, and a tool for promoting assessment as an integral part of management and regulation. A truly international capability of this kind is a *sine qua non* for international environmental action, conflict resolution and management in the years to come.[7]

As noted in the previous chapter, GRID represents a unique, international global data management nucleus, whose function it is to encourage the standardization of data and to store, process and use the data generated by various monitoring networks and activities, with a potential for build-up and diversification of functions. Its purpose is to cope with the data explosion, to establish links between different data sets, to make usable assessments and to relate these with management by preparing country-, problem- and locality-specific assessment packages based on geographically referenced data sets extracted from global and sectoral databases.

As a facility which transcends the sectoral domains and performs functions performed by no other part of the system, GRID should not be objected to by the specialized agencies. Thanks to new technologies and the networking approach, it does not need a large staff at the centre, thus meeting one of the constitutional requirements of UNEP, i.e. that of smallness. Moreover, as the GRID Processor is located in Geneva, GEMS has gained a foothold in the North, an important operational and practical consideration in linking up with the emerging national and regional information systems of global significance. Accordingly, it would have the potential to act as the focus and base for a global system, giving to UNEP a tool to link up and utilize its various components and outputs. And it should give another lever to UNEP, beyond project-funding, to claim a role in guiding GEMS. Indeed, as experience has shown, one of the key premises of a global system of monitoring and scientific coordination is a data-based state-of-the-art communication system linking all organizations and scientists.[8]

---

7 GRID itself has been suspect in the eyes of some because of its close ties with and dependence on data, hardware, software and ideas originating mainly from the United States. How to manage this type of relationship and maximize the value of such links for the international community without giving rise to suspicions and conflicts is a question that requires attention.

8 This point is well captured in the IGBP Action Plan which states: 'the IGBP is by design

The role of other actors in the further evolution of GEMS

The strengthening of GEMS and maintaining its universal character is of great political and practical significance. The existing trend is towards the concentration of global monitoring capabilities in a few countries in the North, and therefore towards their dominance, indeed monopoly, in the interpretation and uses of and access to the data thus acquired. For their part, the developing countries have been largely marginalized, a trend accentuated by their lack of interest and of appropriate organization.

This is not such an innocuous matter as it may appear at first sight. Environment – and in particular the effects of energy policies on future global change – has emerged as a topic of strategic importance in international relations and in national politics, perceived as vital to national security and survival, and for the political and economic well-being of nations and their peoples. Under these conditions, environmental monitoring data will figure as a potent tool with many uses and applications. These include shaping public opinion and influencing domestic politics, and holding countries responsible for their behaviour and performance.

In an age when figures and quantitative data make a powerful social, political and economic impact and have a critical bearing on national and international policy choices and decision-making, it is a matter of the highest policy importance who controls the supply of environmental data and assessment. In a situation where the relevant scientific and technological power is monopolized by some countries and private corporations, 'environmental data dependence' and the control and uses of global environmental monitoring and assessment loom as important issues in the years to come. Illustrative of this is a North–South dispute over their respective contributions to the global carbon emissions, where the figures and assessments generated in the North were challenged by a non-governmental institution in the South.[9]

The time may be ripe to state as a major international policy objective the commitment to fill the gaps in global coverage, and to supplement local capabilities where necessary with international support, funding and technical assistance commensurate with the needs. One element of a truly global international system would be a network of internationally funded, staffed and operated observatories and research centres.

Furthermore, because remote sensing from space may well become the principal method of environmental monitoring, assessment and indeed management, globally, regionally and at the national level, the question arises whether some of these

---

multi-national and multidisciplinary, and coordination must occur at various levels of the network, including across disciplines, between and among tiers of the network, between nations and within a nation' in order to communicate findings, to transmit data stored in different locations but accessible from the data network through a central data directory, to maintain data quality, to confirm observations by other facilities, etc. 'A Plan for Action', *Global Change, op. cit.,* report No.4, 1988, p.107.

9 See A. Agarawal and S. Narain, 'Global Warming in an Unequal World; a case of environmental colonialism', Centre for Science and Environment, New Delhi, 1991. This monograph disputes the World Resources Institute (WRI) data that up to then had been broadly accepted for lack of an alternative source.

capabilities should be internationalized and placed in the service of the international community.

As mentioned in chapter six above, Japan was the first to moot informally the idea of a satellite for GEMS. While this suggestion was not followed up seriously, the idea of 'common good' and 'green' satellites is gaining ground, as truly multilateral undertakings, jointly planned, designed, launched and managed, a global public and independent service, accessible to all. For its part, the USSR declared, at the 43rd session of the UN General Assembly in 1988, its readiness to cooperate in establishing an international space laboratory or a manned orbital station designed exclusively for monitoring the state of the environment. On an earlier occasion, France had suggested the setting up of an international satellite monitoring agency.

In fact, significant cooperation in planning and executing environment-related space missions is already taking place among developed countries. The EEC Earth environment remote sensing project, the US Earth Observing System (EOS), and the Japan satellite, all scheduled for the 1990s, will revive the policy concern and the debate as regards the role, uses and control of remote sensing capabilities and environmental monitoring data. Whether and how these new capabilities will be put in the service of the international community is an important issue on the global environmental agenda and concerns the future and nature of GEMS itself.

Indeed, the possibility that global environmental monitoring systems may be set up and operated by a single country or group of countries from the North is a real one. Moreover, if some of the current trends continue, private actors such as transnational corporations of the North will move forcefully into this interesting and profitable domain, a possibility that would raise novel issues for national and international affairs.

It is essential that the international community should act jointly, before the field is pre-empted unilaterally and the situation becomes more difficult to balance. This is a challenge that has to be taken up in order that the global environmental monitoring effort may be kept within a single framework, that the participation of developing countries may be secured, and that effective linkages may be maintained with the rapid advances taking place in the industrially advanced North.

What is brewing, then, under the umbrella of global environmental monitoring is a major contest between and within countries and regions, and worldwide. In an unequal and unequally endowed world, the weaker are likely once more to get the worst of the deal.

Accordingly, it is in the interest of these countries, mostly in the South, and of global stability and equity, that this evolution should take place within the United Nations, subject to multilateral control, accommodation and negotiations. It is for these countries, then, to take the appropriate policy initiative, including sustained action by the Group of 77 which is backed by solid scientific and technical advice, which in turn has to be based on clear national policies and needs as elaborated by individual countries.

Because of its far-reaching implications for the future of planet Earth and for international relations, this matter requires the timely attention of policy makers and public awareness. This is why it would be important to create an access point to GEMS for the expression of views and interests additional to those traditionally represented and heard in intergovernmental forums.

For example, a body could be established, of the kind already proposed by the World Commission on Environment and Development, whose members would be persons distinguished in public life and in various scientific and humanitarian disciplines.[10]

Its objective would be to keep an eye on what is happening in GEMS, to verify the quality and objectivity of GEMS' products, and to see how they are used. Moreover, either on its own, or through specialized groups, it could undertake independent assessments or initiatives related to GEMS' work and outputs, and thus introduce a new dimension in its work.

Such a body would act as an important safety net and a sounding board, and would offer a degree of continuity and independence, aloof from what happen to be the current interests of institutions or shifting preferences of those directly involved in the operation of GEMS, or of the governments themselves which also tend to change their positions depending on who is in office and controls the power.

Moreover, it could act as an arbiter in certain public controversies and as a counterweight *vis-à-vis* the manipulation of data and particular interests, including those which are able to influence national and world public opinion through their superior resources or access to and control of the media.

GEMS: THE NEED FOR A FRESH START

To sum up: global environmental monitoring is gradually coming of age and becoming a potent and essential tool both for scientific study and for the shaping of national and international policies and decisions. The challenge is how to transform the existing networks and activities into an effective global and truly international system, which would be also responsive and sensitive to the needs of all members of the international community and resistant to potential abuses. The 1992 UN Conference on Environment and Development offers a chance for a comprehensive and critical look at the experience and prospects of GEMS and for a fresh start.

---

10 In its report, and in order to strengthen and complement UNEP's environmental monitoring and assessment function, the Commission proposed the establishment of a Global Risks Assessment Programme (GRAP). Conceived largely as an independent mechanism for cooperation among non-governmental groups and scientific bodies, GRAP was to be composed of eminent individuals representative of major bodies of knowledge, vocations and regions of the world. It was to offer 'timely, objective and authoritative assessments and public reports on critical threats and risks to the world community', and to provide advice and make proposals additional to those emanating from governments, experts under governmental instructions, and intergovernmental organizations on such issues as critical threats to survival, security and well-being of peoples, and their causes and consequences. See *Our Common Future*, Oxford University Press, New York, 1987, pp.321–5.

Part III
What international organization for the twenty-first century?

During a recent debate on institutional reform in one of the major UN specialized agencies, the representative of an important developing country referred to a proposal made by the representative of an important developed country as 'a series of suggestions without adequate justification, a series of remedies without adequate diagnosis, a series of generalizations unsupported by adequate and concrete data, and a case without concrete evidence'. This quote, with reference not identified deliberately, illustrates the controversial and often passionate nature of discussion of institutional reform in the United Nations and a cleavage of views primarily along the North–South divide, reflecting different expectations and perceptions of the role and functions of the organization. It also highlights the need for empirical studies as an essential input for understanding the dynamics and growing complexity of international organizations.

In reviewing the story of GEMS and placing it in a UN perspective, I trust I have made a contribution that would be useful in continuing efforts to improve GEMS specifically, as well as in strengthening international environmental action in the more general sense.

Beyond this and as flagged in the preface, similar to many activities in the fold of the UN, GEMS is a microcosm in a broader scheme of things, reflecting the underlying structural themes and issues common to the system as a whole. I focus on a select few of these in the next chapter, building on the conclusions and observations derived from the empirical analysis of GEMS carried out in the preceding section of this book. In the final chapter, I move still further into the controversial realm of assessing the present trends in the UN system and making a few suggestions regarding the need for its comprehensive reform.

A good deal of what I have to say and how I say it, and the analytical framework that I use, run counter to the dominant orthodoxy and thus may not be to the liking of many. Also, I am aware of the pitfalls of generalizing, and that I tread on tricky ground by using shorthand to address complex reality and by singling out only a few generic issues, which I consider to be at the very root of the dilemmas faced by the United Nations.

These two chapters do not claim to be a treatise, however; they are more in the nature of essays or commentaries, meant to add to the already extensive literature and a wealth of proposals on this highly complex and controversial problématique, and to contribute to a suddenly vigorous debate on the reform of the UN system. Indeed, this moment of transition and likely innovation calls for clarity, plain speaking, and fully elaborated systemic proposals and competing paradigms. Currently, the study of international organizations and the thinking and debate about them are characterized by a good deal of confusion, pretence and double standards, and are highly ideologized.

I have thus no qualms about advancing given objectives and values that I consider not only desirable and worthy in themselves but also essential for peace and development. In this context, I would be glad if my presentation were to sensitize the sceptical readers to a different way of looking at things. Most of all, I would be happy if my arguments are found useful by those seeking a stronger and more effective United Nations, and in particular by the developing countries – for whom the nature of international organizations and international regimes of the future assume vital importance as they struggle for political, economic, and cultural survival and independence.

10 Building blocks for a stronger UN

Many of those who work for the UN and who think about it seriously are bound to harbour mixed and contradictory feelings. On the one hand, they experience satisfaction with the progress made and excitement about the future based on the underlying potential and need for wide-ranging and intensive international cooperation. At the same time, and unavoidably, they are also likely to feel unhappiness with the relative marginality of global international organizations and the gap between the complex demands on their agenda and their powers, capabilities and performance.

To avoid being discouraged by daily frustrations in the life of the United Nations, they need to maintain a global and longer-term vision and to recall that the mere fact that countries do things and learn together – as the case of GEMS also goes to prove – is of intrinsic value and moves international cooperation to a more advanced stage as the years go by. And they can console themselves by the fact that in the final analysis the problems encountered are the result of the nature of the world and of the societies we live in, and reflect the still primitive state-of-the-art of international cooperation and institutions at the global level.

This does not mean, however, becoming resigned to inertia and the course of events. Working continuously for institutional reform and improved institutions is an essential element in the struggle for a stronger and more effective United Nations, as well as for broadening the scope for international cooperation. Indeed, this should be a permanent concern of international organizations and one of their important preoccupations.

The broadly defined economic and social sphere should be the very essence of the UN's task, not only because it contains some of the key challenges on the international agenda, but also because it is the causal nexus of the great majority, if not all, of political and other tensions and conflicts that are manifested in the collective security domain, still seen by some as the *raison d'être* for international organization. This distinction has been at the very heart of a long-standing controversy about the nature and role of the United Nations. Today, the argument that it is necessary to remove the causes of conflict in order to build solid and lasting foundations for peace and cooperation seems to be broadly accepted, at least in principle; the disagreement emerges, however, when these causes have to be identified and evaluated, and solutions devised and implemented.

And this is where the crux of the controversy is to be found: namely, should international organizations be agents of systemic and structural change, or should they buttress the dominant order or at least not question the existing state of affairs? These two visions became so sharply polarized during the cold war that institutional reform in the economic and social domains turned into an unceasing tug-of-war, and amounted to partial and *ad hoc* measures and meagre compromises between the North (whose collective position was based on the lowest common denominator, largely coinciding with the views of the United States which has played a pre-eminent role in these matters) and the South.[1] The developed countries were nervous about collective demands of the developing countries and about anything that might threaten to erode the dominant system. The developing countries, on the other hand, reflecting their collective and individual weaknesses and vulnerabilities, placed too much hope in institutional reform and in the United Nations as a means to obtain the desired changes and concessions from the North without in turn organizing properly and effectively to promote their interests and views. The resulting stalemate prevented the international community from engaging in a serious and sustained process of institution-building and betterment.

In view of the historical events and realignments on the world political scene, some of the important reasons for institutional deep-freeze have been removed and new possibilities opened up; as well, new risks have emerged for the UN on account of the intensified hegemony drive by a small group of dominant nations. Furthermore, the persistence and aggravation of long-standing and stubborn discords in North–South relations rooted in the global development gap, and the surge on the international political scene of complex, new issues that span traditional categories and call for comprehensive, worldwide approaches and long-term strategies in the quest for their management and solution are compelling reasons for rethinking and redesigning international organizations.

It is with this in mind that the following pages are written. Taking into account

1 A word may be in order at this point on terminology and classification of countries. This is an issue over which a lot of ink has been spilt over the decades and where views have been polarized and ideologized. Scholars and governments from the North have argued that there is no such thing as the South, that there is no communality of interests among developing countries, and that their differences and great heterogeneity make nonsense of any claim of solidarity or ability to act together. The self-classification criterion has been denied to developing countries, leading one to believe that this theological debate was in part fuelled by the lack of desire to see developing countries get organized and challenge collectively the dominant powers. Whether the North–South categories will hold, and the boundaries between groupings become less rigid and broader alliances become a reality, is an issue for the future and is becoming a possibility with the geopolitical changes taking place. Ultimately the way to peace and sustainable development is for the divisions between the North and the South to disappear and the world to become one. In this study, the 'South' is used to denote the membership of the Group of 77, which has endured for more than 25 years as a loose coalition of countries bound by common aspirations and grievances and shared traits, while the 'North' refers to OECD member countries. The East, i.e. what used to be referred to as the centrally planned economies, in the past stayed on the margins of this basic divide. Of late, however, these countries have moved into the mainstream of the North, though many of them have in more ways than one come to resemble countries of the South. As regards China, while belonging to and sharing the characteristics and aspirations of the South, it has stood apart. With its permanent membership in the UN Security Council and its size, it could afford to act on its own and not to associate itself with the group action of the developing countries. Naturally, this weakened the collective stance of the South.

both the potentials demonstrated and the major constraints experienced in the build-up of GEMS, four variables are singled out. The first three – concerning the international civil service, money and institutional dynamism – are closely related and have to do with the very character and capabilities of international organization; the fourth relates to the comprehensive policy framework needed to structure the system of international organizations and to inspire and guide their work and that of their member governments in a highly unequal and conflict-ridden world.

The title of this chapter speaks of 'building blocks' in order to convey both the highly selective nature of the discussion, and the generic character of the variables chosen and their key role in any true reform of the UN system. Symptomatically, these issues are seldom admitted when governments broach institutional matters, for such a debate would make surface the basic questions of the nature of national and international political structures and institutions, of the theoretical framework underpinning them, and of their status and role in the contemporary world.[2]

INTERNATIONAL CIVIL SERVICE

One of the underlying themes of this study of GEMS is the critical role of human resources – their qualities and motivation, the opportunities they are given, and how they are utilized – in the work of international organizations. The motivation, professional skill, and ability of those working in and for these organizations is critical for their performance. As the importance and responsibilities of international organizations grow, the stake that the world community has in these people and their profession will increase.

In a growing number of domains of concern both to peoples and governments, the desired outcomes depend on international consensus and action at the global level. Employees of international organizations by the very fact of their institutional locus assume a pivotal role both in helping to define collective interests and in devising and implementing appropriate actions. Yet their roles and expectations from them are ill-defined and controversial, and represent a subject matter that has not been addressed frankly and systematically by governments in their discussion of reform of international organizations. This is not a mere accident.

This neglect is due in part to the historical roots of the international secretariat and the broader geopolitical conditions under which it has had to evolve in the post-World War II period. It is also due to the very ambivalence of governments, which tend to be cautious and reluctant to admit the need for an influential and independent actor on the global scene in the form of secretariats of international organizations. Moreover, governments and their bureaucrats tend to view international civil servants as inferior and subordinate; the more powerful a country, the more inclined it is to claim and exercise its right. Indeed, the very term 'civil servant' is all too often understood literally and hierarchically. It conveys the inferior status of the international official in relation to the representatives of

2 For a trenchant analysis of the issues at stake, placed in the historical and theoretical context, see a seminal little book by M. Bertrand, *The Third Generation World Organization*, Nijhoff, Dordrecht, 1989.

sovereign states. It projects a bureaucratic image of his/her role and executive character of duties, and thus the posture of unquestioning submission to governments and their representatives, who are supposed to be considered as sacred and whose authority and views should not be questioned.[3]

In the broader context of the agenda and evolution of post–1945 international organizations, and leaving aside the effects of the East–West contest, it was the development imbroglio that is most responsible for many of the dilemmas faced today by the UN international civil service in its everyday work and in its search for a role and self-respect. In one way or another, the development issue has become ubiquitous on the agenda of the UN system by the very fact that the mere presence of North and South in any forum brought it on to the centre stage.

Development, as a topic, is highly controversial and highly political, and is more diffuse and pervasive than the relations characterizing the traditional political and security domains. Whether it concerns commodity trade, development assistance, transnationals, national development patterns, environment, or any other of the myriad topics on the international agenda, there were usually two, often polarized views, which mostly pit developed against developing countries. The latter were seeking logistical, technical, and often political and moral support from the UN secretariat in their uneven match with the North, and fought hard for the creation of such institutions as UNCTAD and UNIDO in order to open new negotiating fronts and to secure more committed backing than was possible via the relevant departments in the UN secretariat in New York.

In this they succeeded most prominently in UNCTAD. Indeed, the group action of developing countries was instrumental in the creation of UNCTAD, while the establishment of UNCTAD and in particular the support from its secretariat was of great value to the Group of 77 in its work and in articulating its demands *vis-à-vis* the North. In fact, Raúl Prebisch as the first Secretary General of UNCTAD came to be seen as the champion of the Third World cause on the international scene, and a man with a development mission.

Raúl Prebisch's approach to his role differed from the norm. First at the regional level in ECLA, and then in UNCTAD at the global level, he combined intellectual vibrancy, advocacy, and vitality with an interpretation of his mandate as being to foster the development of the South through change and reform of the dominant global order. His efforts to expose the built-in biases against developing countries within the existing international economic system and to challenge the theory underpinning it, his advocacy of a theoretical cum action framework regarding development and change of the global system, his ability to attract a competent and committed core of secretariat officials, his eloquent speeches, and most of all his openly activist role in organizing and assisting the poor countries to try to gain some countervailing power in an uneven match with the North and to compensate for the lack of an organizational support mechanism of their own all caused visible annoyance to developed countries.

In a sense, rather similar to Dag Hammarskjöld in the political sphere, Prebisch

3 For a series of essays on the international secretariat, see D. Pitt and Th. Weiss, eds, *The Nature of United Nations Bureaucracies*, Croom Helm, London, 1986.

may have done a disservice to the UN secretariat in the economic domain. His activist approach, though tolerated on the periphery in the regional context of Latin America, was strongly objected to by the developed countries at the global level, in particular as it was linked with the Group of 77 and the group system of negotiation. Prebisch combined experience, intellectual strength, professional and technical competence, and charisma with political courage and initiative. He embodied traits of an international civil servant that the North did not want to see challenge it and be active in the development sphere anywhere in the UN system. His behaviour and activities were considered unwelcome and alerted them to potential roles that individuals in these posts could play. They gave additional arguments to those already unhappy about the UN secretariat allegedly 'tilting' to the cause of the South. Also, they helped strengthen developed countries' determination to check the role of international civil servants and to prevent similar occurrences, by demanding 'neutrality' on the part of the staff as well as 'equal treatment' for all groups of countries.

The concept of 'neutrality' of international civil servants, i.e. literally interpreted to mean not taking sides between two or more adversaries in intergovernmental debate or negotiations and originally intended to be observed in political and security affairs, has thus been substituted for 'objectivity' and 'independence', and has been applied across the board, including research and analysis of development-related issues. What this meant in fact was that studies could not express any controversial judgements, or present data that would in any way offend the sensitivity or interests of groups of individual states, regardless of the merit of the case and the need for full and comprehensive analysis and evaluation of facts.[4]

In a world of unequals, where developing countries have failed to organize themselves and put in place an institutional support for the promotion of common interests, this insistence on the UN secretariat's 'neutrality' in development matters has meant an additional advantage for the well-organized developed countries in the defence and promotion of their views, and in ignoring the demands for change. It has also meant that the UN secretariat was not able to carry out fully and with the necessary flexibility its functions in the context of the development strategy in its more universal role and in reflecting the problems faced by and aspirations of the great majority of humankind. Nor could it probe systematically and in depth the underlying structural and systemic determinants, in particular those that govern and shape international economic relations. Effective pressure

---

4 For a discussion of attempts in some specialized agencies to discourage research and studies that challenge the existing order and the insistence that they limit themselves to operational activities, see J. Harrod, 'United Nations Specialized Agencies: From Functionalist Intervention to International Co-operation?', in J. Harrod and N. Schrijver, eds, *The UN Under Attack*, Gower, Aldershot, 1988, pp. 139–40. This is a common occurrence in the daily life of international organizations; as these paragraphs were being written, the Group B of developed countries in UNCTAD blocked the proposal that the secretariat should undertake an analysis of barriers to market access for commodities if and when the Uruguay Round is completed. 'SUNS, South-North Development Monitor', no. 2491, 16 Nov. 1990. Indeed, of late it has become part of the common wisdom within the UN that studies and documents should be 'US-friendly' so as not to incur the displeasure of this important country.

was brought to bear on those who strayed beyond the perimeter of what was considered permissible.[5]

International civil servants, at least those working in politically sensitive and controversial development-related areas, thus find themselves in a permanent dilemma. In a situation where a basic difference of view exists among blocs of nations and where the expression of their 'collective will' is either a meagre compromise or a *de facto* disagreement, how are they to behave? Having several 'masters' may provide some breathing and manœuvring space in some instances. In a stalemate that has characterized North–South relations, the steadfast refusal of the North to make any significant structural concessions, the lopsided power relationships between states and their uneven influence in international organizations has constrained their role as either brokers or advocates. Indeed, in a world of power politics, where each government would like to influence international civil servants and shape the actions of international secretariats in its own image, it is those with power, resources, and ability that find it easier and indeed possible to do so.[6]

What this amounts to is a struggle between different visions of what the development outlook and role of the secretariat should be, with the end result that the countries of the North, through their influence and the power of the purse, have been quite successful in their efforts to neutralize and weaken whatever challenge or dissent was coming from secretariats of those international organizations that follow the 'one state, one vote' principle, while shaping the secretariats of Bretton Woods institutions, which have weighted voting, in their own image and into *de*

---

5 To illustrate with the case of UNEP, at an early point in that organization's history the United States objected to a broad development-oriented approach fostered at that time by the UNEP secretariat, linking the environmental issues and situation in the South to the overall context of development, including the role of international economic relations, policies and actions of developed countries, the unfulfilled international development agenda, the patterns of development and lifestyles, and the very rationale of economic thought. Over a period of time, the US views and pressures carried weight. As the major contributor to the Environment Fund, it was also in a position to express its displeasure and exert pressure quite effectively. This complex story eventually resulted in the change of course by the UNEP secretariat and thus of the organization as a whole, pulling attention away from the structural causes of environmental degradation and moving into the less controversial and more technical area of methodology, environmental impact assessment, and cost-benefit analysis. It was small consolation to those involved that the events in the years that followed amply vindicated the initial arguments and the need for a comprehensive approach, and that the World Commission on Environment and Development (WCED) replicated some of the early arguments. At the time when they needed it, political support and protection were hard to come by and mobilize effectively against the unilateral pressures of the major donor, in part because of the disarray in the Group of 77 and its marginal role on the UNEP scene, as well as because of the general reluctance of virtually all governments to face in earnest some of the difficult underlying issues.

6 Shirley Hazzard has written extensively on the history of the formative years of the UN and the practice of direct intervention by the United States in its recruitment and personnel policy. According to her, one of the objectives was to weed out and exclude confirmed internationalists from the international civil service, and even more so to keep its outlook and behaviour in line with US policies and preferences. For example, see 'Breaking Faith', *The New Yorker*, 25 Sept. 1989. These influences, whether from the US or from other governments, continue in their subtle and not-so-subtle forms. Observers have pointed out frequently that the system of selection and hiring of personnel is in fact discriminatory against those not born and educated within the prevalent culture of the trasnational system. V. Sánchez and H. Santa Cruz, 'Política en los Organismos Mundiales Multilaterales', ILET, Santiago, 1989, p. 11.

facto instruments of their own policies and views of how to deal with the global development challenges.

These underlying tensions and constraints have tended to dampen the initiative and the élan of international civil servants and to limit their freedom of analysis and advocacy; sometimes they are used as a pretext by the officials themselves for following the comfortable path of least resistance and minimum effort.

Nor was the objective of building an authoritative secretariat favoured by the personnel policies and practices common in the UN system. The distribution of posts and appointments has become like a global bazaar, with all governments fighting to secure posts for their nationals and protégés and to maintain or expand their entitlements (e.g., monopolizing certain key posts at the high level), seemingly more concerned with this aspect of the UN than with anything else. On the whole, this has worked against leadership, independence, quality, or optimal distribution of talent.[7]

Within the secretariats, the narrowing down of scope for effecting the desired changes in the world and fulfilling the mission of the UN has meant less job satisfaction and often a sense of futility. This is compounded by the usually rigid bureaucratic and administrative straitjacket and the resulting frustrations about the obstacles to recognizing and promoting adequately quality and excellence, both in daily work and in the opportunities for the best to move up in the ranks or, indeed, in renewing staff in a dynamic manner. This situation has bred mediocrity and has contributed to apathy among staff, and to their being preoccupied with intramural politics and security of their own tenures often at the expense of broader organizational goals.[8]

At the highest levels, the need to survive and thus not to incur the displeasure of important governments has tended to reduce boldness and risk-taking among management and has undermined leadership and innovation.[9]

---

7 For an insightful overview of the current practices and problems of UN personnel policies and their historical roots, see the background paper on organizing the UN system for leadership in the 1990s, prepared in the context of a study jointly sponsored by the Dag Hammarskjöld Foundation and the Ford Foundation and written by Brian Urquhart and Erskine Childers, entitled 'A World in Need of Leadership: Tomorrow's United Nations'. This study and the background papers were printed as a special issue of *Development Dialogue*, 90: 1–2. It appeared when this chapter was virtually finalized. It coincides in many of its diagnoses and observations with my own analysis. Its recommendations are aimed primarily at the process of selection of the Secretary-General of the UN so as to increase the likelihood of selecting the person who would be well-equipped to exercise the leadership function effectively. The very fact that the study speaks of the leadership function and advocates a central and much greater role for the Secretary-General in the socio-economic domain highlights the needs and unfulfilled potential in this post.

8 A seldom noted fact concerning the 'pecking order' among international organizations refers to the higher salaries and benefits enjoyed by staffs of the Bretton Woods institutions as compared with the rest of the UN system. In a highly competitive world, where the paycheck has become one of the most important considerations, the UN has lagged behind. This differentiation has not always been there. It seems to have its origins during Hammarskjöld's tenure and the policy of major countries to put stricter control on the initiative and roles of the Secretary-General. Today, it also reflects the favoured and pre-eminent status of the Bretton Woods institutions.

9 In some cases, however, boldness and risk-taking did not prevent certain executive heads from being re-elected contrary to the will of some important governments. Indeed, they showed that skilled and politically astute players can hold the upper hand over governments and create a power base of their own. The personality traits and abilities of the candidates and incumbents for the top executive posts

On the whole, then, the international civil service in the UN has lagged behind in fulfilling a dynamic and leadership role on the global scene commensurate with the challenges, or even with the limited expectations that one might have. As argued above, this erosion of the role and influence has been a consequence of many factors at work, embedded in the stalemate between opposing positions on some fundamental issues facing the world community today, and the quest for influence or dominion over the secretariat by governments. It is this political stalemate that has boosted the position of those who do not recognize the need for a strong, independent, articulate actor on the world scene, in the form of staffs of global international organizations. This has been reflected, for example, in their favouring for top jobs 'safe' personalities, who would not show excessive independence of spirit and thought, display unwanted initiative, or create trouble. Undoubtedly, this has not helped international organizations to win influence and a leadership role in the international arena.

A slight variation on the above theme, which only goes to confirm the basic argument, is to be found in the Bretton Woods institutions. As the establishment institutions that buttress rather than challenge the status quo, and that also perform significant functions as far as the major countries are concerned, and where the developing countries largely go along with the system, the developed countries have shown an interest in both maintaining the quality and numbers of staff and in placing at their disposal the necessary resources and facilities. As will be discussed in the next chapter, combined with their mandate, the rewards they can bestow, and the deprivations they can impose on many governments, this has been an important factor in their being key international organizations and eclipsing the UN proper in the economic policy sphere. However, when it comes to initiative, freedom, and independence of action, the staffs and heads of these organizations are kept on a very tight ideological and action leash of strict adherence to the dominant system and the prevailing body of economic theory that they are to defend, and of need to respect the policies and wishes of a few dominant nations on their boards of directors.[10]

Turning to GEMS, while its evolution and fortunes were strongly influenced by the geopolitics of environment and development, those international civil servants

thus emerge as a significant variable in the performance and trajectory of international organizations. One of these concerns the ability to function in the high pressure environment and to cope simultaneously with the multiplicity of demands and policy games, as well as with the administrative, paper and meetings overload common to such high level jobs in the UN.

10 Paradoxically, then, the staffs of various other organizations within the UN system may in reality have a greater latitude of thought and action than the staffs of IMF or the World Bank, by the very fact of greater diversity of thinking among their member governments, which is allowed to surface precisely because of the more democratic modes of decision-making and expression. If this logic is followed, then one could argue that a secretariat like that of OECD has an even lesser margin of manœuvre than those of the Bank or the Fund, and that, having an ideologically tight-knit group of states as its members, it resembles much more a national bureaucracy which is under the command and control of those in seats of political power. To what extent an international bureaucracy is capable of sensing and capturing this inherent advantage of leadership and innovation is closely related to its qualities, vision and independence of thought, all of which are the subject of discussion in this section. This may also explain the open reluctance of dominant nations to allow institution-and quality-building in secretariats of those organizations where the reigning paradigm might be doubted or challenged in any serious manner.

working on its implementation were largely spared many of these underlying tensions. For the major part, the programme moved in the relatively non-controversial domain of system build-up and improvement. Combined with specific institutional advantages, this made it possible for them to act with a degree of creativity and latitude that is usually not encountered in the UN, and though only a few in number, to carry on with a global activity with a significant degree of success, highlighting the role of inspired and qualified individuals in these posts and of the facilitating conditions for their work.

However, with assessments based on the monitoring data becoming a major preoccupation, both opportunities and the need for a more active and committed stand will grow, giving rise to political and scientific controversies and demanding from the international civil servants involved greater activism and political exposure.

Indeed, a plausible argument can be made that, in more general terms, the topic of the environment and the need for and possibility of arriving at a definition of a common planetary interest on the basis of some objective criteria and indicators, using environmental monitoring data-based assessments, will open up greater scope for a more independent and assertive international secretariat, drawing its major support and inspiration from an agreed global vision and world public opinion and civil society.

If the secretariats of international organizations are to fulfil their functions in the global quest for sustainable development, the governments will eventually have to agree that they should be given a greater and more independent voice, by definition a 'non-neutral' one in the sense that in defence of the planetary goals they might have to act and argue against particular interests and views of individual countries or groups of countries, which includes such sensitive matters as national economic policy and the structures governing international economic relations.

The recognition of the fact that the international civil service – which in a way reflects its global constituency – should be more than a simple bureaucracy and should have a recognized role to play in formulating, implementing and continuously adapting policy to evolving requirements and lessons is an important building block for international organizations of the future.

As this scenario largely corresponds to the evolving state of the public mind in the North, the environmental compulsion, and the overarching global consensus that it might yield, it may yet turn out to be responsible for eroding the long-held national views and for opening a crack in the door towards stronger and more independent secretariats of international organizations to fulfil the necessary functions and exercise a growing leadership role on the world scene.[11]

11 The other side of the medal needs to be recognized also. Precisely because of the increasing stakes, the desire of the governments to control the secretariat and to use it for their own purposes will also increase. This will sharpen the existing dialectic between the dominant few nations and the rest, and in particular between the North and the South. The environmental issues – being as they are heavily science- and technology-biased when it comes to monitoring, assessment as well as management – will give a major advantage to those countries advanced in these domains. The dependence on these countries for data, assessments and ultimately many of the solutions, will make it easy for them to wield the desired influence, without much effective challenge from others. These underlying trends are becoming evident, as seen in the discussion of GEMS in the preceding chapters.

While governments may find these trends uncomfortable and disturbing in terms of institutional and policy implications, they will need to accept the fact that international civil servants have a special and independent role to play and that they should be encouraged in this, given the necessary means, and placed in a politically favourable and facilitating organizational milieu if international organizations are to measure up fully to their task.

These people work in the global context and are exposed to impulses and views of the broad constituency, which places them in a position where they can evolve a posture and perception that transcends particular interests. This is essential for dealing with increasingly complex global problems. And their function gives them a degree of influence and initiative and provides them with a unique institutional platform to help define shared interests and to point the way, as a service to the international community.

The human factor thus emerges as a key variable for institution-building and functioning of international organizations. The task of these organizations is challenging and they are bursting with unfulfilled potential. This places a premium on excellence, on highly motivated and capable people, able to overcome constraints, create new and maximize existing opportunities, and infuse their organizations with vigour and the spirit of continuing renewal and improvement. It is these people who should build and defend the respectability of international organizations, fight against the bureaucratization of the secretariats, build and defend their own independence and roles, and wield influence on the international scene through political skill, integrity and courage, intellectual rigour and excellence, professional and technical competence, and a global vision.

This is why such concerns as improved recruitment and personnel policies, career development, and reducing the proportion of time spent on bureaucratic chores could play a vital role in building international organizations of the future. In particular, attention needs to be devoted to careful selection of the top echelon of the UN system, drawing on the rank and file working in international organizations, as well as bringing from outside people with necessary merit and qualifications.[12]

This does not mean simply transplanting the administrative, technocratic, managerial or scientific qualities and professionalism common in the complex organizations of the modern post-industrial society, although these are quite important and essential ingredients for successful and dynamic institutions. It also means talent, inspiration, courage, dedication, enthusiasm, experience, and learning the job in the international setting.

12 The Urquhart–Childers study is thus a welcome event, as it opens up a debate on some important issues concerning the performance of international organizations. It draws attention not only to the role of the Secretary-General and by implication that of the executive heads of other organizations, but also to all the top positions in the system, whose occupants largely determine both the posture and the performance of the international secretariat. It is these individuals who act as transmission belts of leadership and the key nodes in the management structure, and as such inspire, lead, and organize their staffs. The study uses the term 'cascade effect' to denote the impact of individuals in key posts on the rest of the staff, calling it either positive, negative or indifferent. See 'A World in Need of Leadership', *op. cit.*, p. 86.

And it means transcending the ethnocentrisms and frequently the arrogance of national and regional backgrounds, imagery, and languages. It calls for a broad knowledge of cultures and languages backed by an internationalist ethic, a global vision, humility in one's own job, and empathy with all strata of humanity, especially the weaker and underprivileged ones, who have little chance to promote their interests and aspirations in a harsh world of power realities. The ethics and the conceptual framework inspiring and guiding those working in the international civil service are of paramount importance for the evolution of international organizations.

All this calls for a purposeful and longer-term build-up of international cadres and a careful selection and special education – prior to joining and during their tenure – of a pool of talented international civil servants of top intellectual quality, professional excellence, and resourcefulness, who would dedicate themselves to work in international organizations and find this not only materially but also professionally attractive and politically, intellectually, and emotionally rewarding.[13] They should grow in their jobs. They should learn and continuously develop their tools of trade in the international environment and through multilateral processes, as well as in the field, to be able to rise to the challenge of important posts in these organizations. A deliberate effort will need to be made to keep alive the motivation and dynamic spirit among the staff, for example, through recurrent training and through exposure and circulation of the staff between and within organizations.

They should be given high public recognition and attracted to this work. Their task should be facilitated and encouraged, while they should be required to fulfil the necessary standards of excellence in their performance, with a premium placed on innovation and skill in working in the international context. Ultimately, the public respect and support will be gained only through professional quality of work, the value of ideas they generate, and the services that they perform.

Although fully protected against political pressures and the vindictiveness of governments and indeed of their own administration, they should be subjected to periodic, independent evaluations of performance and professional competence. Indeed, a judicious balance should be found between the need to protect the security and tenure of staff from political pressure and the need to obtain appropriate levels of performance.

Parenthetically, a view commonly held in the North, and propagated by its media, is that the erosion in the quality of UN staff is due to political appointments and to the national quota system that has both reduced the proportion of, by definition, qualified Northerners and brought into the secretariat less qualified, usually 'non-white' individuals who come predominantly from the developing countries. This has projected an image of the South as being to blame for these

---

13 A graduate school of the United Nations could be established that would admit only the best and motivated candidates to be provided with advanced training for international civil service. (The Ecole Nationale d'Administration in France could provide some useful lessons on how to proceed.) Many governments would not look with favour on such an idea. Yet this would contribute to building gradually top-notch cadres and solidarity among them, as a way to institutionalize excellence at the very base of international organizations and secure for them public respect and confidence.

trends in the UN secretariat and for not possessing human resources of adequate quality for work in international organizations, while the North abounds in talents and professionalism that cannot find their way into the UN because of the developing countries and their claims for representation.

This certainly would be a matter worth some empirical research and comparative analysis, especially because it has been used to attack and disparage the efforts to democratize the UN and make it more representative of its broad constituency. For those who have been in the ranks, the alleged intellectual and professional superiority of the Northerners is not all that evident – in particular as many of the developed countries have a deliberate policy of not releasing their best people for service with the UN organizations. Moreover, the nationals of developed countries still account for the bulk of the UN staff, especially at the higher levels, reflecting national quotas based primarily on budgetary contributions.[14]

A long-term, systematic build-up of cadres for the UN within the newly elaborated and adjustable quota system, which is aimed to reflect broad groupings and interests – rather than countries' paying ability – based on merit, experience, and training, could help overcome some of these drawbacks. Moreover, staff should be allowed to move through the ranks to the very top positions in their organizations, posts that have become a virtual monopoly for outside appointments, thus contributing to the stratification of the staff and lack of leadership perspective for those working their way up through the ranks.[15]

The high-quality staff, however, should not be limited only to those working

14 As concerns the top leadership posts in the UN system, during the period 1946–90, of the 136 Executive Heads, 57 or 42 per cent were from Western Europe and 40 or 29 per cent from North America. In other words, the North accounted for 97 heads of organization, or 71 per cent of the total (79 per cent during the period 1946–70, and 62 per cent in the period 1970–90). See table 5 in 'Statistical Analysis of UN System Leadership Posts', Annex in 'A World in Need of Leadership', *op. cit.*, p. 100. This pattern is even more pronounced in the case of 17 major organizations of the UN system, where of 83 appointees in the period 1946–90, 62 or 75 per cent were from the North and 21 or 25 per cent from the South (Table 7, *ibid.*, p.103). The fact remains, however, that international organizations are largely modelled on Anglo-Saxon institutions, with the English language being dominant, which places at an advantage those coming from given cultural and geographic backgrounds and makes it easier for them to fulfil their role. Individuals who do not speak English could not aspire to leadership posts in the UN system; were they to get there, they would be severely handicapped in the performance of their duties and in their effectiveness within the secretariat. A person who speaks only English, in contrast, would qualify automatically for these jobs, in spite of the fact that he or she may not know a single word of another language; those who have English as a mother tongue enjoy an added advantage of ease of writing and communicating, a domain where others have to use significant additional time and energy. This problem seems to have been overlooked in the Urquhart–Childers study when they propose that individuals who have been prime ministers in national governments should be considered for the post of the Secretary-General. The language requirement would severely limit the number of former prime ministers who would be eligible.

15 Only a few international civil servants move through the ranks beyond the director level, except in administrative posts (e.g., finance, personnel, general services). This practice deprives the organization of the experience and talent of their own staff in leadership positions, and increases the chances of inadequate appointments to top posts. This is an important issue for any reform of the UN system. Urquhart and Childers touch on this question in passing. Their criteria for selection of individuals for top posts in the UN system imply that recruitment nets will have to be cast more widely to include also those who have grown and matured within the organization and acquired international skills and perceptions essential for the leadership roles.

directly and on a full-time basis in the international organizations: an ever broadening network approach should be utilized to harness talents, energies, and the creative drive of interested and committed individuals in all parts of the world and in all walks of life in joint endeavours on behalf of the international community and in the service of the public interest.

This approach would help expand and diversify the base and thus give greater dynamism and induce a continuous flow of ideas to secretariats that tend to be drawn into the bureaucratic rut and insularity of large organizations. This would also be a manner of creating backward and forward linkages between global international organizations and the social and political base in national societies and in regional and other groupings. Only such involvement of people and back-up from public opinion could help secure and maintain broadly the support and institutional base for strong, articulate, truly internationalist secretariats of international organizations. Modern communication and information processing technologies can play a vital role in making such networking effective, efficient, and comprehensive, and in expanding the notion of international civil servants beyond the traditional meaning referring to those engaged on a full-time basis or occupying a permanent place within the four walls housing their organization.[16]

In summary, excellence and distinction from within at all levels, especially at the very top, but also at middle and lower echelons, and the related *esprit de corps* is a vital building block for the international organizations of tomorrow. The post of the UN Secretary-General merits special attention as it provides the tone and direction for the whole UN system, and plays multiple roles, including those of leadership in domains of policy and ideas. For the Secretary-General to fulfil these roles individual qualities are essential, combined with institutional support and the general acceptance of the fact that the person is cast in the position of the number one citizen of the planet, speaking on behalf of and promoting the common interest.[17]

Excellence without purpose and a vision is not sufficient, however. It is this purpose, drawn from the global constituency, and the ethic underlying it that will

---

16 Along these lines, the Urquhart–Childers study has proposed the creation of 'special task forces' and 'short-term intensive units' as a means to secure the necessary talents, leadership, and institutional back-up in dealing with given issues. 'A World in Need of Leadership', *op. cit.*, p. 77.

17 Urquhart and Childers characterize the Secretary-General's job as the most difficult job on Earth, requiring both the mastery of an immensely complex and broad-ranging subject matter and the ability to keep above the tide of events, multiple pressures and daily chores, which are overwhelming at the top of the pyramid where he sits virtually all alone. As a consequence of this overload which is magnified by management and protocol detail, they note that in fact the incumbents of the post have largely ignored social and economic matters, having assigned priority to peace-keeping and political functions. According to the authors, one of the ways to help overcome some of the dilemmas of this office, improve the Secretary-General's chain of command, and increase the leadership potential for his post would be to create posts of three deputies (*ibid.*, pp. 21–2). (This proposal is further elaborated in a paper by the same authors entitled 'Reorganization of the United Nations Secretariat; a Suggested Outline of Needed Reforms', New York, Feb. 1991. It suggested a fourth deputy, on humanitarian and human rights affairs, to the three proposed earlier, namely: political, security and peace affairs; economic, social, development and environment affairs; and administrative, management and conference services. The choice of the Secretary-General turns out to be a critical link, especially so as the changing global political constellation may present the incumbent with greater opportunities for action and for exercising the leadership potential.)

orient the secretariat and help it stay on course in its daily dealings with international politics and myriad, often contradictory, demands from member states and other actors on the international scene. The purpose should also inspire it in its chief objective, that of a leading force for a better world, which also casts the secretariat in the role of a challenger and advocate, a gadfly, and a voice of the collective conscience of humankind, and may often set it on a collision course with some of the established interests. Governments will have to learn to live with and value a dynamic and independent international civil service, which if it is to execute its many roles effectively will need to be accorded greater powers, institutionalized protection, latitude of action, and confidence by the world community as a recognized actor on the world scene.

Excellence, however, will not be able to flourish in an institutional environment and within structures that stifle initiative and do not provide enabling conditions for work; nor can the secretariats of international organizations be cast fully in their global roles as long as some powerful countries aspire to control or neutralize them, or even to turn them into obedient instruments of their national policy. In this context, one of the key variables has to do with financing of international organizations.

FINANCING INTERNATIONAL ORGANIZATIONS

One of the refrains in the chronicle of GEMS concerns the critical role of the financial resources that were available for its build-up.

Money is needed to pay for communications and travel, to bring people together to meet, discuss, and work, all of which are the lifeblood of international organizations. Money is also needed to pay for administrative costs and to hire staff and consultants, to commission research, and to execute various operational activities.

Money is required to involve developing countries in various activities and to assist them in their national efforts. Indeed, it is the North–South disparities that are responsible for at least a part of the dilemma faced by global international organizations. The needs of the South substantially increase the total bill that has to be paid for their operation; yet what is available is a drop in a bucket when compared to the real needs. Also, the unequal ability of countries to contribute gives rise to tensions and undermines the democratic character of these institutions.

Money means power, greater outreach, and greater independence, freedom, and latitude of action for the staff of the secretariats of international organizations.

The role of money is understood by governments all too well. This is why those who foot the greatest portion of the multilateral bill – i.e., the regular budget derived from countries' assessed contributions and extra-budgetary funds (voluntary funding and *ad hoc* funding), which together make the total cash flow of any given organization – feel that they are entitled to call the tune and use their powers of the purse in shaping policy and programmatic orientation of these organizations and their secretariats. This is at the very root of the secretariats' dependence on governments and of their vulnerability to governmental pressures.

One of the principal weaknesses of international organizations, and indeed their

Achilles heel, is the lack of independent, adequate, assured sources of finance. This is coupled to the general impression that such organizations cost too much in relation to the benefits that they yield, and are inefficient and wasteful, with the major share of the organizational dollar eaten up by administrative costs.

The money available, however, is usually not adequate for optimal functioning of various secretariats; it tends to be dramatically insufficient relative to programmatic needs and to the high costs of many activities, were these to be carried out properly.[18]

The story of GEMS amply demonstrates a common theme in the UN, namely inadequate funds or underfunding. It is for lack of money – a situation aggravated by its frequently inefficient uses – that activities are not undertaken, or are executed superficially. For the same reason, countries fight unending and bitter battles over priorities, in the end spreading far too thin what is available, thus undermining important and ambitious programmes and activities that have to be designed and executed in a marginal and often improvised manner on account of the limited resources available.

Likewise, it is because of the paucity of resources that international bureaucracies compete intensely for mandates and operating space, within and among organizations, thus intensifying the underlying jurisdictional, sectoral, and administrative tensions and jealousies, and diverting time, attention, and creative energy of international civil servants from the pursuit of the organizations' primary objectives. Rationing of money is also used by some governments to exercise pressure – to stop or hamper given activities and organizations while promoting others, thus introducing unilateral influence in their work.

One reason why the money factor has been singled out in this study is to illustrate how the tight-fisted attitude of governments towards international organizations subverts their own decisions and declared intentions. Indeed, it is because of this attitude that many of the UN programmes can so easily be the subject of caricature and are vulnerable to criticisms, especially from those already sceptical and hostile to the UN system. This leads to a vicious circle of declining support and disillusionment with outputs on account of initially inadequate financial inputs, and to a serious imbalance between objectives and the means available for their achievement.[19]

18 This is not to say, however, that the money available is well and effectively spent, or to deny that there is much inbuilt waste in procedures and practices, or to argue that the quality of performance or the effort of the UN staff is on a par with what it costs. This has been one of the flaws in the posture of the UN secretariat whenever it tried to defend itself against criticism, budgetary and personnel cuts, and other streamlining demands made by the developed countries. In fact, inefficiencies, waste and excessive bureaucratization are inadmissible and harmful, especially in a situation of financial stringency. It should be added, however, that these are inevitably also a product of large and unwieldy structures, of sometimes chaotic and undisciplined intergovernmental process, of lack of coordination within and between governments and/or organizations, and of procedures, practices, and decisions imposed by governments, and should not be attributed simply to the secretariat.

19 This vicious circle has been characterized as 'no money – no performance – no money'. See P.S. Mistry and P. Thyness, 'Financing the Multilateral System in the 1990s', The Nordic UN project, Report No.13: 1990, p. 15. While this study deals with the development and technical assistance side of financing the multilateral system, it offers a number of insights and incisive comments with

As shown in chapter nine, for the first decade of GEMS, the Earthwatch budget line of the Environment Fund made it possible to allocate to GEMS-related activities a total amount of about US $15 million. While this may appear a large amount and though it denotes what was primarily meant to be seed money that would attract additional and significant resources to the implementation of GEMS, this allocation over ten years is in fact less than the cost of a single advanced fighter plane, or the earnings of a heavyweight boxing champion from a single match, and surely is evidence of gross imbalance in the scale of societal priorities and values. If one takes into account that a large proportion of these US $15 million unavoidably had to be spent on lubricating the machinery, including the running costs of GEMS PAC, the high costs of various activities, and the many unfulfilled or partly fulfilled objectives on account of inadequate finance, then it is even more striking how frugal was the international community in allocating monies for GEMS.[20]

Preparing for the future of the planet on a shoestring budget should be to the credit of international organizations and their inherent potential. It also goes to the credit of the founding fathers of UNEP who conceived the institutional scheme in Stockholm, which made it possible to mobilize much greater resources than were likely to be available centrally. Yet, on the whole, the money factor was at the root of most dilemmas, difficulties, and shortfalls experienced by GEMS.

GEMS is but an instance of the broader malaise affecting UNEP, discussed in chapter one, that of minuscule finances available in the Environment Fund relative to the needs, which had to be dispersed among many deserving programmes. Not surprisingly, this contributed to UNEP's weak performance and inability to meet expectations, especially as concerns development–environment needs where a major gap emerged with serious consequences for the state of the global environment. In turn, this made it easier for key governments to confirm their scepticism about the UN and to write off UNEP, and for other institutions, most prominently the World Bank, to muscle in to fill the vacuum. This was illustrated vividly when the lacuna was finally recognized some twenty years after UNCHE, and the Global Environmental Facility (GEF) – projected for $1.2–2.4 billion for the 1991–3 period – was created and, with some UNEP and UNDP participation, *de facto* entrusted to the World Bank as the institution in which Northern contributors have confidence. This was done in spite of doubts that, being dominated by the major developed countries, the Bank is the proper forum to handle the sensitive and politically explosive interface between the North and the South on environment and development.[21]

---

broader implications. In relation to this vicious circle, the authors note that the combination of inter-organizational wrangling and competition and the 'one country, one vote' system prevalent in the UN has led to growing frustration among donor countries, who have come to perceive the World Bank as a 'multilateral oasis' to which to transfer mandates and activities (*ibid.*, p. 56).

20 For the breakdown of expenditure see chapter nine, footnote 6, on p. 213. In absolute terms, the situation improved somewhat in the period that followed, with close to $2 million spent per year on GEMS-related programme activities. The total figure for the 25-year period (1974–1988) thus stood at $28 million. See 'GEMS: Global Environment Monitoring System', UNEP, Nairobi, 1990, p.7.

21 Mistry and Thyness comment on these trends (*op. cit.*, pp.14–15, 26). They note that even the Brundtland Report failed to face squarely the issue of inadequate funding for environmental action.

In a way this was a belated recognition of the deliberate oversight at UNCHE and in the original conception of the Environment Fund. However, while the GEF endowment may appear a major advance in comparison with that of the Environment Fund, the financing approach practised by the Bank is likely to exclude many of the key issues and needs, while the sums available remain quite inadequate compared to true global needs. This leads one to conclude that governments still have not drawn appropriate conclusions from the past experience.

Indeed, governments tend to rely primarily on their own national priorities and criteria when assessing the financial needs of international organizations. The question is seldom asked how much is truly required to carry out a given function properly or to fulfil the organizational task; the dominant concern is with the bottom line, i.e., how much it will cost and what is their own contribution, often to the exclusion of substantive goals or without worry about content, structure, and method of work.

True, the financing needs of international organizations may appear large in absolute terms, and are continuously growing and diversifying. This undoubtedly makes for short tempers in finance ministries and legislatures, especially of those countries that are major contributors or donors to international organizations and that have to allocate the resources to pay assessments or contributions to special purpose funds. In general, they are not eager to pay for structures and activities where the results are often ephemeral and difficult to pinpoint and prove, and are keen to see a direct link between their 'investment' and the end result, often judged on the basis of narrow national interest criteria and policy preferences.

Their patience is even thinner if they have little sympathy for multilateral objectives and/or global international organizations, or if they perceive the UN as being inefficient, staffed by an 'overpaid, oversized, unqualified and underworked bureaucracy', where an 'irresponsible multitude' of developing countries who contribute little financially each have an equal vote when budgets are decided upon, and where a 'subversive or alien ideology' not consistent with their world view is being advocated.

The unequal financial and economic power of members of the UN, further accentuated by North–South cleavages, and thus their different abilities to contribute are important structural dilemmas of international organizations and weaken their democratic character and independence.

Significant changes in relative strengths and abilities of countries are not to be expected in the foreseeable future. Nor is it likely that the prevailing attitudes will change quickly or easily and that the financing of international organizations or programmes via traditional modes of governmental contributions will become any smoother or more generous in the time to come. Indeed, if the US precedent is any guide, unless the trend is checked forcefully it should not be suprising to see other major financial contributors following its example and feeling it legitimate to use their position as a lever to impose conditionality and thus to influence international organizations – their ideology, programmes, policies, management, staffing – unilaterally. Moreover, hostile public opinion and pressure groups, conditioned by the media, could make their influence felt in legislatures and at the stage of budget formulation and approval.

However, in contrast to the rest of the UN, multilateral development banks and financial institutions will continue to earn handsomely for their own upkeep and activities and to use these assests for task expansion, to fund their operations and extend their influence worldwide, increasing their power and marginalizing other organizations of the UN system, an issue discussed in the next chapter. The fact that these institutions generate their administrative budgetary resources internally – in the case of the World Bank, the annual budget exceeds $1 billion – 'gives them far greater flexibility in programme innovation, development and expansion than is the case with the UN organizations and agencies'.[22]

It seems, therefore, that for international organizations not engaged in banking and finance the only way to raise adequate resources without strings attached will be through non-traditional, independent, and/or automatic means of financing, intended to generate resources for given international programmes and for permanent activities. Among the many schemes that have been mentioned are international taxation and income collection pegged to the growth of world product, special purpose levies, non-profit economic ventures including exploitation of global commons, royalties on new products and technological processes derived from global commons and biodiversity, taxes on pollution and energy intensity, taxes on TNCs, private initiatives, and even global lotteries.[23]

These modes of financing would help overcome the undesirable side-effects of 'the major donor syndrome', by increasing the money available and diversifying its supply base away from governments, parliaments, and budgetary cycles and by making financing of international organizations immune to and independent of political changes and moods in particular countries. It would also help shield organizations from daily pressures, whether on behalf of national interests or of specific interests, which tend to throw off balance the collective will and put at a disadvantage those governments or interests not in a position to pull their weight.

It would strengthen international organizations and help them pursue their activities in a more ambitious, steady, and independent manner. And, by spreading the costs, it would make the whole process less painful for governments. It would also make it possible to revise the personnel quota system in the UN, whereby the big contributors have reserved for themselves a major share of posts, including some key ones, and would thus make possible a more equitable, independent, and quality-oriented recruitment process.

Naturally, to argue for automatic and adequate financing is not to deny the vital importance of accountability, transparency, effective financial control, and monitoring of the use of financial resources, which at present is lacking. Indeed, the combination is essential for the political credibility and sustainability of the

---

22 *Ibid.*, p.63.
23 The lottery idea might be the easiest to implement and for sure would be welcomed by common people, who would see it both as a means to contribute to a shared cause and a possibility to win a prize. Governments, on the other hand, may not be that enthusiastic about the idea which, admittedly, goes a step further than the widely accepted UNICEF cards and calendars approach.

proposed arrangements. It is also needed to overcome the dilemma flagged above, namely that 'of a system being underfunded yet unable to utilize the scarce funds that are provided to it and through it with sufficient efficiency, effectiveness, transparency and accountability to satisfy the providers of those resources'.[24]

The idea of non-traditional financing is not novel. It was mooted at the intergovernmental level as early as 1972, at the Stockholm Conference, which recommended that it be explored. Based on this, UNEP commissioned a study on the subject, with the idea being eventually endorsed by both the Brandt and Brundtland Commissions.[25]

While many vested interests oppose this notion, and while it goes against some long-established principles of national sovereignty, the time has come when it will have to be given serious consideration by the international community. In this domain, as well, the global environmental interests and objectives are likely to provide the necessary impulse for an initial breakthrough in international attitudes, perceptions, and cooperation. In view of the usual difficulties of obtaining consensus and the iron law of the 'lowest common denominator', those countries willing and interested to participate could begin to operate limited pilot schemes, also as a way to get the idea better established and to gain the necessary practical experience.[26]

UNEP's catalytic approach – by enlisting others to spend on their own activities in support and as a part of international programmes – has shown another possible way to obtain additional funding supplementing that available in regular budgets and special funds. As seen, this approach has worked better in the North, where the experience and the habit already exist, where resources are plentiful and can be mobilized more readily by attracting contributions in kind, including those from the non-governmental sector. Possibilities of this nature, however, also exist in the South, and give added scope for contributions to developing countries strapped for scarce foreign exchange.[27]

---

24 *Ibid.*, p. 2. The differentiation made between the World Bank and the UN type of organizations by the donors is that in the case of the latter they claim that they cannot hold the staff accountable for poor performance or lack of performance. In the former, the Board of Governors meets continually and is supposed to be in full control (*ibid.*, p. 49).

25 For the UNEP-commissioned study, see E. B. Steinberg and J. A. Yager, *New Means of Financing International Needs*, The Brookings Institution, Washington, D. C., 1978. See also Independent Commission on International Development Issues, *North–South: a Programme for Survival*, Pan, London, 1983, and World Commission on Environment and Development, *Our Common Future*, Oxford University Press, New York, 1987.

26 In this context, it is interesting to note that the first annual summit of the Permanent Summit for South–South Consultation and Co-operation, composed of 15 important developing countries, adopted as one of its guiding principles the idea that when some of the members decide to proceed with a given action, others who may not be ready to join will not object or block them from doing so. If and when they change their mind, they are free to join. See 'The Communiqué of the Group of Fifteen on South–South Co-operation and Consultation', Kuala Lumpur, 1–3 July, 1990. This represents an attempt to soften the blocking effect of the requirement for full consensus in the fold of the Group of 77 and Non-Aligned Movement, where many useful initiatives could not be followed up on account of opposition by one or a few countries.

27 As travel and communications costs are often a burden on a tight budget, this could be an area to seek some relief for the regular budget through contributions in kind. For example, employees of IATA, civil airlines, and travel agents and their families travel by air on personal business or for

242 The quest for world environmental cooperation

In concluding these comments, the point to be stressed is that political leaders, governments and public opinion need to be educated as to the vital importance of financing adequately the work and functioning of international organizations. They should also accept that international organizations need to be free of unilateral financial pressure, claims and interference of the large contributors, and that they cannot be held hostage to their political preferences and national interests. After all, international organizations are not business corporations to be owned and directed by those who pay most for their upkeep and programmes, i.e., their major shareholders.

The politicians, governments, and the public must be also made more aware of the nature of processes and outputs, and the structure of the costs of international organizations, what the money is spent on, and what is the total of the bill paid each year.[28] This is essential in order to help demystify the issue, to rationalize the use of resources, and also to free some resources by reducing the irrationalities and in-built waste that exist, including excessive bureaucratization. One of the issues that often escapes attention has to do with the effort and time invested by staffs of international organizations to manage financial resources, which includes fund raising, budgeting, securing approval of governing bodies, managing projects, reporting on uses of money. Very often this gains the upper hand, and becomes the major preoccupation at the expense of policy and substantive work, leadership, and innovation, not to mention the huge bureaucratic apparatus and overhead that it generates, eating up a major portion of each dollar obtained

Ultimately, the governments and the public must learn to live and deal with the fact that complex global organizations require a given institutional mass to be viable and effective, and that they carry a price tag – which though high at first sight is in fact very low in relative terms – and that the price will increase if a higher degree of performance is desired, and that they will simply have to be financed, fully and adequately, as an essential service for humankind. For this, appropriate

---

pleasure, using the ample empty space available on the aircraft and paying only 10 per cent of the regular fare. Were some form of special treatment, based on this model, to be extended by the airlines to those travelling on UN-related business (e.g., meetings, missions), considerable savings could be effected by the UN system. (Swissair has set a useful precedent by granting a 20 per cent discount on its flights to Red Cross and Swiss organizations and churches engaged in humanitarian and development activities.) While the airlines and a few appointed travel agents would be deprived of some business, the budgets of international organizations would benefit and greater mobility and interaction would be introduced within the UN system. A similar approach could be tried in the communications and data processing field, for example, with free or special rate access to satellite channels, including those already leased for other purposes, to commercial global databases, to computer networks, etc. Perhaps the way to start, similar to the suggestion for automatic finance made earlier, would be to invite volunteers to initiate experimental schemes of this kind, and to 'pay them back' through favourable publicity in the global media, giving annual awards and honourable mention to airlines, companies, postal services, etc.

28 An informative table has been compiled by M. Bertrand, containing the regular budget and extrabudgetary resources and manning tables of the main entities and organizations of the UN system for the year 1985 (*op. cit.*, pp.200–1). Were such a table to be updated annually and expanded to include the Bretton Woods institutions and GATT, as well as some regional organizations from the North (e.g. European Commission and OECD), and even a few transnational corporations to help place the issue in a better perspective, it would serve a useful purpose in intergovernmental dialogue.

modes of financing universal international organizations will need to be devised and implemented.

By their nature, these will require new international mechanisms, to raise, allocate and manage funds between many objectives and organizations, and corresponding norms, relationships and practices in national and international discourse and action. These should be derived starting from current and future needs as the first consideration, while taking into account past practices and experiences.

INSTITUTIONAL BETTERMENT AS A PERMANENT OBJECTIVE

This study has offered glimpses of both the inertia-prone intergovernmental process and bureaucratic obstacles to more effective action. Indeed, international organizations find themselves in a double-bind, that of the standard bureaucratization of large and complex organizations, which is accentuated by the international status and the absence of clear-cut lines of authority and responsibility, and the tensions and irrationalities at the intergovernmental level imposed by the opposing interests and the lack of consensus among governments pulling in different directions, diffusing authority, watering down or blocking action.

In the UN, where there is a continuing flux of people, usually with different backgrounds and preferences, changes in policies and programmes, and a large number of governments – each one at least in theory claiming for itself part of the role of a master of the organization – continuity and stability are anchored in the administration and the administrative rules, which begin to live a life of their own, as an end in themselves.

The primary purpose of the administration becomes to follow the rules and to play it safe within the established framework. This, together with control of the finance, becomes the source of power, or at least blocking–enabling power. Rules and caution are put ahead of innovation, initiative, or action, even when the latter are of obvious importance for fulfilling the organization's objectives. Administrative and mechanical performance becomes the measure of institutional performance; punctilious observance of regulations, the criterion of excellence.

As a result, those trying to build international programmes often find themselves obstructed and discouraged by the machineries and procedures of their own organizations, and frustrated by the behaviour and attitudes of their administrations. They have to spend a great deal of time, energy, and nerves to work their way through the administrative undergrowth and to secure administrative action and support. All too often the administration becomes the *de facto* master rather than the servant of the organization, and a major factor that affects not only how a given activity is executed but also how the organization behaves.

Much the same phenomena are known in national bureaucracies. Internationally, the difficulty is aggravated by the complexity of rules and requirements, which are based on and reflect the demands of a multifaceted intergovernmental constituency (e.g., as concerns distribution of posts and the conditions that limit possibilities to fill a given vacancy with the most appropriate candidate), and by the lack of superior authority and of direct responsibility and transparency in the work of the administration. Moreover, top-level substantive staff, who often come and go and have not moved up through the ranks, experience difficulties in mastering the UN administrative maze and in gaining the upper hand over the administration. To make things more complex, there are also multiple pressures exerted from outside via administrative and financial channels, where certain countries have traditionally cultivated their right to exercise influence. Different backgrounds of the staff can also contribute to administrative tensions and difficulties, as they can introduce the divisions and conflicts from the outside world into the secretariat, or bring in contrasting styles and types of organizational behaviour.

A recent study of the UN reform notes as a characteristic of a good leader 'the ability to go beyond the limits of tradition and custom' and 'seek to redefine the parameters of what is possible'.[29] This might apply equally well to any international civil servant, who must strive to innovate and continually try to expand the outer limits of action.

One of the important constraints the international official has to face is institutional inertia, both in maintaining given practices and continuing given activities past their optimal lifespan. Indeed, innovation and going beyond established practice is resisted even in those instances where the need for change is self-evident.

This inertia is a reflection of a pronounced rigidity of the structures of international organizations and of their mandates, resulting from the unwillingness and/or inability of governments to modify these and to respond readily to lessons of experience, once they have negotiated a consensus and agreed on an institutional construct, which tends to acquire virtually a sacrosanct status. It feeds on the political tensions and lack of basic consensus among, and often within, governments. It is magnified by the tendency towards immobilism of the international bureaucracy, which prefers safety and continuity, and by the centrifugal tendencies within the inherently polycentric and overlapping system of organizations.

Yet, if there is one single lesson to be learnt from the experience of GEMS, and for that matter of UNEP, it is the critical importance both of sound institutional design and of flexibility and adaptiveness in making it work. The latter presupposes glasnost in taking into account fully the experience, the organizational performance and related criticisms, and changes in the overall political environment, for the purpose of modifying and improving the approaches and responding to new and changing needs.

In addition, it is also important to assess critically the performance and contribution of governments, and to work out more advanced institutional and legal

29 See P. Fromuth, ed., *A Successor Vision: the United Nations of Tomorrow*, University Press of America, Lanham, Md, 1988, p.26, a Report of an International Panel convened by the UN Association of the United States of America in the context of its United Nations Management and Decision-Making Project, 1985–7. This definition echoes a similar one made some twenty years earlier by the Secretary General of the Stockholm Conference. See Maurice Strong, 'One Year After Stockholm: An Ecological Approach to Management', *Foreign Affairs*, July 1973. Having also been appointed as the Secretary General of the 1992 UN Conference on Environment and Development. Strong will be in a good position, once this event is over, to comment on the applicability of this motto in the everyday life of the UN, on the principal constraints experienced by those who may attempt to follow it, and on the strategies for leadership in the UN system of tomorrow.

Building blocks for a stronger UN 245

frameworks and approaches for involving them, committing them to action in international undertakings, and holding them effectively responsible and accountable for their actions, or the lack thereof, in the process of implementation.

GEMS has been affected by the built-in constraints and standard practices of the UN and its intergovernmental process; at the same time, it has benefited precisely because it had a certain amount of leverage and flexibility created through the institutional design of UNEP. As shown, a degree of financial and administrative autonomy of GEMS PAC and decentralization, backed by new technologies, have been important elements in enabling such a small unit to carry on with its task. In fact, this example illustrates one of the challenges for the UN, namely, the need to loosen the bureaucratic constraints and to encourage and enable its staff to pursue the organization's objectives more freely and effectively, and to apply and develop continuously its talents and skills in the international setting. New information and communication technologies could make this easier, at least in part by reducing the paperload, bureaucratic and other essentially physical chores, and the number of staff, and in making communication within the system faster and more efficient.

The break that GEMS PAC enjoyed, as discussed in chapter one, was due to the foresight of those who took part in conceiving and setting up UNEP and who understood the importance of financial resources, of administrative independence, of flexibility and initiative, of networking and harnessing outside resources and people in a common undertaking, and of trying to bridge the sectoral and disciplinary gaps common in the UN system. Indeed, the model approach inherent to UNEP made it possible to go a considerable way with limited resources and to build the foundations of a global earthwatch literally on a shoestring.

The UNEP model, however, remains largely an exception to the rule in the UN system. Governments continue to be rigid and hardly prone to innovation and adjustment in the institutional domain. While this conservatism and the desire not to 'rock the boat' may spare them some altercations, and does protect the bureaucracies from shocks and discontinuities, on balance the international community stands to lose, and the pursuit of substantive goals is hampered by this attitude.

Pragmatism and overall design are not mutually exclusive. Moreover, it ought to be possible to seek and approximate a judicious balance between continuity, organizational self-protection and structural stability, and the need to innovate, adjust, fulfil functions, and meet challenges more effectively and with greater dynamism. Initiative, flexibility, and innovation should be encouraged in pursuing commonly agreed objectives.

In sum, cultivation of institutional memory and dynamic institutional review, including collective self-questioning, exposure to critical scrutiny of inner workings and procedures, adaptation, experimentation, and redesign, combined with greater transparency of the system and access to the system by actors from the civil society, needs to be legitimated as a major objective at the very centre of multilateral action. Both the governments and the international secretariats need to be kept on their toes; together they share the responsibility for the management and performance of international organizations, in particular the implementation of the agreed objectives and not abandoning these largely to inertia, as is often the case.

An honest and well-meant scrutiny of these bodies is essential for their evolution and, by the way, is also the best antidote to the half-truths and ideologically motivated animosity to which they have been subjected over the years.[30]

THE CHALLENGE OF INTERRELATEDNESS

The preceding pages have argued for a set of institutional qualities that, for the most part, the UN system is at present still deficient in: adequate financial resources, greater confidence from governments in the international secretariat and more scope of action and flexibility for it to pursue organizational objectives, human material of the highest professional quality and dedication, greater institutional adaptability and dynamism combined with greater commitment, involvement, and responsibility of countries in multilateral processes.

The emphasis has been on a positive approach to institution-building, on preparing for the future, and on developing tools, institutions and approaches that will be needed by the international community in order to confront and manage shared challenges collectively.

These challenges are emerging in a world in which physical and economic distances are shrinking in relative terms and in which political boundaries can no longer assure isolation and are losing their traditional meaning in a number of domains. Peoples, countries, national economies, societies, and the biosphere are ever more tightly bound through multiple global feedback loops created by rapid scientific and technological change, the globalization of the world economy, closer and more complex economic, social, and political relations between countries, and humanity's growing impact on the planetary environment.

This newly discovered complexity combined with the traditional agenda of international relations, which are set in an unstable and volatile global, political and development mix, has moved humankind a long way, not only from the initial conditions that gave rise to the creation of the UN system after World War II but also from the conditions that prevailed for decades and that have largely shaped its present physiognomy and practices.

The standard approach in the UN system, reflecting the preference of some powerful developed countries and the paradigm that guides their thinking and action, has been to pursue narrowly a given functional goal, programme or project, and often an operational activity, without recognizing and establishing a link with the broader realities that often hold the key to its successful implementation.

Shrugging off the arguments for an integrated approach with the platitude that

---

30 As implied in the introduction, for lack of staff, resources, and concern, and for fear of self-criticism, 'institutional memory' has not been paid adequate attention in the UN system. As people come and go, they take bits and pieces of this memory with them, while the rest remains buried in the files, a fact that is also partly responsible for occasional abrupt changes of course, experienced when staff in key positions change. For sure, it should be possible to create an independent niche in the system, entrusted with the task of collecting and analysing relevant materials, and keeping the analytical history of organizational evolution up to date. This would provide an essential input for institutional review and betterment, and expand the relevant knowledge beyond the occasional memoirs of international civil servants who find enough energy and inspiration to write after retirement.

'everything is related to everything else' and avoiding the key issue of the nature of these relationships, the sectoralized and 'case-by-case' approach has been put forth as the only and pragmatic way for attaining results. This has been accompanied by the critique of the UN on account of the political and ideological nature of its proceedings, often juxtaposed with the 'business-like', technocratic, i.e. as the 'as it should be' style of the Bretton Woods institutions. In other words, anything critical of the system is disparaged and labelled as 'ideological' and 'politicized', and thus by definition to be considered as bad and undesirable. At the same time, the profoundly ideological foundations of the technocratic mode are overlooked and denied.

This, however, amounts to a political stance that discourages the questioning of the status quo and the recognition by the international community of the systemic relationships that shape the operational context and should inform and guide decisions and action. Also, it reinforces one of the major lacunae in the UN system, namely a forced exclusion from its agenda of those matters that concern national and regional policies of the countries of the North.[31]

The GEMS story offers many illustrations of interrelationships of issues, processes, institutions, and solutions, and of how they were related to and influenced by the North–South and/or East–West differences and tensions. Indeed, the underlying logic points to the conclusion that in order to be fully understood and dealt with effectively all the major issues arising out of international economic and political relations, including externalization of the costs and effects of national actions on others and on the environment, should wind up on the same negotiating table as part of a single basket of collective goods. For this to happen, it will be necessary to accept that there is a need for a higher order of international cooperation and for corresponding international organizations to make it work.

Yet the UN system and the approaches practised within it are hardly conceived or geared to meet these challenges. In the words of someone who has been associated with the UN from its very early days: 'Governments have in many respects structured the UN system *against* a cross-sectoral, integrated approach to world problems. The system is also structured severely to limit the degree to which the Secretary-General can take such an approach.'[32]

Indeed, the basic approach to the creation of the UN was rather similar to that practised when GEMS was conceived and formed, as described in chapter two. In the case of GEMS, projects and monitoring activities were decided on the basis of the existing agencies and their claims, before GEMS as a whole could be launched and consolidated. In the case of the UN, the sectoral thinking and structure were established before its own structure and functions were decided on. The sectoral thinking that was dominant in those days – and that has largely continued since, reflecting and reinforced by structures, lack of coordination, and jurisdictional tensions at the national level – worked to strengthen the separate agencies or offices that had operated parallel to the League of Nations, and to create new ones. Thus,

31 For an early warning about this trend in the UN system, see M. Elmandjra, *The United Nations system: an Analysis*, Faber and Faber, London 1973.
32 Urquhart and Childers, 'A World in Need of Leadership', *op. cit.*, p. 73.

the basics of ILO, UNESCO, FAO, the World Bank, and the IMF were largely decided upon before the UN, so that 'by the time the mandate of the UN's Economic and Social Council (ECOSOC) was considered, the structural die was already cast', ECOSOC itself being almost an afterthought and 'a quite late addition to a UN initially conceived as primarily for peace and security purposes'.[33]

Thus, while the UN Charter refers to interdependence of peace and security with economic and social progress, to borrow the words of the same authors, 'in practice the post-war international structure was not designed to recognize that interdependence or to achieve integration of the work of its different parts' – which *de facto* are largely 'sovereign international bodies' with 'intergovernmental legislature largely representative of the respective national sectoral authorities'.[34] The appearance of a number of issues that cut across the traditional sectoral boundaries of these organizations, e.g., environment, has led to increasing competition between them and institutional claims, especially where such an issue can become a source of funding on account of the interest of key countries. At the same time, the need to overcome sectoralization has become more urgent than ever.[35]

In the years to come, and once more keeping in mind the GEMS experience, the objective should be not to attain some sort of bureaucratic, pro forma or hierarchical coordination as a first essential step, but rather, to agree on the basic policy framework and policy goals that only governments can ultimately decide upon and impose on the UN system and its member organizations.

The global nature of the world's major problems calls for integrated analysis, understanding, and actions. It calls for a modernized, fully reorganized and forward looking UN system, equipped to help the international community meet these challenges. Also, it calls for significant adjustment and change in national behaviour, in how governments perceive the UN, and in how they act in global international organizations.

---

33 *Ibid.*, p. 70.
34 *Ibid.*
35 One of the proposals for alleviating this situation concerns the establishment of 'interdisciplinary development agencies' at the regional and sub regional level, which would deal simultaneously with a series of sectoral concerns. This decentralization cum integration could help overcome many of the present difficulties, and could serve as a conceptual lever for a major institutional reform. M. Bertrand, *op. cit.* , p.91.

11 UN system at a crossroads

Because of the lack of political consensus and collective motivation on the part of governments, the UN system as a whole has not been subjected to a comprehensive multilateral review and even less so to a systemic restructuring exercise. However, while on the surface things may appear static because of the lack of agreement among states and the absence of a blueprint for reform, the quest for a reform has been a continuing one, for the most part in the economic and social sphere, and the system has been undergoing incremental change and adaptation, including that given rise to by unilateral initiatives and actions of the major developed countries, within the UN and elsewhere.

Indeed, especially of late, their policies and actions have set in motion processes that are modifying the map of international organizations. The UN system has been undergoing a mutation and has arrived at a very important juncture in its evolution, while the public, and indeed some governments, are not fully aware of what is taking place. A recognition and assessment of this state of affairs is essential to any consideration of the future of the UN.

NON-REFORM AND CHANGE THROUGH UNILATERAL ACTION

The UN system grew and organizations proliferated and evolved without an agreed vision, and often at cross purposes. Its basics were hardly touched since World War II. This institutional and political deep-freeze complicated an inherently difficult task of the UN system, that of helping deal with the interrelated problems of peace, development, economy and, of late, environment. It resulted in a disarticulation and often a marked gap between a structure, on the one hand, and the functions, objectives and aspirations, on the other hand. An effective link between the two was difficult to forge on account of cleavages and different objectives of countries grouped on either side of the North–South and East–West divides.

The UN has been a target of reform initiatives, by governments at least since the mid–1960s, and by the Secretary-General at an even earlier date. These attempts at reform, by the very nature of the intergovernmental debate and the positions taken by the major groups of countries – who understood it differently and usually had diametrically opposed expectations – were limited in scope and failed to come to grips with some of the key issues. Nor did they address the UN system as a whole, at least in part for lack of will to confront the sectoral and jurisdictional barriers. It is, however, easier to criticize and notice the shortcomings, and to produce proposals for reform, than to reach some sort of intergovernmental consensus. In spite of the wealth of comments and proposals, including those of the UN's own Joint Inspection Unit (JIU), the system and its organizations were thus largely left to adapt and respond to emerging needs in an incremental and *ad hoc* manner.[1]

The well-known Jackson 'capacity study' appeared in 1969. It produced some valuable conclusions and recommendations of a general nature, which on the whole were ignored by the governments. However, its main objective was UNDP, where it led to some changes and adaptations.[2] Soon thereafter, in 1975, the Non-Aligned Movement and the Group of 77 spearheaded the effort to strengthen and rationalize the UN's role in economic and social sectors in order to provide more effective support for bringing to life the New International Economic Order (NIEO). The initial proposals, however, were watered down in the course of negotiations. And even what could be agreed was further eroded at the stage of implementation where vested interests interfered. In the end, only some relatively minor modifications of little practical significance were introduced.[3]

The reforming drive of the 1980s was initiated by the North. It was triggered by the financial crisis originating in the US actions, *inter alia*, an attempt within the Congress to cut the US contribution to the UN budget by 20 per cent unless weighted voting was established for decisions concerning finance. It was not inspired by a genuine wish to improve and strengthen the UN, for it was part of a political offensive mounted by some major developed countries against the organization and the notion of multilateralism that it embodied. The supposed objective of the advocated reforms was to improve management, organizational efficiency and cost-effectiveness. However, as it turned out, the 'reforms' were meant primarily to cut down costs – mostly by reducing the number of posts and through a freeze on recruitment which caused a disturbance in the work of the organization – and to reduce the share of the major contributor to the regular budget of the UN, as well as to place additional constraints on the independence, and on the analytical,

---

1 For an insightful analysis and overview of the attempts to reform the UN and of its adaptation to change, see A. Donini, 'Resilience and Reform: Some Thoughts on the Processes of Change in the United Nations', *International Relations*, Vol. IX, Number 4, Nov. 1986, pp. 289–315. Also see M. Bertrand, 'Can the United Nations be Reformed?' in A. Roberts and B. Kingsbury, eds, *United Nations, Divided World*, Clarendon Press, Oxford, 1988, pp. 193–208. The various attempts to reform the UN were called 'harmless tinkering with organizational charts'. See M. Elmandjra, 'UN Organizations: Way to their Reactivation', Sept. 1986, paper submitted to the UNU Tokyo International Round Table 1986 on 'Future of International Cooperation: Prospects for the Twenty-first Century'. The paper was reprinted in *IFDA Dossier*, No. 67, Sept.–Oct. 1988.
2 'A Study of the Capacity of the United Nations Development System' (UN publication, Sales No. E.70. I.10).
3 For the proposals of a group of experts appointed by the Secretary-General, see 'A New United Nations Structure for Global Economic Co-operation', (UN publication, Sales No. E.75. II. A.7). For the decisions of the General Assembly, see resolution UN GA 32/197. From the same period, see 'Towards a New United Nations Development and International Cooperation System', a chapter in 'What Now', *Development Dialogue*, 1975, No. 1/2.

the advocacy and challenger functions of the UN secretariat in the sphere of development.[4]

This approach was inspired by an underlying strategy, which although not explicitly articulated, could be readily inferred from the policies, actions and statements of these countries. The three principal objectives of the strategy can be summed up as follows:

- To keep their mutual relations as much as possible out of the UN system and to do business elsewhere, e.g., in their regional institutions and through other mechanisms. In this manner, not only did they remove from international scrutiny in universal forums a series of critical issues and simplify the decision-making process by limiting it to a few like-minded countries with similar backgrounds and levels of economic growth, they also avoided getting these issues embroiled with the demands of the developing countries and global development politics.
- To keep the UN, particularly in its development roles, in check and to 'reform' it in a restrictive manner, which was largely at variance with the strengthening and task expansion advocated by the developing countries. They were annoyed by the latter's tendency to use the UN as a forum and vehicle for mounting populist pressures and, through group action, for demanding structural changes in the world economy and trying to obtain concessions from the North. They also began to view some secretariats with suspicion when these started, by their research and studies, to challenge the dominant system and conventional thinking.
- To shift the centre of gravity away from traditional strongholds of the 'one country–one vote philosophy' to Bretton Woods institutions, and of late to GATT, by moving them to the forefront of action, giving them marked policy preference, vesting in them increased powers and roles mostly as concerns Third World-related issues, and encouraging migration of development-related tasks and activities from the UN proper and other agencies.

While these were long-standing objectives, they became fully manifest during the 1980s. Several factors contributed to this. Foremost, it was the conservative *Weltanschauung*, which swept the United States and some other developed countries, that singled out the UN and the internationalist values underpinning it as a

---

4 The General Assembly established a high-level group of intergovernmental experts to review the problems. For the recommendations of the Group, see 'Report of the Group of High-Level Intergovernmental Experts to Review the Efficiency of the Administrative and Financial Functioning of the United Nations', New York, 1986, UN doc. A/41/49. See also UN General Assembly resolution 41/213 approving the Group's recommendations, and UN doc. A/42/234, containing the progress report on the above subject prepared by the UN Secretary-General. (For a caustic critique of the work of the Group of 18, see a comment by one of its members, M. Bertrand, *The Third Generation World Organization*, Nijhoff, Dordrecht, 1989, pp. 107–17.) For a non-governmental attempt at analysis and prescription formulated during this same crisis period, see the report of a Panel convened by the UN Association of the United States of America, P. Fromuth, ed., *A Successor Vision: the United Nations of Tomorrow*, University Press of America, Lanham, Md, 1988. See also a report entitled 'United Nations Financial Emergency: Crisis and Opportunity', New York, 1986, prepared under the auspices of an informal group chaired by Sadruddin Aga Khan and Maurice Strong. For an iconoclastic view from the same period, see J. Galtung, *The United Nations Today: Problems and some Proposals*, Centre of International Studies, Princeton University, November 1986.

target of attack. In a sense, the UN was seen as the counterpart of the public sector and welfare state on the international scene, and thus to be remade in the image of the new coalition in power.[5] The shifting balance of power, and the disarray of progressive and liberal forces, made it easier to pay lip service to some key premises of multilateralism and universality and to downgrade them in international discourse. These new conditions were reflected in the fact that the formation of the Group of Seven – which in some ways acts as a self-appointed super-directorate for the global economy and takes decisions that affect vital interests of all countries and peoples – has been objected to only sporadically and usually *sotto voce*. Yet in most areas, because of the importance of the developed countries, what they decide and how they act domestically or as a group has implications for and concerns the whole international community and the well-being of all its members, and should normally be the subject of consideration also in global forums.

In practical terms, it can be said that the policy of the North *vis-à-vis* the UN system has been three-pronged. First, the objective was to confine the role of the UN essentially to debates and airing of views about matters related to development, to neutralize its ability to question the very basis of the dominant world economic system and to minimize its practical role and interference in matters having to do with policy aspects and management of global economic affairs.[6]

The renewed emphasis on the peace-keeping function of the UN and the favourable press given to the UN in western media in the broader context of the end of the cold war and changes in the East do not detract from this underlying strategy. In fact, it fits quite well with the idea of using the UN – in its role as the universal security organization – to provide the legal and political umbrella for intervention by the North aimed at controlling and moderating the situation in the increasingly unstable Third World.[7]

5 Pierre de Senarclens has noted the parallel between the dismantling of the New Deal legacy in the United States, and the questioning of the liberal values that inspired the creation and development of the United Nations. This has been accompanied by the abandonment almost everywhere of the internationalist values, the return to local and regional preoccupations, and an increase in nationalist, xenophobic and racist attitudes. See his thought-provoking *La crise des Nations Unies*, Presses Universitaires de France, Paris, 1988, p.223.

6 One analyst has argued that the accusations and criticisms levelled at the UN – wastefulness, corruption, inefficiency, high salaries – could very well be directed at any large international bureaucracy, but that they were used as an excuse to attack the UN and were motivated by the allegation that the UN was challenging the status quo and that its staff began to include an ever greater number of people from the South who brought with them a world view and sensitivities different from those of the North. He also argues that the purpose of this kind of reform is 'to silence the even weak international voices which are questioning the veracity, justice and viability of current world structures and practices'. See J. Harrod, 'United Nations Specialized Agencies: From Functionalist Intervention to International Co-operation?', in J. Harrod and N. Schrijver, eds, *The UN Under Attack*, Gower, Aldershot, 1988, p. 141.

7 This point, illustrated by the handling of the Gulf crisis within the UN, was anticipated by an earlier comment that 'the new UN agenda' discussed in the North can be summed up under four headings, i.e., 'narcotics, environment, terrorism and refugees', all four seen to emanate from the South and considered as a threat to the North's comfort and prosperity. The author goes on to say that in a world of the nineties, and as the distinction between the East and the West fades, the North–South divisions will become a key concern in the North and the tendency may be to cast the UN in a role of a mechanism for 'riot control' in the South. E. Mortimer, 'East–West to North–South', *Financial Times*, 19 Dec. 1989.

Closely related to above is the second objective of the strategy, namely to weaken, neutralize, and demoralize UNCTAD. This organization has been seen, ever since its creation in 1964, as the traditional stronghold of the South in the UN, closely reflecting its outlook on the world economy, the principal global forum for advocacy of systemic change, a source of data and analysis differing from and challenging those originating in more traditional institutions, and an organization providing technical support for an alternative vision to the dominant order. Indeed, UNCTAD in particular has been considered by the developed countries as an ideological nuisance, a hangover from what they felt was a discredited past still clinging to the assumption that both the state and intergovernmental action have a key role to play in development and in managing the international economy, an assumption out of step with the dominant thinking radiating in particular from the IMF–World Bank–GATT troika.

Thus, benefiting from weakness and vacillation in the ranks of the developing countries, hesitancy of the UNCTAD secretariat, and the melting away of the Soviet bloc, an effort was mounted by the developed countries to bring UNCTAD's philosophy and outputs more into line with the OECD vision of the South's problems. This was done in part by reducing the scope and freedom of action of the secretariat, including its technical support for the Group of 77, through tighter administrative, personnel, and budgetary constraints, and through direct political pressures. It was also done by modifying the agenda of the organization to include as a major preoccupation the national policies and responsibilities of the developing countries, and thus dilute its primary concern with the role and impact of international economic structures and processes in their national development. By frustrating the Group of 77 through lack of progress and by unwillingness to engage in any serious negotiations in this forum, while boosting the role and responsibilities of GATT, including proposals to transform it into a multilateral trade organization, they sought to diminish the role of UNCTAD in the evolving scenario and indeed to sow doubts as to the very reason for its existence.

This last point is connected with the third and most significant element in the strategy of the developed countries, namely the objective to evolve a system of global economic organizations, based on the World Bank–IMF–GATT axis. It is in these institutions that the developed countries can use their economic preponderance, including the weighted vote in the financial institutions, to influence and control the policy and management. It is these institutions that they consider as more competent, trustworthy, accountable and responsive to their concerns.[8] It is

---

8 P. S. Mistry and P. Thyness note that the US Congress and the Senate in particular have resisted US contributions, even when this is required by international treaty to which the United States subscribes, to those organizations where it is not allowed 'sufficient control and influence of how budgets are determined and how funds are spent'. Their attitude is different in the World Bank, for example, where 'weighted voting permits the US, along with one or two large cohorts, to exercise virtually unilateral control over management, orientation and budgetary decisions'. See 'Financing the Multilateral System in the 1990s', the Nordic Project Report No. 13: 1990, p.8. The same authors also note that the perceptions of international organizations by the major donors reflect more their concern with the vulnerability of these to political influence of developing countries and with 'control over voting power, budgets, management, staffing and policy direction of multilateral institutions than with actual institutional effectiveness and quality of output' (*ibid.*, p. ii).

these institutions that have emerged as a major multilateral tool, controlled by the North, for the purpose of influencing policies and actions of developing countries. Thus, the immunity of the developed countries to international scrutiny and responsibility came to be contrasted with the growing exposure of many developing countries to direct intervention in the running of their domestic affairs by the staffs of the IMF and the World Bank. Benefiting from the weaknesses and pressing needs of the developing countries, and using debt management as a pretext, these two institutions, and in particular the IMF, have come to act as a kind of a 'supranational' authority, or a 'gendarme' in a large part of the South, shaping the destiny of whole countries and their populations, changing social and political structures, undermining the public support for governments, and rewarding or punishing certain social strata.

Not suprisingly, these bodies have not been known to challenge or question the status quo and have been its defenders all along. Moreover, in these organizations the developing countries have not really tried to mount effective group action and have tended to accommodate themselves to the dominant system. One reason is that no country wishes to prejudice its chances of securing tangible benefits that can be obtained by playing according to the rules, or to risk being disciplined for not doing so. Another reason is that the pursuit of specific national objectives made them relegate to the background collective and structural goals, a trend accentuated by the crisis of the 1980s. Last but not the least, the nature of the agenda, the method of work and negotiations in these bodies, as well as the fact that the national representatives and technocrats come from traditionally cautious or conservative institutions, such as ministries of finance or trade, or treasuries and central banks, contributed further to the posture of the developing countries which made it easier for the North to achieve its goals.

This third aspect of the North's strategy – the progressive concentration of power in the Bretton Woods institutions and GATT and the systematic undermining and erosion of the traditional role and influence of General Assembly and ECOSOC-based system – merit additional comments.

Nominally, the Bretton Woods institutions and GATT belong to the UN system. Yet, from the very beginning they have been kept and treated as something special and apart from the rest of the UN.

These institutions are in fact under the strong influence of a few developed countries. Over the years they have become closely linked to, dependent on, and responsive to the governmental, political, trade, and financial centres in the North. Their executive heads have all been nationals of developed countries; in the case of the World Bank, the Presidents have all been American citizens.[9] Their secre-

---

9 UNDP, of course, has also had as its Administrators American citizens, handpicked by the US President. As the major donor to UNDP the US has always claimed and exercised this privilege. However, UNDP is based on the 'one country, one vote' principle, and operates within the framework of the UN. It was thus not as easily amenable to full Northern control as the Bretton Woods institutions. Still, it remains an organization heavily influenced by the donors and their preferences, as reflected at present in the emphasis it places on promoting privatization in recipient countries. For the story of the origins of UNDP in the unsuccessful bid of the developing countries to extend the role of the UN into capital funding of development by creating the Special United

tariats are tightly steered and operate within the basic ideological and management framework that these countries espouse. They are characterized by lack of pluralism, by strict ideological discipline and intolerance for job applicants or staff members who do not conform with the accepted mould or express doubts about the existing scheme of things. In general, their ability to punish or reward staff is much greater than elsewhere in the UN system.

The three organizations are also known for tight security, secretiveness, and lack of transparency, not only towards the outside world, but also towards the majority of their member governments, with developing countries being *de facto* denied access to their inner sanctums.

The above factors surely account for their successful institutional performance and it should not be surprising that the IMF and the World Bank, and indeed GATT, as the institutions of the establishment and the pillars of the status quo in international economic relations, especially in the case of the first two, have been allowed to attain the necessary critical mass of staff and resources, and have been encouraged to follow a path of task expansion and institutional adjustment.

Any suggestions for their reform advanced by the developing countries, especially as concerns democratization of the decision-making process and management in the IMF and the World Bank, have been dismissed outright by the developed countries. They have held these institutions beyond reproach, portraying them, with ample support from the media, as paragons of efficiency and quality, technical excellence, impartiality, and administrative competence. And while the UN proper was scolded for its high salaries and wastefulness, many of the same critics from the North – with the exception of some members of the US Congress upset by the greater affluence of their neighbours from international civil service in the District of Columbia and its environs – seemed to overlook the significantly higher salaries and benefits enjoyed by the staffs of the Bank and the Fund, including the controversial and costly first class air travel which puts them in a category apart, an 'international nomenklatura' of sorts.

The rare criticisms that reach the public eye from time to time – which are usually countered effectively or are simply ignored – show that these organizations merit careful and critical scrutiny by the international community.[10]

Nations Fund for Economic Development (SUNFED), see J. G. Hadwen and J. Kaufmann, *How United Nations Decisions Are Made*, Sijthoff, Leyden, 1960. The compromise, namely task expansion for the UN into technical assistance, initially the Expanded Programme of Technical Assistance (EPTA), was very much in line with the thinking of the western powers, who saw technical assistance as 'legitimate and palatable channels with which to continue to influence, if not govern, the ex-colonies from a distance', and to mould new nations into 'permeable States open to all influences and penetrations which by the nature of the world would originate in the ex-metropolitan and more powerful States'. Harrod, *op. cit.*, p. 136.

10 See M. Irwin, 'Patience Runs Out With the World Bank', *Wall Street Journal*, 11 March 1990, and 'UN's World Bank condemned for "immoral" profits', *Sunday Observer*, 22 April 1990. See also a monograph by D. L. Budhoo, 'Enough is Enough, Open Letter of Resignation to the Managing Director of the International Monetary Fund', New Horizons Press, New York, 1990. A personal and intense style is a handicap of this exposé, which offers a glimpse into the inner workings of the IMF. *Inter alia*, the letter depicts IMF as a bastion of the North imbued with the 'white man's burden ethos', questions its impartiality, and voices the exasperation of a Third Worlder working for and familiar with practices of this organization in its handling of debt and structural adjustment in the

The IMF and the World Bank have acted subtly as instruments in the hands of the North to discourage collective demands and actions of the South, and in general, South–South cooperation. Through macro-economic moves and direct intervention in the domestic affairs and surveillance of individual developing countries, they have been used to extend a net of economic and political control over the majority of developing countries.[11] The weakness of most developing countries and their discouragement after years of fruitless development dialogue, a mounting political and development crisis at home, a hostile international economic environment, institutionalized political and economic bondage on account of debt and haemorrhage due to the debt service, bilateral pressures and the spreading use of conditionality by the North, and last but not least the impacts of the collapse of the Soviet model, have all contributed to the achievement of the avowed or unavowed purpose of the leading developed countries.

What has also been taking place is duplication and often a simple take-over of development-related functions from the 'NIEO-tainted' and 'one country, one vote' UN forums by the World Bank and the IMF. Starting from their strong financial base and political support in the North, with a well-equipped and adequately staffed infrastructure, a strong research capability, good quality and attractive publications that have gained virtually unchallenged acceptance as sources of facts and wisdom, and a friendly and attentive global media coverage, the two Bretton Woods institutions have become the true centres of institutional power on the world development scene and the extension of national policies and ideology of their major shareholders. Together with GATT, they draw their strength and relevance from direct links with those national institutions that have a strong power base and which really matter when it comes to economic policy and development.

The Bretton Woods institutions, in particular, promote given development models, theories, and methodologies; they generate statistical data and choose how to present and interpret them; under the façade of technical impartiality, they play a profoundly political role in the national development of developing countries, devise fashions and fads of development thinking and foster their translation into practice, and influence public opinion, decision-makers, and action worldwide.

---

developing countries. It is true, however, that criticisms voiced by insiders are usually easy to dismiss as 'sniping by disgruntled ex-officials', and tend to lose in credibility by the very fact that they often expose, or base their arguments on, what is considered as confidential information and thus act contrary to their loyalty oath. Criticisms come also from undoubtedly bona fide analysts. Thus, one of the studies of the Nordic UN project has criticized the World Bank 'for an almost frivolous waste of scarce budget dollars', singling out 'unfocused, operationally unrelated intellectual pursuits through poorly managed and expensive internal research programmes; an extensive and growing range of publications attractive in appearance but poor in content, produced more for staff gratification and public relations than to meet genuine needs; a series of navel-gazing retreats, seminars, conferences and training programmes, which do less to develop its human resources than to provide convenient outlets for staff frustration with growing bureaucratization and mismanagement'. Mistry and Thyness, *op. cit.*, p. 64.

11 For an analysis of the historical background and issues at stake, see S. Dell, 'Relations between the United Nations and the Bretton Woods Institutions', paper presented to the North–South Roundtable on the Future Role of the UN, Uppsala, 6–8 Sept. 1989, and also his 'Reforming the World Bank for the Tasks of the 1990s', the Fifth Export–Import Bank of India Commencement Day Annual Lecture, 5 May 1990.

They also act to counter the South's own perception and analysis of its problems, armed with a well-established theory, methodology, and a few select indicators, plus a powerful institutional presence.

This trend has been further accentuated by the effort to strengthen and expand the roles of GATT manifested in the Uruguay Round. By broadening its agenda to include such matters as trade-related aspects of services, intellectual property, and investment, and by the attempts to bolster its constitutional foundations through what has been labelled the 'Functioning of the GATT System' (or FOGS) exercise, the developed countries have taken the initiative to associate GATT with the broader economic agenda and international regimes build up and to pave the way for its conversion into a multilateral trade organization. In itself, this would be a laudable initiative, recalling the stillbirth of the International Trade Organization in 1948 and the major gap that resulted in the UN system. However, the North seems to feel that such an organization should be kept at arm's length from the development agenda, overlooking or effectively denying whatever limited advances were made in this domain over the decades to help the developing countries and to recognize their special status and needs in the international trading system.[12]

By formally recognizing the linkage between money, finance, and trade – something they had steadfastly refused to do during decades of debates in UNCTAD and other UN forums – the developed countries have opened the way towards creating, parallel to the UN, an integrated system based on the IMF–World Bank–GATT trio for the management of those aspects of the world economy that require universal participation and that they cannot regulate or control fully in their own narrow circle. In line with this, and arguing that these issues are dealt with elsewhere in 'competent' organizations, they now object to and block in-depth consideration of international financial and monetary issues in the UN proper and its various forums, a practice that in the past had yielded many new ideas and generated political pressure that led to the adoption of various measures by the initially reluctant Bretton Woods institutions.

Quietly and almost imperceptibly for the broad public, significant changes have been thus taking place in the system of international organizations. Because the changes have been incremental and because there has been no institutional opportunity for analysing and debating all the issues and trends or their interaction, it has been difficult to grasp their significance and to respond to them comprehensively.

The response of the developing countries was hesitant. Many complained, but their reactions seldom went beyond speech-making and did not lead to elaborating a joint position, backed by collective action. The changes in the East, and the political and psychological impacts of the Gulf War, accentuated the feeling of helplessness *vis à vis* the massive display of power and self-assurance by the North. With few defences left, and under the impact of a political, media and intellectual offensive from the developed countries, they engaged at best in a holding and damage-limiting operation.

They came to realize that the end of the cold war had reduced their traditional

---

12 For a critical assessment of the Uruguay Round, see C. Raghavan, *Recolonization – GATT, the Uruguay Round and the Third World*, Third World Network, Penang, 1990.

influence and margin of manoeuvre in the UN as they witnessed the countries of the former Eastern bloc move solidly behind their Western partners and express indifference or open hostility to traditional demands of the South, including those on institutional matters.[13]

Also, the developing countries sensed that the UN Security Council which, for all practical purposes, was under full control of the North led by the only remaining 'superpower', would erode the democratic character of the organization in the political sphere. It would turn the Security Council into an instrument for the North to handle the situation in the South, reinforcing and complementing the pattern already in place in the Bretton Woods institutions. On both accounts, the very basis of the UN and the place and role of the countries of the South in the organization were being challenged.

The foundations of the 'New World Order' were being cast. It was quite clear that the North intended to maintain the initiative and try to reform the UN to its liking, benefiting from the new geopolitical constellation and the apparent inability of the developing countries to formulate and defend a strategy of their own.

In this situation, a set of policy objectives concerning the full-fledged reform of the UN system was becoming essential for the South in order to initiate a debate, to slow down changes set in motion through unilateral initiatives of the developed countries, and to negotiate with them on the basic features of a reformed and democratic organization adapted and equipped to deal with the new realities and challenges, reflecting and protecting in a balanced manner the interests and aspirations of its broad constituency.

THE NEED FOR A COMPREHENSIVE REFORM OF THE UN SYSTEM

After half a century of existence, the UN system of international organization is well embarked upon a period of transition and discontinuity, when a comprehensive re-examination of its purpose and structures, similar to that carried out at the time of its formation in the closing stages of World War II, is required.

The international community is at a critical point in history when effective global institutions are increasingly needed and indispensable in order to help in managing ever more complex international affairs.

The globalization of the international economy, the transnationalization of culture, politics, power, and influence, thanks to modern means of communications and rapid advances in science and technology, and the novel linkages and environmental interdependence through the biosphere are all contributing to blur the

---

13 Interestingly, it was the address by the President of the USSR to the 43rd session of the UN General Assembly on 7 December 1988, that put forward a positive vision of the reform of the UN system and the strengthened role of the UN Secretary-General. The core of this reform was to be in the concept of 'collective economic security', based on a systemic vision of challenges facing the international community, an approach which gave hope to the South. This initiative was gradually whittled away as the USSR's superpower status went into decline. For a while, however, bilateral consultations and studies on the future of the UN were carried out by joint US-USSR groups composed of academics and government officials.

traditional distinctions and separation between domestic, international, and planetary affairs.

Because human-based actions have been superimposed and are making impacts on the 'natural order' of the planet Earth with far-reaching effects on the very nature and stability of the earth system, which in turn has momentous implications for the future of humankind and civilization, the international community and in particular the governments will have little choice but to respond and act, individually and collectively, and co-ordinate their actions. In this, they need to be aware of the critical issues and causal and systemic relationships that define and shape the challenge, most prominently the long-festering conflict between the North and the South. This conflict is rooted in the structurally biased global economic system, the centre–periphery syndrome, and the development gap that separates these countries, their respective population numbers, what they need and what they consume now and will consume in the future, and how this relates to the natural resource base and to the global outer limits.

If one were to try to define the very essence of the task facing the international community, one might say that many of the issues that in different forms have preoccupied states on the domestic level, and of late the EEC, as the most advanced specimen of regional integration, are increasingly moving to the global level. They will need the international community's collective response. Because of a lack of useful precedents and on account of the scale, numbers, complexity, socio-economic and political heterogeneity, and levels of development of the participants that will need to be involved in such a collective undertaking, this will require advanced forms of international cooperation, buttressed by corresponding institutions, norms, and ethical values.

In an unequal, fragmented, and polarized international community, the approaches favoured by different actors are in broad terms likely to be similar to those that have been known at the national level. Among the 'haves' these have ranged across a broad political spectrum, from ignoring the challenges and demands for equity and perpetuating divisions, privileges, and hegemonies to responding in a constructive manner, seeking equitable solutions and promoting the collective good through negotiation and accommodation as a 'positive sum game', guided by a set of humanistic norms and values for ensuring the common well-being.

Trends indicate that those now in control of economic and political power in the North, where a well-off and self-confident minority of the world's people live, are likely to prefer the perpetuation of the structural status quo, combined with carefully dosed concessions on their own terms. They will seek to boost their dominion on the world scene and the control of access to key natural resources, relying on their superior economic, technological and military power, knowledge, resources, and control of the global media.

In this scenario, the political, business, technocratic, and military élites of the South would be firmly wedded to and benefit from the links with the North, in return for not questioning the system and the unequal relationships, including those likely to arise from global environmental and resource constraints. Those developing countries whose ability to develop on their own is doubted would be kept in some form of an institutionalized second-class status, under the benevolent tutelage of the powerful North and with the tacit acquiescence of all.[14] It is assumed that in dualistic societies, people would accept their status in return for the benefits of the trickle-down process from association with and opening up of their countries to the North, seen as the only alternative and a solution preferable to the poverty and deprivation and externally imposed hardships that have come to be associated in the popular mind with attempts to pursue development paths independent from the dominant system, to challenge the latter, or indeed to show zeal for self-reliance.

This confluence of interests and needs could lead the South and its peoples deeper into dependency *vis-à-vis* the North and the North-based transnational actors, which could effectively come to dominate their national development and political scenes from within, through what will amount to a twenty-first century variant of 'neo-colonialism', and *de facto* recolonization of many countries. This could be achieved through more subtle and more efficient means than the direct political and territorial control of yesteryear, including the influence over the way people feel, think and perceive the world around them.

The mixture of payoffs and benefits, and the generalized feeling of helplessness against the hegemonic powers and the dominant system, could weaken the individual and collective resolve of developing countries and their efforts to resist and change the systemic bias, in spite of the fact that they would remain frustrated, unhappy second-class members of the international community. Instead, they would be co-opted by the system, engaging in an individual scramble for upward mobility on the global ladder, towards maturity and 'graduation' and even membership of the OECD as the ultimate status and recognition that a country has 'made it'.

A collective renunciation by the South of its long-standing claims and its acceptance of the terms demanded by the North, as some hope and expect will happen because of the current weakness and disarray of developing countries, would not be a sustainable proposition. The heritage of the age of imperialism, the systemic biases in the world economy, the North–South gaps, the widespread poverty and destitution, the growing population numbers, and global environmental issues carry the potential of intensifying strife and conflict, endangering the fragile and sensitive international economic relations and the environment. This was illustrated by the Gulf events, which first alerted world public opinion to an uncertain future after the end of the cold war and which, beyond the immediate crisis, exposed the lopsided relationships and systemic issues linking the South and the North in a potentially tragic embrace of interdependence.

Which way the course of history will move and whether there will be enough time, patience, and enlightenment to arrive at some form of *modus vivendi* and equitable and stable world situation can only be conjectured. It is certain, however, that a great deal of tension and conflict will mark this interregnum, and that if a

---

14 For an argument in favour of a 'declared, internationalized colonialism in Africa' and 'installing a frankly paternalist international authority in Africa and continent-wide development structure and programme' as the only way out of the current impasse facing African states, see W. Pfaff, 'For Distressed Africa, What About International Colonialism?', *International Herald Tribune*, 24 April 1990.

positive vision of international cooperation, with the UN system of international organizations as its main pillar, is to have a chance, it will need to be articulated and defended forcefully, and those who support it will need to get well organized for a protracted and difficult process of argumentation, persuasion, negotiation, and political struggle.

The two broad scenarios sketched above embody different visions of international organizations. In view of the existing trends and the bifurcation that is taking place in the UN system, an essential and first step in the direction of positive change at this watershed in history would be to revise and update the basic paradigm of the UN system of international organizations and to reaffirm its democratic orientation. This revision would have to recognize the changes that have swept the world in the fifty years since the UN system began, the expanded and closely interrelated global agenda, and the existence of a broad and diverse constituency of interests. And it should be based on the recognition of the central role of economic and social issues – and in particular of resolving and managing the North–South conflicts, and eliminating global poverty – both for world peace and for sustainable development, two goals of concern to the North.

Is such a fresh departure possible in view of the development gap, lack of consensus between the North and the South and the *dialogue des sourds* that has persisted for decades? Is it likely in view of the wish of some countries to dominate the UN and of unequal power relationships, now accentuated by changes in the former Eastern bloc and its member countries emerging both as competitors with the South for economic space in the industrially advanced North, and as newly converted and eager supporters of world economic and political structures and relationships? And is it possible in view of the traditional lack of interest and pronounced antipathy toward the UN on the part of some key sectors of political and public opinion in the North?

In the past, world wars have acted as triggers or catalysts for a fundamental re-examination of the rationale and structure of international organizations. It is to be hoped that the international community will find it possible to embark on such a comprehensive reform without waiting for a world war to take place.[15] The discontinuities and changes that the world is undergoing, the accumulated problems of yesterday and today and the prospective ones of tomorrow, are in many ways more serious, complex, and challenging than those that inspired the institution-building spurt at the close of World War II, and call for serious thought on what has been called 'the Third-Generation World Organization'.[16]

Three principal groups of actors are likely to figure in the debate about the future of international organization:

---

15 For a notion of not waiting for a world war as the only stimulus to rearrange the system of international organizations, see H. Cleveland and L. P. Bloomfield, 'The Future of International Governance: Post-war Planning without Having the War First', in 'Multilateralism and the United Nations', *Journal of Development Planning*, No.17, 1987, pp. 5–17.

16 See M. Bertrand, 'Some Reflections on Reform of the United Nations', written as a report of the Joint Inspection Unit for the occasion of the fortieth anniversary of the UN (UN doc A/40/988), as well as his *The Third Generation World Organization, op. cit.*

- a differentiated North, with the USA as the paramount power trying to steer the process, with the EC and Japan also to the fore, an uncertain role for the successor state(s) to the USSR, and with a special part for smaller, traditionally internationalist European countries, whose basic instincts often coincide with those of the developing countries and who feel bothered by hegemony and unilateralism in international organizations;
- the South – the institutional domain being one where developing countries could iron out their differences *inter se* and assume a common stance, which is a *sine qua non* if they wish to exercise influence – with a very important role for China as the only developing country with the veto power in the Security Council; and
- the non-governmental actors, which include among others the public opinion representing the amorphous political base which those engaged in the debate will have to address,[17] the many organizations and groups that belong to the civil society, powerful national and transnational organizations and economic interests that have a vital and growing stake in international organizations and regimes, the media, the academics and those working in international organizations, especially the UN Secretary-General, who are well placed to help in thinking through the issues and in devising the blueprints for change.

Central and special responsibility for promoting a new and democratic vision of international organization rests on the developing countries as a collectivity, i.e., the South. They are the party vitally interested in a stronger and more democratic UN system. Collectively, well organized as the South, with a clear set of objectives and a strategy backed by solid analysis, the developing countries have enough political and economic weight, and authority, to advocate and press for required changes.

Yet all these years it has proved impossible for the South to organize properly for dealing with the North at the global level. At present, the prospects of such organization may appear even more remote in the context of discouragement and loss of faith in South–South cooperation on the part of many leaders and strata of opinion in the developing countries. And while the new geopolitical realities may be acting to discourage such global action by the South, they also speak more loudly and urgently than ever in favour of better and fuller organization of the developing countries in defence of their shared objectives.

Having their own technical support organization would be of help in this undertaking and in taking part in the proceedings of the UN system in general. It would make it easier for them to flesh out their arguments, to articulate their

---

17 Since most of the guiding principles for a reformed UN correspond to values and norms prized by the developed countries in their domestic politics and their mutual relations, it should be possible to seek support in the North by re-educating and enlightening the public opinion. The chances of success for democratic and effective international organizations are small as long as parochialism, narrow nationalism, or even exclusive regionalism hold sway over domestic political scenes and voting publics, in particular in those few countries that can project their power globally. In the case of the United States, congressional, local and interest group politics, as well as swings in intellectual fashion, have had a major impact on the proceedings and fortunes of international organizations.

UN system at a crossroads 263

positions, and to negotiate more flexibly in multilateral forums.[18] It would help them to analyse critically their external environment.[19]

Beyond this, and more ambitiously, the developing countries could also envisage setting up a global intergovernmental organization of their own – a United Nations of the South – as a way to establish a greater degree of balance, equity and order, and to initiate true negotiations in those forums where the North and the South meet, including Bretton Woods institutions. As it is, the developed countries have built up over the years an elaborate system of multilateral organizations and mechanisms of their own, where the like-minded meet, work together, negotiate, sort out their differences, and coordinate their positions including those *vis-à-vis* the South.[20]

In their quest for a reformed UN, as an essential step, the developing countries will need to press for the convening of a world conference to negotiate and approve the design of the global system of international organization for the twenty-first

18 For lack of such organizational support, the coordination of developing countries' policies and action at the global level has been inadequate, and generally limited to declaratory statements and periodic gatherings. It is this policy and organizational vacuum that gave rise to the initiative to create the South Commission, chaired by Julius K. Nyerere. Individuals from all corners of the Third World, with different backgrounds and ranging across the political spectrum, acting in their personal capacity and independent from the governments of their countries, worked for three years to produce a report that takes a global view of the South – its status, as well as its options and prospects for the future. See *The Challenge to the South*, Oxford University Press, Oxford, 1990. One of its major recommendations addressed to developing countries collectively concerns the setting up of a 'South Secretariat', as the first, limited step in a process possibly leading to a fully fledged organization of the South at the global level (*ibid.*, pp. 200–5). Addressed to the South and pleading for national and collective self-reliance of the developing countries, the Report has struck a responsive chord especially among the readers in the Third World, but also in the North. However, there has been some hostility to it among the more traditional establishment in the North, in part because of dislike for the very idea of a South, especially an organized and self-reliant South, and the denial of its existence and viability, and also because the Report argues cogently a case that does not toe the current development wisdom, including that promoted via international financial institutions. Interestingly, most governments and commentators from the North, rather than criticize what they disprove of have tended to applaud the Report for addressing the issue of national development in the South and have lifted out of context those arguments and statements coinciding with their own views (e.g. on corruption, militarization, ineffectiveness of the state), while largely ignoring most other issues and the integrated and comprehensive character of the analysis.

19 This includes analysing critically the outputs and policy advice which are emerging from what has been dubbed as the IMF/World Bank 'duopoly', and which are seldom challenged by countries or by other international organizations. Mistry and Thyness speak of the recipients being 'vulnerable to inappropriate (sometimes plain wrong) policy advice doled out by a duopoly with no credible countervailing voices to argue against bad advice', and argue for 'policy analysis and options evaluation capacity to be built up within developing countries' (*op. cit.*, p. 25). While this advice pertains to individual countries, it is applicable as well to the collectivity of developing countries.

20 In view of the long-standing inability and unwillingness of developing countries to organize themselves in the required manner, and on account of their expectation that this support function be performed by UNCTAD and other UN secretariats, one of the ideas mooted concerns the 'secondment', or a simple transfer, of UNCTAD to the South by the international community, to service their needs, including their mutual cooperation and their negotiations with the North. It has been suggested, for example, that, as part of a global North–South arrangement, the original functions of UNCTAD be maintained by splitting it into two organizations, one to service the developing countries, while the other would merge and expand the UNCTAD and GATT functions into a global International Trade and Development Organization. See J. Pronk, 'Towards a New International Trade Organization?', in J. Harrod and N. Schrijver, eds, *The UN Under Attack, op. cit.*, pp. 90–2.

264 *The quest for world environmental cooperation*

century. Similar in scope to the institution-building conferences in the 1940s (starting with UNESCO in 1942, ending with ITO/GATT in 1948), and in view of the complexity of the issues and the long time needed to arrive at an agreed solution, this should be more of a process similar to the model used by the Helsinki Conference on Security and Cooperation in Europe, which could last for years, if necessary, and which would set up permanent mechanisms for the review and monitoring of the functioning and performance of the system.

Many ideas are already on the table. The Stockholm Initiative on Global Security and Governance, entitled 'Common Responsibility in the 1990s', adopted in April 1991 at a meeting under the auspices of the four independent commissions which worked during the preceding decade, namely the Brandt, Palme, Brundtland and South Commissions, has as one of its main messages the need for a profound reform of the UN system. It calls for a World Summit on Global Governance, similar in intent to the San Francisco and Bretton Woods conferences, to be convened on the 50th anniversary of the UN in 1995; and it argues that the best way to prepare for this event would be to establish an independent, non-governmental 'International Commission on Global Governance'. Such a commission would undertake 'dispassionate and enlightened examination' of issues before these reach the stage of intergovernmental deliberations.[21] At about the same time as the launch of the Stockholm Initiative, the Nordic UN Project produced its final report after three years of research and deliberation. This represented a significant contribution by a group of interested governments to the UN reform drive.[22]

The Group of 77, however, cannot be content to sit back and only react to the initiatives of others.[23] It will need to work out a global, coordinated strategy of institutional change, which should encompass the whole UN system, and in particular the Bretton Woods institutions and GATT where a great deal could be

21 See 'Common Responsibility in the 1990s', The Stockholm Initiative on Global Security and Governance, 22 April 1991, Prime Minister's office, Stockholm, Sweden.

22 See *The United Nations in Development – Reform Issues in the Economic and Social Fields; a Nordic Perspective*, Nordic UN Project, Final Report 1991, Almqvist & Wiksell, Stockholm 1991. Note that various Nordic UN Project studies cited in this text are reproduced in a single volume under the title *The United Nations – Issues and Options*, Almqvist & Wiksell, Stockholm, 1991.

23 The attitude of the Group of 77 in the spring 1991 discussions on the reform of the UN in its economic and social sectors was one of reserve and uncertainty. Having taken the initiative, the Group of 77 pulled back, fearing that the developed countries would use this opportunity to weaken further and erode the UN and neutralize the political leverage in the organization enjoyed by the developing countries. Thus, the resumed session of the General Assembly on this subject was largely devoted to procedural matters of marginal importance and an opportunity was lost. (See General Assembly resolution 45/177 on restructuring and revitalization of the United Nations in the economic and social sectors, which was sponsored by the Group of 77. See also the formal position paper of the 77 on this subject [doc. A/45/991] and the resolution adopted by the General General Assembly [doc. A/45/L.49]). With this attitude, however, the developing countries both cede the initiative to the North, and also undermine their own stand and flexibility by often acting as uncritical defenders of organizational status quo. The developing countries will need to devote serious attention as to how to overcome this situation where both the analysis of the UN and proposals for its reform originate almost exclusively in the North. On this last point, see Donini who highlights the dominance of Anglo-Saxon analysts and the 'monocultural pervasiveness' of the study of UN (*op. cit.*, p. 320). For a review of the essentially US-based study of international organization, see F. Kratochwil and J. G. Ruggie, 'International Organization: A State of the Art on the Art of State', in *International Organization*, 40 (Autumn 1986).

done were the developing countries to pull their act together. The Group of 77 will also need to fashion and propose a paradigm and a corresponding institutional model in keeping with a vision of a democratic UN equipped to lead the international community into an age of cooperation and peace. This would entail putting on the negotiating table the taboo theme of the revision of the UN Charter. This, in turn, would mean reopening some of the key issues which concern the foundations of the United Nations, this time in the presence of many countries that joined the organization since the Charter was conceived and adopted. Indeed, the end of East–West confrontation makes it possible to envisage such steps.

The contemporary approach to international organization-building should extend and apply to the global level and to the UN system the counterpart norms, structures and practices that are derived from the democratic and human rights values that have been forcefully advocated for application at the national level. The public and governments of the developed countries that have been in the forefront of arguing the case for democratization in the domestic affairs of developing countries should be made aware of the inconsistencies and double standards, and be held accountable for the undemocratic model of international organization and of management of international affairs that they practise and defend.

Indeed, international organizations should be a driving force and catalytic agents for the kind of change that would help bring about a better and just international community and shared prosperity, rather than act as bulwarks of the status quo and instruments in the hands of already powerful countries, or be marginalized and reduced to arenas of perpetual stalemate and futility.

An up-to-date, forward-looking UN system should represent the centre-piece in a system of international relations capable both of dealing with and overcoming the *damnosa hereditas* of the past, and of confronting and managing the new generation of problems facing society in the late twentieth and early twenty-first centuries.

In creating such a system, and in part because of the large number of states, and the lack of consensus and great inequalities among them,[24] two basic challenges, among others, need to be dealt with:

- *how to secure the democratic character of international organizations*, both as

---

24 One of the contributing reasons for the differentiation, stratification, and erosion of the democratic character of international organizations is the different ability of states to keep up with the multitude of organizations and the related processes and negotiations, continually increasing in numbers, importance, and complexity. For many developing countries, their participation has become largely representational. Only a handful of them can mobilize the necessary staff, technical effort, and resources. For example, in the Uruguay Round of negotiations a series of factors turned many developing countries into observers with no other choice but to follow the tide. Without much to offer in terms of concessions, and faced with the massive economic power combined with high technical preparedness of the negotiators from the North and a clear set of objectives *vis-à-vis* the economic space in the South, these countries willy-nilly are engaged in a process that will greatly influence and in some cases determine their economic and political fortunes and futures, and that they seem hardly able to influence. Were it not for a few developing countries that can mount a credible negotiating posture and technical presence, and for the fact that, as the process unfolded, a greater degree of solidarity and consultation among developing countries materialized, the South would have been in a much more difficult position in these negotiations.

a normative value subscribed to fully and sincerely by the international community and all of its members, and practically, i.e., in terms of collective decision-making, negotiations, organizational processes, freedom of thought, analysis, and inquiry, the rights and duties of international civil service, etc. ; and
- *how to make international organizations into effective and responsive instruments* for settling disputes or conflicts, defining and managing problems of common interest, negotiating agreements, and pursuing shared goals, and how to rationalize and make more efficient their organizational processes.

The institutional cum normative package should have as its major component and binding force the twin objectives of protecting world peace and securing the well-being of all humanity. The concept of peace and collective security will need to go beyond the firefighting, military, peace-keeping, policing, and curative approach conceived in the 1940s and still favoured by the North. It should be defined broadly and in systemic terms as 'peace-building', to encompass the removal of the causes of conflict and of conditions that are the source of such contemporary ills as the poverty trap in which billions live, exploitation, inequality, humiliation, and the frustration and alienation of individuals, groups, and entire nations.

Neither a Darwinian approach to global problems nor self-interest based on narrowly conceived group or national cost–benefit analysis as the paramount and ultimate criterion, both of which are implicit in the currently dominant thinking and practice, is a promising way to build solid and lasting foundations and conditions for peace and cooperation in the decades to come. It also places the poor and the countries already behind in the global race at a tremendous disadvantage.

The systemic approach to global challenges should have among its principal components the following:

- Defining and implementing rules and regimes of international economic relations and global economic macro-management, with the overarching objective of achieving a degree of fairness and equity, and benefits for all, and thus the willingness among the key actors to accept the system as legitimate and to work within it. There can be no enduring peace and stability in an international system where more than 1 billion people live in poverty and human misery. Nor can there be stability in a system where the majority of nations are relegated to second-class status and feel disenfranchised, powerless, exploited – a system dominated by those with power, money, technology and resources.
- Launching of global solidarity and mutual confidence-building programmes intended to narrow and bridge the North–South gaps, to assist the South in dealing with and overcoming some of its acute problems and chronic deficiencies, and to help the developing countries build the essential take-off capabilities in critical domains, such as human resources, science and technology, and infrastructure, in order to be better equipped to meet challenges and fulfil their

needs in the decades to come.[25] An approach of this kind gains in relevance when looking at events in Eastern Europe and the USSR and the ability of the advanced industrial countries to muster an uncommon degree of solidarity and generosity to assist the former socialist countries so that these can change their political and economic systems, and maintain a minimum of social stability. There can be no peace and stability in a world where the countries of the South are not able to confront and resolve their problems and meet the aspirations of their populations.

- Moving the environmental issue to the very centre of the international agenda, both as an urgent issue for peace, global health, survival, and economic prosperity for all, and as an interconnecting topic that cuts across all political, social, and economic domains and that provides a foundation for global package deals or multisectoral, multilevel, and time-spaced quid pro quo bargains involving all actors on the world scene. The Earth's environment is indivisible and can be managed and protected only by means of a global and comprehensive approach, and with the participation of all countries and peoples in accordance with their responsibilities and abilities to contribute to the common objective.

The international organizations of the future should, of necessity, be quite different from anything known in the past or acceptable to governments at present. The age of transition, structural change, and new international regime-building requires global policy guidance and management that are democratically inspired, and collective efforts and appropriate institutions to channel social and economic forces and change in the direction of shared goals, international help and preferential treatment for the disdvantaged countries and relationships of mutual trust in a joint undertaking.

Institutionally, the recognition of the integrated character of the global agenda would call for integrating effectively all institutions of the UN system, including the three splinter organizations of IMF, World Bank and GATT, appropriately reformed and democratized, in a single policy, conceptual, and institutional framework. It is essential to devise a comprehensive framework, with all groups of countries being fully and effectively represented, to guide and direct this large system. To attain this objective, one of the proposals pertains to creating an Economic Security Council at the intergovernmental level, as the centrepiece of the 'economic UN', parallel and equal to the 'political UN', and backed by an interdisciplinary, central secretariat, highly qualified and large, and able to perform its functions in a critical and independent manner.[26] Likewise, it is essential to

---

25 A Global Infrastructure Fund (GIF) was proposed in 1972 – as an expression of a 'Global New Deal' – to carry out multilateral projects in areas that would contribute to the development of the Third World (e.g., energy and resource development, improvement and use of swamplands, greening of deserts, large-scale construction of transportation systems, canals, and tunnels using state-of-the-art industrial and engineering technologies, global telecommunications, global network of superports). See M. Nakajima, 'A New "Global Deal"', in *Journal of Development Planning*, No. 17, 1987, pp. 191–7.

26 See M. Bertrand, 'Some Reflections on Reform of the United Nations', *op. cit.*, and *The Third Generation World Organization*, *op. cit.*, pp.83–90. As a possible model for the secretariat, the author notes the Commission of European Communities, which is mandated to state 'the Community point of view', and to study and propose compromise solutions.

redefine the role and priorities of the Secretary-General, which should evolve away from the predominant concern with narrowly conceived political and security domains, and towards the more comprehensive leadership role that is required.[27]

Ideally, and as the experience of GEMS shows, in the quest for a new system one should not start from the existing pieces, but rather from a global design and the very purpose of the system as the first step. Would the governments have enough fortitude and imagination to engage in such an exercise, and would this not be too much to expect in a situation where the so-called political realities do not even allow a discussion of some basic issues? Yet, it is certain that the governments would find it easier to negotiate and arrive at a consensus and a global vision without getting bogged down in secondary institutional details and bureaucratic quagmire of an awesomely complex system, without having to worry about jurisdictional preserves or about protecting jobs for their nationals or, in the case of some, about the incomes and other benefits that they derive from organizations located on their soil.[28]

As already noted, in any serious review of the future of the UN system, inevitably the issue of democratization of international organizations, in the sense of decision-making and control of their work, policies, and orientation, as well as of people participation, will be one of the central issues.

In regard to decision-making, some form of compromise and *modus operandi* will need to be reached in a polarized debate between, on the one hand, the rich and powerful nations who insist that the procedure of decision-making must

---

27 Urquhart and Childers speak of the need for an 'international cabinet', with the heads of key UN agencies joining the Secretary-General, in order to evolve a coordinated and energetic approach to global problems. B. Urquhart and E. Childers, 'A World in Need of Leadership: Tomorrow's United Nations', *Development Dialogue*, 90: 1–2, pp. 75–6.

28 A built-in constitutional provision for periodic dismantling and renewal may be the best way to re-energize international organizations, get rid of dead wood and break down bureaucratic strongholds, and transcend obsolescent ideas, unwanted structures and practices, while maintaing the essence of institutional memory and desired continuity. In this context, the location of international organizations and impacts of location on their functioning and physiognomy need to be given serious consideration. This study has gone to some length in discussing the controversial issue of UNEP's location and its impacts. This was done not to argue for or against the merits of Nairobi but to show that location has significant influence on the nature and functioning of an international organization. Some circles still have not given up the idea of moving UNEP from Nairobi and can make a plausible technical argument in favour of such a move. Ultimately, however, on the wings of technological change many of the practical disadvantages will disappear, leaving as the key issue the political and other influences and pressures on international organizations emanating from the local environment. In this context, another taboo theme, that of the location of the UN Headquarters and of the Bretton Woods institutions, deserves special attention by the international community. For the evolution of the UN system in the twenty-first century, it would be highly beneficial in political terms were these headquarters to be moved to more neutral environments. Now that the cold war is over, this is no longer an impossibility and such a move would set the stage for a new phase in the life and evolution of international organizations, tempering their dependence on and role of supplicant *vis-à-vis* this global power. In practical terms, the proposed move could cause some inconveniences and costs, which are not insurmountable, however. They would certainly be worth the effort in relation to long-term benefits for the UN system and to removing it physically from a world power with clearly defined aspirations to dominate international organizations on its own terms, and from the direct influences of its highly volatile and vocal domestic politics usually hostile to the UN. Indeed, it is not by accident that the North's own organizations involving cooperation among developed countries are not located in the United States.

UN system at a crossroads 269

conform to their terms and confer on those with money and power a special status, as in Bretton Woods institutions or in the UN Security Council, and who would like to see it extended throughout the system, and, on the other hand, those who cling to the principle of 'one country, one vote' as their only chance to maintain a modicum of influence and protection vis-à-vis the North's juggernaut. Certainly, more sophisticated forms of decision-making and representation can be negotiated, which would accommodate the power realities, protect the basic democratic inspirations of the UN, and the need for action and for overcoming in given circumstances the paralysing effects of the need for full consensus.

The new approaches would also need to take fully into account a series of new factors that have come to play a part in world affairs since World War II when the current structures were devised, including the group system, regional groupings, and the environmental dimension. They need also to recognize different categories of decisions and the need to vary or adapt decision-making processes and criteria according to particular situations and types of decisions.[29] New approaches to decision-making could contribute to a reinvigorated UN system. To be fashioned adequately, they will require among other things some give and take, generosity, and enlightment on the part of all governments, especially their bureaucracies locked in traditional thinking, and of public opinion. In view of increasing unipolarity of power and of 'international authoritarianism' on the global scene – demonstrated, for example, by the new situation in the UN Security Council – the issue of decision-making assumes special importance for the developing countries and for small and medium-sized nations in general.

A further issue that will need to be recognized and dealt with concerns the dichotomy within international organizations between the requirement to continue to rely on officialdom, i.e., the formal structures of government and establishment, and the requirement to endow these organizations with a broader and more independent spirit, beyond the immediate concerns of governments or what they can agree to at a given time.

It is the former dimension of international organizations that has been dominant, and has been hitherto the only one that the governments would accept. Yet the two-track approach is essential for a UN of the future. This is so both because of the need to allow a greater access to and role for the civil society in its proceedings, and because of the need to provide a counterweight to the official sphere, which is mostly bureaucratic and rigid in character and increasingly monochromatic, and which, to say the least, does not fully reflect the rich diversity of human concerns

---

29 See M. Bertrand, *The Third Generation World Organization, op. cit.*, pp.85–7. He distinguishes between 'weighted voting' on such questions as budgets or amending programmes and 'weighted representation', in other types of negotiations, when the issue is how to represent basic interests by a limited number of negotiators. The Group of 77 would thus trade off some of its numerical advantages enjoyed in UN forums for gaining access to and influence at a negotiating table where significant issues are dealt with and from which they are at present excluded. (This indirect representation was tried once in the North–South negotiations in Paris, which followed early actions of OPEC and the adoption by the UN General Assembly of the resolutions on the New International Economic Order. It caused a good deal of rancour among those developing countries that were excluded, and a lack of confidence that their interests would be properly represented.)

and pluralism, and is prone to domination by those with superior power and ability to organize and muster the resources.

The role of the civil society in educating public opinion and influencing governments assumes added importance with regard to increasing homogenization of views and hegemonic trends in analysis of basic social processes, aided by massive concentration on the global scale of information and communication capabilities. A large number of issues transcend national borders and concern directly and visibly entire populations, or given strata and groups, making possible bridges and coalitions between them; these do not necessarily find expression through national politics, individual governments, and standard modes of multilateral discourse in international forums.[30]

In conclusion, the forward-looking, progressive and democratic world view should have as one of its chief pillars an up-to-date vision of the UN system that would embody shared strivings and promote the satisfaction of the needs of all peoples. This vision should represent the basic reference in a period of growing conflict and turmoil. Its objective should be to usher in a new age of cooperation and mutual respect. For this to take root an educational effort is called for to enlighten the public, governments, policy-makers, and decision-makers of the potential benefits of such a transformation and to improve the public image of international organizations.[31]

More than ever, therefore, the required vision is predicated on a correct perception of global problems and institutions, and on adequate information and

30 Provided the basic principle of direct popular participation in the work of the UN is adopted, the practical modalities could be worked out. One such idea could be an Assembly of People, as a global policy forum to which delegates could be elected or appointed on a regional, national or even professional or group basis, and which would function in parallel with other major organs of the UN system. A proposal for a 'citizen chamber' has been made. See M. Nerfin, 'The Future of the United Nations System', in *IFDA Dossier*, No. 45, Jan.–Feb. 1985. A similar idea was mentioned by M. Gorbachev in his article 'The Reality and Guarantees of a Secure World', *Pravda*, 17 Sept. 1987. He speaks of a 'World Consultative Council... uniting the world's intellectual élite', including 'prominent scientists, political and public figures, representatives of international public organizations, cultural workers, people in literature and the arts, including the laureates of the Nobel prize and other international prizes of worldwide significance, eminent representatives of churches', who would contribute to 'enrich the spiritual and ethical potential of contemporary world politics'.

31 It must be kept in mind, however, that public opinion has proven to be an unreliable ally, easily swayed and influenced by negative information and views hostile to international organizations. In some instances, what is needed is the targeting of whole nations that have persisted in exhibiting traits and ambivalence inimical to international organization. Research has thus been suggested into cultural and psychological barriers to international cooperation, into the 'psychopathological personality of nations', and into the 'large power complex' that makes them indifferent to the needs and perceptions of smaller countries. See J. Kaufmann, 'Developments in Decision-Making in United Nations', in J. Harrod and N. Schrijver, eds, *The UN Under Attack, op. cit.*, p. 30. For certain, these deeply entrenched prejudices feed on and are reinforced by facts from the life of international organizations, often taken out of context and anecdotal in character; a view of the UN commonly encountered in the North has thus been summed up as 'misguided priorities, lack of transparency and accountability, absence of overall directional thrust and vision, lack of coordination, quality control and consistency, being overstaffed and much too politicized (i.e. influenced too much by developing countries who do not directly bear the cost of their follies than by donors, who do)'. Mistry and Thyness, *op. cit.*, p. 15. This imagery has been reinforced by the world view of the ascendant conservative and neo-liberal forces, and their anti-UN bias and stress on unilateralism of power, the hard bargain, and interest-dominated political process.

knowledge among the political leadership, the public, and the opinion makers including academics, many of whom, despite all the new tools of analysis and information at their fingertips, persist in their relative ignorance or misperception of the real world, hold a narrow and parochial vision, and are confused about the problems, potentials and achievements of the UN.

As recent experience has shown, effective communication and publicizing of ideas will be crucial to a reform and expansion of functions of the UN. The media are thus potentially a very important ally in this undertaking. However, the global media are concentrated in the hands of a few corporate and private actors in the North, and are closely identified with the establishment and the market. Mostly, the media have been reserved towards multilateralism and reflect closely the views of some powerful, conservative political and economic interests. Therefore, the media can also act as the most formidable opponent and a tool of those opposed to a stronger and reformed UN system. The influence over and control of people's minds via communications and media has become one of the most effective, global means for depicting and interpreting reality, and for promoting political and economic objectives and values in the modern world. Who controls the media and what they say thus emerge as issues of special importance also for the future of the UN system and for the future of international organizations in general.

BY WAY OF A CONCLUSION

This study was an attempt to reflect only some of the processes and facts from the complex daily life of international organizations. These should inform intelligent debate and decision-making concerning institutional approaches to international cooperation at the global level, a subject which for too long was bound by political inertia but which has entered into a stage where significant innovation appears possible.

The nature and direction of such innovation is of vital importance. Among this study's many conclusions, those singled out in the last two chapters therefore bear repeating. These are:

- the need for a high professional quality, independent-minded, motivated and inspired, truly international secretariat;
- the need for adequate and independent finance for international organizations, and reducing their dependence on the major contributors and/or donors;
- the need for institutional dynamism and continuous adaptation and improvement;
- the need for an integrated approach to understand and deal with both the causes and effects of the interrelated, systemic issues facing the international community (security based not on a 'balance of power' but on a 'balance of development', to borrow the words of a statesman from the North);
- the need for full and committed participation of governments in international undertakings and their readiness to transcend narrowly defined national or group interests in favour of a common perspective; and

272   *The quest for world environmental cooperation*

- the need to secure the democratic character of international organization.

These considerations should be at the very base of any sincere effort to reform and redesign the UN system into an effective and principal instrument for managing international affairs, securing global prosperity, equity, sustainable development, democracy and peace for all, and combining inevitable confrontation of views and interests with greater problem solving and accommodation. Only a truly world organization can provide the necessary leadership and help to define and manage with the global, transnational, or supranational dimensions of international politics, which have come to stay and will increasingly dominate human existence. Likewise, only a truly world organization that is democratic and accepted by all major actors or groups of actors can help deal with and resolve the challenges of North–South divisions and of South's development.

The future of international community and of nations and peoples will be greatly affected by how these two sets of interrelated issues are handled and sorted out. This challenge also offers a chance for revitalizing the United Nations and placing it at the very centre of collective policy and action, an undertaking which calls for mobilizing public sentiment and for the political courage and internationalist vision of national leaders, especially those at the helm of major world powers.

How this broader set-up evolves will be of importance for the future of GEMS. Likewise, emerging global instruments such as GEMS will play a major role in shaping the world of tomorrow and its international organizations. A sustained effort will need to be deployed to secure GEMS' truly international character, since the information and knowledge that it is supposed to provide will affect fundamentally national and international life and the future thinking, action, and choices of humankind. This study has tried to depict the historical and policy context in which GEMS was born and how it has evolved in the early stages of its development, as a small contribution towards thinking through its future options and those of the UN system of which it is an integral part.

Name index

n after page nos indicates material in notes

Aga Khan, S. 251n
Agarwal, A. 218n
Albritton, D.L. 121n
Andersen, N. xxi, 149n
Arrhenius, S. 92

Barth, D.S. 167n
Baylocq, J-F. xxii
Becker, F. 165n
Berthoud, P. xxii
Bertrand, M. xiiin, xxii, 225n, 242n, 248n, 250n, 251n, 261n, 267n, 269n
Bie, S. 194n
Bloomfield, L.P. 261n
Bojkov, R. xxi
Bolin, B. 177n
Bolle, H-J. 165n
Brown, K.W. 171n
Brundtland, G.H. 30
Budhoo, D.L. 255n
Budyko, M. 94n

Childers, E. 229n, 232n, 234n, 235n, 247n, 268n
Cleveland, H. 261n
Cordovez, D. 10n
Clark, W.C. 72n
Croze, H. xxi, 157n, 164n

Dell, S. 256n
Deudney, D. 162n
Donini, A. 250n, 264n
Dovland, H. xxi, 189n, 191n

Ehrhardt, M. 127n
Elmandjra, M. 247n, 250n

Farrar, J.F. 171n
Fehsenfeld, F.C. 121n
Finkelstein, P.L. 110n
Fromuth, P. 244n, 251n
Frosch, R. xxii

Galtung, J. 251n
Georgii, H.W. 102n
Goldberg, E.D. 144n, 146n
Goodman, G.T. xxi, 51n, 171n
Gorbachev, M. 270n
Graisse, J.J. xxii
Granat, L. 177n
Gwynne, M. xxi, 157n, 164n, 210n

Haas, E.B. xxi, xxii
Hadwen, J.G. 255n
Haji, I. xxii
Hakkarinen, C. 102n
Hammarskjold, D. 226, 229n
Hansen, P. xxii
Harrod, J. 227n, 252n, 255n, 263n, 270n
Hazzard, S. 228n
Helmer, R. xxi
Holben, B.N. 156n
Houghton, J.T. 113n

Ingelstam, L. 177n
Izrael, Yu.A. 72n, 167n
Irwin, M. 255n

Johannesson, M. 177n
Justice, C.O. 156n

Kaufmann, J. 255n, 270n
Keckes, S. xxi
Keeling, C.D. 95n
Kennan, G.F. 15n
Khosla, A. xxii

Kingsbury, B. 250n
Kohler, A. xxi, 64n, 93n
Komar, L. xxii
Kohnke, D. 127n
Koning, H.W. de 64n, 93n
Kratochwil, F. 264n
Kreisel, W. xxi
Kronebach, G.W. 93n
Kullenberg, G. xxi

Lamp, J. 194n
Lampe, R.L. 101n
Lanly, J-P. 156n
Lehmann, A. xxii
Levy, E.M. 127n
Lundholm, B. 167n

McElroy, M. 118n
Manning, M.R. 102n
Martin, B. 93n
Mattson, E. 177n
Miller, J.M. 110n
Mistry, P.S. 237n, 238n, 253n, 256n, 263n, 270n
Mitchell, A.W. 155n
Mooneyhan, W. xxi
Morgan, G.B. 167n
Mortimer, E. 252n
Munn, R.E. 72n, 93n, 167n

Nakajima, M. 267n
Narain, S. 218n
Nerfin, M. xxii, 270n
Newell, R.E. 113n
Norton-Griffiths, M. 157n
Nyerere, J.K. 263n

Oden, S. 177n

Pfaff, W. 260n
Pietila, H. 30n
Pitt, D. 226n
Pravdic, V. 141n
Prebisch, R. 226–7
Pronk, J. 263n
Preston, A. 128n, 130n, 149n
Pueschel, R.F. 92n, 93n, 102n, 107n, 112n, 177n
Puzak, J.C. 101n

Raghavan, C. 257n
Randolph, K. 173n
Reichle, H.G. 113n
Roberts, A. 250n

Rodgers, C.D. 113n
Rodhe, N. 177n
Roll, H.U. 126n, 129n
Rovinsky, F.Ya. 170n
Rowntree, P.R. 165n
Ruggie, J.G. xxi, xxii, 264n

Sand, P. xxi, 186n
Sanchez, V. xxii, 228n
Santa Cruz, H. 228n
Senarclens, P. de 252n
Schrijver, N. 227n, 252n, 263n, 270n
Seiler, W. 113n
Sella, F. xxi, 93n
Smith, R.A. 177n
Sobtchenko, E. 127n
Sors, A.I. 51n
Steinberg, E.B. 241n
Strong, M.F. ix, xxi, xxii, 19n, 244n, 251n
Suzuoki, T. 127n
Synnott, T.J. 155n
Szekiada, K.H. 149n

Tamin, C.O. 177n
Taylor, F.W. 113n
Temple Black, S. 42n
Thacher, P. xxii, 161n, 194n
Thompson, J.R. 171n
Thyness, P. 237n, 238n, 252n, 256n, 263n, 270n
Tokuhiro, A. 127n
Tolba, M.K. 118n
Townshend, J.R.G. 156n
Trainer, T. 30n
Tuck, A.F. 121n
Tucker, C.J. 156n

Urquhart, B. 229n, 232n, 234n, 235n, 247n, 268n

Vahter, M. 71n

Waldrop, M.M. 118n
Wallen, C.C. xxi, 93n, 102n, 191n
Webber, J.O. 200n
Weiss, Th. 226n
White, H.H. 147n
Wiersma, G.B. xxi, 167n, 170n, 171n
Wruble, D.T. 112n

Yager, J.A. 241n

Zammit Cutajar, M. xxii

Subject index

n after page nos. indicates material in notes

For meanings of abbreviations see pages xxv–xxviii

acid rain 171, 175–6, 177, 187; at UNCHE 177–8,
Acidification of the Environment Conference (1982) 181
Action Plan (UNCHE) xviin, 9, 12, 177–8; IOC involvement 129; monitoring-related recommendations 38–41; sector-related recommendations 41
Administrative Committee on Coordination 54
aerial photography 155
aerosols 97, 112n, 112
aflatoxin monitoring 65
agency–UNEP partnerships 20
agriculture, land converted to 91
air pollutants 182–3; long-range transport and deposition of 105
air pollution: background levels 93–4, 100–1; human exposure to 66, 72; monitoring of 41, 182; *see also* transboundary air pollution
air-quality monitoring 64, 69
alkalization 154
animal feed monitoring 65
anthropogenic impacts 105, 259; on climate 92–9, 113; and the marine environment 148
assessment–collective action linkage 138
Assistance in the Case of Nuclear Accident or Radiological Emergency Convention 87
atmosphere systems, monitoring of heat budget system 97

background monitoring 169
BAPMoN 36, 64, 93, 114–15, 121, 122, 179; adaptability of 105; build-up of 99–105; criticized by WMO Working Group on Environmental Pollution and Atmospheric Chemistry 115–17; and GAW 211; main objective 94–5; monitoring tools and capabilities 112–13; stations to be used for other monitoring 167, 171; voluntary nature of 110–11; *see also* baseline stations; Canary Islands baseline station; Kenya, Mt, baseline station
baseline monitoring 36n
baseline stations 92, 101–2n, 101, 107–8, 109, 121, 145n; disparities in 116–17; work of 94–5
bioaccumulation in the environment 65
biological monitoring 63, 66, 69n, 70n, 70, 72
biota–atmosphere interaction 173
Brandt Commission 241, 264
Bretton Woods institutions 229n, 247, 251; developed countries' interest in staff 230; and GATT, concentration of power in 254–8; political role in development of developing countries 256–7; *see also* IMF; UN system; World Bank
Brundtland Commission 241, 264; *see also* WCED
bureaucracy, effects of 243–4

cadmium (+ compounds) 45, 65; monitoring of 62n, 66
Canary Islands, baseline station 108n, 108
carbon dioxide 92n, 93, 104, 112n, 122; as a heat trap 91–2; sources and sinks 96, 116

carbon emissions, North–South dispute 218
carbon monoxide 62n, 64, 113n
catalytic approach (UNEP) 16, 17, 25, 97, 159n, 178–9, 192, 198, 211, 214, 241
cause–effect chain, transboundary air pollution 187–8
CFCs 99
chemical pollutants 43
Chernobyl accident 45, 67, 209; effects on monitoring 86–90; usefulness of EMEP monitoring network 188; and WMO 116
civil society, role of 270
climate change 91, 94, 117, 118; human induced 92–9, 113
climate-related monitoring 53, 91–123; beginnings in GEMS 92–9; costs of 107–10; early indicators of progress 99–105; factors at play 105–13; voluntary nature 110–11; see also BAPMoN; global atmosphere monitoring; World Climate Programme
coastal areas 143
coastal and territorial waters, marine pollution monitoring 131–2
Cocoyoc Declaration 26n
Codex Alimentarius 71
complexity, of modern world 246
compliance: with EMEP targets, monitoring of 182; more formal approach for 80
compliance checking 37n
computer hardware 200, 201
computer software packages, usefulness of 194n, 194
Conference on Security and Cooperation in Europe 178
contributions in kind 241–2n
coordination 54; in GEMS HEALTH 80–1; issues 111; role of UNEP 17–18, 20; in UN system 18–19
countries, terminology and classification of 224n
cryosphere monitoring 97

data: integrating and linking 199–200; GRID a global nucleus for 200–1; location-specific 69; see also monitoring data
data quality, EMEP 190–1
DDT 45, 62n
debt and development crisis (1980s) 29
decision-making 268–9

deforestation 156, 165
degradation, environmental 30, 44
desertification 44, 165
desertification monitoring 151, 154
Designated Officials on Environmental Matters (DOEM) 20, 54, 209
developed countries 13, 224, 227; alarmed, lost interest in UNEP 14; attitude to GEMS 208; effects of conservative governments 28–9; financial support to WHO 84; global economic organizations based on World Bank–IMF–GATT 253–4; single issue campaigns 34; see also industrialized countries
developing countries 4, 8, 12, 13, 33, 41, 77, 132n, 224; acting as South 262; attitude to GEMS 208; baseline stations in 107n, 107–8, 108n; climate-related monitoring 106–8; discouraged by cost of health-related monitoring 78; Environment Fund as additional finance 25–6; favoured natural resource monitoring 151, 158; felt Security Council under control of North 258; and GEMS global network 74–6; and international colonialism 259–60, 260n; more interested in regional seas approach 141–2, 143–4; participation in MED POL 135, 136–7; pollution in 61; priorities put forward for GEMS 44; setting up own intergovernmental organization 263; support for 109–10; and UN reform 264n; and UNCHE 6, 38; vital natural resources and need for information 164–5
development issue 226; neutrality of international civil servants 227
development process, environmental problems of 43
donor pressure 26, 239–40, 250
drifting buoys 149
drinking water monitoring 83n

Early Notification of a Nuclear Accident Convention 87, 89
Earth, as a system 149
Earth Observation Satellite Company (EOSAT) 163
Earth Observing System (EOS) 118–19, 120, 149, 163, 201–2, 219, 231
Earth Resources Technology Satellite (ERTS) 161

Earth System Science research and observation programme 119
Earthwatch xvi, xvii–xviii, 14, 199, 200, 204; and GEMS 238; global programmes of xviii, 23; *see also* environmental assessment
Earthwatch Working Group 54
East–West divide 249
ECE 53; work on transboundary air pollution 178–81
ecological monitoring 168; *see also* integrated monitoring
economic downturn, effects of 29
Economic Security Council 267
ECOSOC 248
El Nino/South Oscillation 98n
EMEP 121, 180, 214; Chemical Component (MMA and EMA) 189n; financing of 186–7; implications of 191; and North–South syndrome 189; phases of 180–1; Protocol on Long-Term Financing 186–7; significance of for GEMS 184–92; slow progress 189–90
endangered species monitoring 158
energy production and use, environmental impacts 41
environment 6; safe/low-risk 62
Environment Coordination Board (ECB) 19–20, 20n, 47
Environment Fund 8, 17, 20, 23–7, 48, 51–2, 55, 214, 215, 238–9; argument over 80; key role of 23–5; major donors 26; management of 26–7; money for natural resource monitoring 158–9; paid for additional secretariat posts 22; pays for staff/operation of GEMS PAC 55; rules of the Fund 24–5; use of in developing countries 74–5
environment–development link 32
environmental assessment xvi–xvii, 5, 50–1; in EMEP 214; in GEMS HEALTH 85; using GRID 203–4
environmental conditionality 34
environmental costs, externalized 176
environmental data sets 193–4
Environmental Health Criteria 71
environmental issues 3, 205; a key concern 121; tension in xix; and the UN xvi, 18–20
environmental management xvi–xvii
environmental measurement, harmonization of 202

environmental monitoring: climate-related 53, 91–123; defined 36–8; health-related 52, 61–90, 214; at UNCHE 35–42
environmental politics, global 33
environmental quality monitoring 63
environmental stress indicators 45
environmental trends 37
Europe, monitoring of long-range transboundary air pollution 177–83
European Space Agency 163
excellence: in UN secretariats 235–6; for UNEP posts 21, 22–3
Executive Director, UNEP 15, 24–5
expert groups 65–6, 80; for GEMS 49–50, 54–5
experts, funded by UNEP 78

FAO 53, 63, 134; soil degradation and tropical forest monitoring 152; World Forest Inventory 154
finance 159; for BAPMoN 107–8; for environmental monitoring 215; further resources needed 84; for global climate- related monitoring 109–10; for international organizations 236–43; lack of 140, 143; non-traditional 240–1; used as pressure 237
Finland 183
fluorides 62n
FOGS exercise 257
food monitoring 65
forests 169–70; effects of air pollution, programme 185–6; forest death 175; tropical, monitoring of 152, 154–7, 166, 173–4
Founex report 5
France, SPOT satellite 162
freshwater monitoring 64–5
freshwater pollution, long-term trends in 65

GATT 251, 253, 254–8; Uruguay Round 257; *see also* UN system
GAW 120–1, 211
GEEP 145n
GEMS xv–xvi, xviii–xix, xx, 1, 23, 34, 42n, 50–7, 89, 121–2, 221, 237, 272; assessment of 205–11; and climate change 118, 120; creativity and latitude of staff 230–1; definition of 51; dilemmas in implementation 116; dual function of natural resources remote sensing 161; financing of 215n, 215,

236, 238; at first UNEP Governing Council 41–3, flawed by uneven participation of member states 206–7; future of 211–20; and GRID 199–204; and human-induced climatic change 92–9; institutional questions 46–50; and integrated monitoring 169, 170–1; and integrated ocean monitoring 146, 148; intergovernmental body suggested 49–50, 54–5, 216; interrelatedness in 247; and *Level Three* 55; and *Level Two* 52, 55–6; marine pollution monitoring in 140–3; nature of 213–14; position of radiation monitoring 87–8; priority list 45; programme goals 44–6; a role rationalized and redefined 122; role of UNEP 214–17; setting up of 42–50; significance of EMEP 184–92; structure and linkage missing 209–10; and UN practices 245; *see also* specialized agencies
GEMS clusters 52–3, 59, 78, 202, 210; environmental monitoring 93; natural resource monitoring cluster 152; status and progress indicators 67–8
GEMS HEALTH 171; assessment and prospects of 82–90; background documents 68n, 71, 86; coordination 80; diversification and movement into research and monitoring 76–7; indicators of progress 67–73; international funding for 77–8; link to integrated monitoring network 171; logistics 81–2; nature and constituents of 62–7; North–South differences 74–6; voluntary nature of 78–80; *see also* radiation monitoring
GEMS office 48, 49, 53–4
GEMS PAC 53, 81, 88, 131, 166, 209, 214, 216–17; analysis of forest data 156n; and GRID 193–4, 196, 199; integrating core of system 54; less involved with evolving EMEP 192; impacts of location 210n; proposals for Latin America 160n; secretariat too small 210; some financial and administrative autonomy 245
GEMS/Air 63, 64, 69–70, 71, 86
GEMS/Food 63, 65, 69n, 70
GEMS/IBM 170–1
GEMS/Water 63, 64–5, 69n, 69, 73n, 81, 83; linked to other GEMS clusters 84; use of existing national networks 70

GEMSI 145n
geographic information systems (GIS) methodologies and procedures 196
Geosphere–Biosphere Observatories 120n, 120
Germany, Federal Republic of 108n, 160n, 179, 185, 202
GERMON 89n, 89
GESAMP 141n, 144n, 144, 145, 146; open ocean report 141; Working Group on Integrated Global Ocean Monitoring 147, 150
GESREM 145n
GIPME 124n, 126–7, 129, 133, 144, 145
glaciers 95
global atmosphere monitoring, an integrated approach 113–23
global challenges, systematic approach to 266–7
global climate, basic actors 91n
global environment 61–2, 214
Global Environmental Facility (GEF) 238–9
global environmental monitoring 37–8, 212, 218
global environmental monitoring system 122, 219
Global Information Resources System (NASA) 194–5
global integrated background monitoring network 170–1
global interactive system 31
global interdependence xv
global monitoring networks 35–6, 50; BAPMoN and WGI 104
global systems build-up, logistics and requirements 81–2
global trends 93
global warming 92, 94, 113, 205; politics of 117–18
GO$_3$OS 98–9, 121
Governing Council (UNEP) 11, 16, 19n; central coordinating body for GEMS 54–5; confrontation between developed and developing countries 42–3; first session 13, 15, 42–3; searching for role 27–8; third session, issue of governments' role 54–5
governments: commitment to system build-up sought 84; contribution to GEMS' operational costs 56; and international organizations 236, 237, 239; and MED POL 138–9; no wish for UN environmental body 6–7;

performance and contribution to international undertakings 244–5; and role of independent civil servants 231; squabble over secretariat posts 229; unwilling to assist GEMS HEALTH 79; wary of environmental assessment 203–4
GRAP 220n
greenhouse effect 92, 94n, 113
greenhouse gases 104, 105, 116, 117–18
GRID 53, 59, 158, 165, 166, 172, 192, 193–204, 212, 216–7; commercialization and communication issues 198n; and GEMS 199–204; global network 198, 203; financing of 195, 197n, 199; implementation phase 197–9; influence on GEMS PAC and UNEP 217; origins of 193–6; pilot phase 196–7; for tropical forest data 157
GRID control, Nairobi 196, 203
GRID processor, Geneva 196, 202–3, 217
Group of 77 10, 11, 12, 13n, 13, 24, 32, 45, 208, 219, 224n, 228, 250; actions needed 264–5; and Environment Fund 24; helped by UNCTAD 226, 227
Group of Seven 252
Gulf War 257, 260

HEALs 63, 66, 70, 76–7, 81, 83; barriers to be overcome 74; coupling of research with monitoring 73; in GEMS HEALTH evolution 84; proposal for single location monitoring effort 66; sites 72
health-related monitoring 52, 61–90, 214; cf. climate-related monitoring 93–4; factors at work 73–82; *see also* GEMS HEALTH
human factor, key variable of international organizations 232; *see also* international civil service
human health 214; acids in air 175–6; nuclear radiation 88; *see also* air pollution; pollutants/pollution
human resources, in international organizations 225–6
human settlements 8

IAEA 87–8, 134
IBM, equipment donation to GRID 199n, 199
ICES 35, 124n, 126, 128n, 133, 134n, 144
ICSU 98, 120

IGBP 118–19, 120, 171, 173, 210, 211, 213; Action Plan 201, 217–18n
IGOM 128n, 145n, 145–6, 147–8, 150
IGOSS 125–6, 127, 129, 132, 144
IMF 253, 254–8; *see also* UN system
impact monitoring 64, 69
impact-effect time lag 62
India 163
individuals, role of xix–xx
Indonesia, Tropical Earth Resources Satellite (TERS) 163n
industrialized countries 4; interest in GEMS reduced 43; pollutants in 61; unhappy with location of UNEP 10–11; *see also* developed countries
inertia in GEMS 207–8; institutional 244
information exchange xvii
information-gathering function 16
INFOTERRA xviii, 23
institutional betterment, a permanent objective 243–6
institutionalization, in EMEP 185, 186
integrated approach 96; to global atmosphere monitoring 113–23; in work of UN system 246–8
Integrated Global Monitoring, Riga Symposium 128n
integrated monitoring 53, 153, 166–72, 173–4, 187–8; of air pollutants 182–3; of climate-related activities 96–7; global ocean 145–8; official definition 168; pilot project 169–70; regional network of CMEA countries 170; use of new capabilities 213–14
integrated monitoring network 121, 122; global, for atmospheric pollution 211
Interagency Committee for Coordinated Planning and Implementation of Responses to Accidental Releases of Radiation Substances 88–9
interagency tensions 81n
Interagency Working Group, view of GEMS excluded UNEP 47–8
intergovernmental deliberations, rigidity and inertia of 207–8
Intergovernmental Meeting on Monitoring (1974) 44–50, 54, 124–5; decided priorities for GEMS 44; favoured natural resource monitoring 151–2; looked into institutional questions 46–50; and radiation monitoring 45–6, 86–7
Intergovernmental Working Groups: on

Marine Pollution 37n, 124n; on Monitoring 37, 38n, 38
international civil service 225–36; erosion of role and influence 230; selection and education of staff 232–3, 234–6
international colonialism 259–60
international community, task of 259
international cooperation xv, 271–2
International Drinking Water Supply and Sanitation Decade 71
international economy, globalization of 258
international institutions, democratization of 268
International Meteorological Organization 92
international monitoring systems, necessity for 88
international organizations xii, 223, 224, 257; actors in future of 261–2; basic challenges 265–6; financing of 236–43; future 232–3; need to be free of donor pressure 242
International Space Year 122
internationalism, attacked 251–2
interrelatedness 246–8
IOC 53, 124n, 127, 134, 142–3n, 145, 148; founding of 128–9n, 128–9; inadequate funding 140; Training, Education and Mutual Assistance (TEMA) component 130; weak institutional position of 142
IOC–UNEP tension 130–2, 140, 142
IRPTC xviii, 23
ISLSCP 165
IUCN 158

Jackson 'capacity study' 250
Japan 163; satellite for GEMS 219
Joint Inspection Unit 250
joint (thematic) programming 20

Kenya, Mt, BAPMoN baseline station 95, 103n, 108n, 108, 109n
KREMU 157n, 193

lake acidification 175
LANDSAT data 161–4; commercialization of 162; doubts 162, 163
lead 64, 65
Level One 16, 96
Level Two 16–17, 28, 55, 56, 98, 109, 114, 131, 209; governments will assume and discharge obligations 73–4
Level Three (Fund Programme) 17, 26n, 55–6, 100, 109, 209; EMEP 188
LIDAR systems 112, 200
Long-Range Transboundary Air Pollution Convention 179–80, 185–6, 187, 214

MAB 152–3n, 152–3, 167
MAB biosphere reserves 153, 170
major donor syndrome 240
MAPS 113n
MARC 51n, 71n, 217; broad data base proposal 194
marine pollution monitoring 53, 124–50; factors at play 140–3; first steps and controversies 125–32; global, movement towards 143–50; North–South differences 141–2; in regional seas: Mediterranean model 132–40
MARPOLMON 127
MED POL 134–40
media, reservations towards multilateralism 271
media interaction 167
Mediterranean Action Plan 133–40; *see also* regional seas
Mediterranean initiative 131
Mediterranean Trust Fund 135
mercury 62n, 65
mercury monitoring 137n, 137
Mexico City, blood lead level 61
microbial contamination 62n
Minimata Bay disease 124
monitoring: costs determined by detail/quality required 160; as part of comprehensive scheme (EMEP) 184–5; regional approach revived 43
monitoring and assessment techniques 83
monitoring data 50–1; and GEMS HEALTH 85; organized in GRID 193; a potential tool 218
monitoring data–assessment–management linkage 200
monitoring programmes, local, important role in GEMS 56
Montreal Protocol on Substances that Deplete the Ozone Layer 99
multilateralism xi, 250, 252, 271
'mussel watch' programme 146–7n, 146, 148n, 148
mycotoxins 62n

Subject index 281

Nairobi: home to GRID control 196, 203; as location for UNEP 10–11; *see also* UNEP, location of
NASA: Global Habitability programme 118, 195–6; and GRID 195, 196; *see also* LANDSAT
national authorities, use of monitoring data 71
natural disasters 113
natural resource base monitoring 41, 151
natural resources 5, 8; exploitation of 151; macro-view of 172–3
natural resources monitoring 53, 153–66; costs and financing of 158–60; factors at work 158–60; prospects for 172–4; tool for management and decision-making 173; remote sensing 160–5
nature protection/conservation 4–5
neutrality, of international civil servants 227–8
New International Economic Order (NIEO) 26n, 250
NGOs 3, 9
nitrogen compounds 62n, 185; reduction in emissions 182
nitrogen oxides 64, 176n, 176
Non-Aligned Movement 250
non-governmental actors 262
Nordic countries 183, 184, 187
Nordic UN Project 264
North 228; differentiated 262; policy regarding UN system 250–8; UN reforming drive of the 1980s 250–1; and the South 33–4; view of UN staff 233–4
North–South conflict 24, 32–4, 259
North–South consensus politics 15
North–South development gap xi, 164, 206, 224, 259
North–South differences: carbon emissions 218n, 218; climate-related monitoring 106–8; consequences for GEMS 43–5, 49–50; on eve of UNCED 31–4; Environment Fund 24–5; health-related monitoring 74–6; impact on marine pollution monitoring in GEMS 141–2; remote sensing of natural resources 161–5; role of UN 224, 226
North–South divide 221, 249, 266; financial and economic 239; neocolonialism 260

North–South environment/ development interface 238
North–South relations, stalemate in 228
North–South tensions xix, 4, 5, 259; concerning GEMS 44, 228, 231n
Norway 177, 178, 191; study on acid rain effects 178, 179
nuclear releases, accidental 86–7, 88

OCA PAC 140n, 144n, 145, 166; *see also* RS PAC
ocean baseline stations 125, 126
oceans, degrading and dispersing waste materials 149–50n, 149–50
OECD 177, 178, 230n
oil pollution *see* petroleum pollution
open ocean monitoring 126, 141, 144–5; and UNEP 127, 130–1
organochlorine compounds: in breast milk 66; monitoring of 45, 62n, 65
ozone 64
ozone layer, depletion of 113, 214
ozone monitoring 36, 45, 99

Palme Commission 264
passivity, of North in GEMS 208
peace and collective security 266
peace-keeping function of UN 252
personal exposure monitoring 63
petroleum pollution: effects of 124; monitoring of 126, 127–8; visual observation 127–8n
Pilot Programme on Integrated Monitoring (Europe) 183
pollutant impacts 151
pollutant pathways 41, 47
pollutants 61, 73, 169, 214; air-borne, effects of 175–6; long-range air transport monitored 53; monitoring list expanded (EMEP) 181; monitoring of 44, 84; priority pollutants 45
pollution 4, 5; monitoring 42–3
polychlorinated biphenyls (PCBs) 65
population growth, rapid, implications of 5
Prebisch, Raul, Secretary General, UNCTAD 226–7
priority areas, of UNEP 14
Programme Activity Centres (PACs) 23, 131; *see also* GEMS PAC; OCA PAC; RS PAC
programme goals, GEMS 44–6
programmes implementation, GEMS HEALTH 82–3
Protection of the Mediterranean Sea

Against Pollution from Land-based Sources, Protocol 138
public opinion 187, 231, 270

quality assurance 71, 83

radiation/radionuclide monitoring 35, 45, 62n, 67, 86–90, 89n, 121n, 121, 191, 214
radioactive waste disposal 67
radioactivity, accidental release monitoring 191
rangelands monitoring 152, 157–8
Red Data Books 158
Reduction of Sulphur Emissions Convention (1979), Protocol 181–2
regional organizations 14–15
regional seas monitoring, logistics of 133
Regional Seas Programme 130, 131, 141; Mediterranean model 132–40
regional seas programmes 143, 144
remote sensing 112–13, 148–9, 172, 201–2, 213, 218–19; of natural resources 160–5; North–South differences 161–5; pressure for commercialization 163–4; for tropical forest monitoring 155n, 155, 156
remote sensing data, in GRID 197
renewable resource base 5
river pollution, transfrontier, Europe 191n
RS PAC 139–40, 140n, 166; *see also* OCA PAC

salinization 154
satellites: for climate-related monitoring 97; for GEMS 165, 219; oceanographic 148; for research and monitoring 165; *see also* remote sensing
scientific objective–environmental monitoring dichotomy xix
SCOPE reports 38, 46, 167
sea level rise 94
seawater quality, coastal 124
secretariat (UNEP) 21–33; defining GEMS 51; and the Environment Fund 22, 26n, 26–7; high-level posts for 21; limitations of smallness 15–16; non-operational/non-executive nature a problem 17, 23; small 7–8, 21–2; three-level approach 16–17; tried to impose new types of coordination 19–20; *see also* UNEP, location of
secretariats 53–4, 229, 231, 234, 235; *see also* international civil service

Secretary General, UNCHE 5–6
soil degradation 44; monitoring of 151, 152, 153–4
solar radiation monitoring 105–6n
South 262; definition 224n; felt disadvantaged by North 257–8; prospect of deeper dependency on North 260; seen as responsible for decline in UN staff quality 233–4
South Commission 263n, 264
space surveillance agency 165
specialized agencies 6, 36, 47–50, 52, 77, 208–9; discouraging research and studies 227–8n; environmental monitoring, UNEP to pay 55–6; funds for monitoring activities needed 51–2; GEMS office alarm 48–9; and MED POL 135n; view of UNEP and Fund 8, 19, 25
SPOT satellite 162
Stockholm Conference *see* UNCHE
Stockholm Initiative on Security and Governance in 1990s 264
Study of Critical Environmental Problems (SCEP) 91n, 124n
Study of Man's Impact on Climate (SMIC) 104
sulphur dioxide 61, 64, 71, 72, 86, 178, 179; monitoring of 45, 62n
sulphur emissions, reduced 181–2
sulphur oxides 176n, 176
supporting measures xvi
suspended particulate matter (SPM) 45, 61, 62n, 64, 71, 86
sustainable development 31, 151
Sweden: national case study for UNCHE 177; national monitoring programme (PMK) 183n, 183
systems coordination, role for UNEP 215–16
Systems-Wide Medium-Term Environment Plan (SWMTEP) 20, 54, 209

target monitoring 37n
technological tools 200, 201
temperate mixed forest ecosystems, integrated monitoring 169–70
Third World: and GEMS HEALTH 83; and pollutants 61
30 per cent Club 181
three-level programmatic approach *see* catalytic approach
toxic rain *see* acid rain

Subject index 283

toxic/radioactive waste dumping 62
transboundary air pollution 105, 171;
  long-range, monitoring of 175–92; *see also* EMEP; Long-Range Transboundary Air Pollution Convention
transfrontier water pollution monitoring 191
trend monitoring 37n
Tropical Earth Resources Satellite (TERS) 163n
tropical forest monitoring 152, 154–7, 166, 173–4

UN Charter, revision of 265
UN Conference on Desertification 154, 157
UN Conference on Environment and Development (1992) 12, 31–3, 122, 205, 220
UN Security Council 258
UN system xix, 222, 245, 249–72; the development issue 226; environment and coordination in 18–20; need for comprehensive reform 258–71; North's strategy 250–8; reform attempts and evolution 249–58; standard approach 246–7; structured against integrated approach 247; UNEP location a policy departure 11; *see also* United Nations
UNCHE xi, xvi, 1, 3–9, 28–9, 41–2, 177, 205; Action Plan 9, 12–13,129, 177–8; after UNCHE 9–12; assessment-management scheme, MED POL 138; call for control of marine pollution sources 133; Declaration 9, 12–13, 177–8; environmental monitoring at 35–42; exposed underlying controversies 3–4; institutional question at UNCHE 6–8; post-Conference period 28–31; preparation for 37–8; recommendations regarding Earthwatch xvii–xviii
UNCTAD 9, 26n, 226, 253, 263n; *see also* UN system
underfunding 159, 237; *see also* international organizations, financing of
UNDP 20, 215, 238, 254n; and health-related monitoring 75n, 75
UNEP xvi, 1, 19n, 21–8, 32, 42, 53, 77, 98, 115, 121–2, 151, 158, 178–9, 209; ambivalent institutional role xviii, 50, 51, 52, 203; an exception to the rule in UN system 245; background to formation of 6–8; changed attitude to open ocean studies 130–1; changes in the broader setting 28–34; a cooperating and supporting organization 20; coordinating organization for the environment 47; expert meeting on forest cover monitoring 156; finance 55, 238; financial support for BAPMoN 93; first Governing Council 12–15; funds and other assistance, MED POL 136, 140–1; institutional model 15–17; and GEMS 48, 50, 114, 212, 214–17; and GEMS HEALTH 80–1, 85–6; and GRID 195, 197–8; location of 10–12, 14, 202–3, 210n, 268n; model and reality 17–34; performing roles 21–8; pressure from USA 228n; problems of institutional model 208, 245; programme priorities for 12–15; pursued adaptive approach 47; sectoral approach 210; sidelined for 1992 Conference 32; support for IOC pilot project 125; *see also* catalytic approach; *Level One; Level Two; Level Three*; Environment Fund; Governing Council (UNEP); secretariat (UNEP)
UNEP fund 7; to supply flexibility 8; *see also* Environment Fund
UNEP–FAO project, interpreting forest data 155–6n, 155
UNESCO 53, 63, 81, 142, 168
UNIDO 226
unilateral action, change through 249–58
UNISPACE 118, 195
United Nations xii, 221–72; 1980s crisis xi, xv, 250; Charter revision 265; controversy over nature and role of 223–4; decision-making 269; and development of global monitoring 219–20; financing of 236–43; institutional betterment 243–6; institutional limitations and rigidities xv; international financial and monetary issues blocked 257; interrelatedness 246–8; role of civil society 270n, 270; sectoral thinking at creation of 247–8; a target for reform 244, 249–50; two-track approach needed for the future 269; USA intervention 228n; *see also* international civil service; international organizations; UN system

UNSCEAR 35, 45–6, 46n, 67
USA 38n, 119; critical of regional seas approach 142; Global Habitability programme 118, 195–6; pressure on United Nations and UNEP 26n, 26, 228n, 250
USSR 207, 219; critical of regional seas approach 142; interest in open ocean pollution monitoring 145–6, 147; kept integrated approach alive 167, 169
UV-B radiation 98–9

vector-borne diseases 43
Vienna Convention for the Protection of the Ozone Layer 99
volcanic activity monitoring 97
Voluntary Cooperation Programme 107n

WCED 30, 33, 220, 228n; report of 30–1
weather observation networks 35
WGI 95–6, 97
WHO 52, 62, 134; on environmental radiation monitoring 67, 88; and GEMS HEALTH 80–1; GEMS/Air, GEMS/Food and GEMS/Water 63; global infrastructure for health-related monitoring 68–9; monitoring methodologies and procedures 68
WHO International Reference Centre of Radioactivity 67n
WMO 35, 36, 53, 63, 92, 93, 96, 97, 98–9, 105–6, 134, 167, 168, 211; and European air pollution monitoring 177–9; not a strong position for BAPMoN 117; responsible members concept 111n, 111; set up GAW 120–1
Working Group on Atmospheric Pollution and Atmospheric Chemistry 93n, 120
World Bank 238–9, 240, 253, 254–8; criticism of 255–6n; *see also* GATT; IMF; UN system
World Climate Programme (WCP) 97–8, 100, 103–5, 114–15, 118
World Climate Research Programme (WCRP) 98, 165
World Conservation Monitoring Centre 158
World Forest Inventory 154
world ocean system, monitoring-assessment approach 148
world problems, global nature of 248
world wars, re-examination of international organizations 261
World Weather Watch (WWW) 35, 50, 103–6, 125
World Wide Fund for Nature (WWF) 158